METHODS IN MOLECULAR BIOLOGY™

AF774283

Series Editor
John M. Walker
School of Life Sciences
University of Hertfordshire
Hatfield, Hertfordshire, AL10 9AB, UK

For other titles published in this series, go to
www.springer.com/series/7651

Plant Developmental Biology

Methods and Protocols

Edited by

Lars Hennig

and

Claudia Köhler

Department of Biology, Swiss Federal Institute of Technology (ETH), Zurich, Switzerland

Humana Press

Editors
Lars Hennig
Department of Biology
Swiss Federal Institute of
Technology (ETH)
Universitätstrasse 2
CH-8092 Zurich
Switzerland
lars.hennig@ipw.biol.ethz.ch

Claudia Köhler
Department of Biology
Swiss Federal Institute of
Technology (ETH)
Universitätstrasse 2
CH-8092 Zurich
Switzerland
claudia.koehler@ipw.biol.ethz.ch

ISSN 1064-3745 e-ISSN 1940-6029
ISBN 978-1-4939-5734-7 ISBN 978-1-60761-765-5 (eBook)
DOI 10.1007/978-1-60761-765-5
Springer New York Dordrecht Heidelberg London

Printed on acid-free paper

Humana Press is part of Springer Science+Business Media (www.springer.com)

Preface

Plants come in myriads of shapes and colors, and the beauty of plants has fascinated mankind for thousands of years. Long before Mendel discovered the laws of heritability and Darwin developed his theory on evolution, the affection for ornamental plants led people to select alleles that establish novel plant forms. Today, plant developmental biology tries to discover the mechanisms that control the establishment of specialized cell types, tissues, and organs from the fertilized egg during a plant's life. Although the underlying processes of cell proliferation and differentiation are similar in plants and animals, plants are different because their development is usually open, and its outcome is not the faithful repetition of a general plan but is strongly influenced by environmental conditions. In the last few decades, plant developmental biology has pinpointed a large number of developmental regulators and their interactions and the mechanisms that govern plant development start to emerge. In part, this progress was enabled by the advance of powerful molecular tools for a few model species, most importantly Arabidopsis.

This volume of the *Methods in Molecular Biology* series provides a collection of protocols for many of the common experimental approaches in plant developmental biology. All chapters are written in the same format as that used in the *Methods in Molecular Biology*™ series. Each chapter opens with a description of the basic theory behind the method being described. The Materials section lists all the chemicals, reagents, buffers, and other materials necessary for carrying out the protocol. Since the principal goal of the book is to provide experimentalists with a full account of the practical steps necessary for carrying out each protocol successfully, the Methods section contains detailed step-by-step descriptions of every protocol that should result in the successful completion of each method. The Notes section complements the Methods material by indicating how best to deal with any problem or difficulty that might arise when using a given technique. Reflecting the current balance in the field, the book is most detailed for Arabidopsis but includes also protocols for other model species such as rice, maize, or Medicago. The book is divided into six major parts: growth protocols, manipulation of gene activity, assaying developmental phenotypes, assaying gene activity, testing protein–protein interactions, and probing chromatin. Presented methods are diverse and range from grafting over bimolecular fluorescence complementation to chromatin immunoprecipitation. In the first place, the book addresses a target audience of plant developmental geneticists and biochemists. In addition, colleagues from other fields such as stress physiology or plant nutrition will find this book helpful. Developmental biology was usually not the prime interest of these colleagues, but when analyzing mutants, which are nowadays so easily available using reverse genetics, many researchers will suddenly be confronted with phenotypes of abnormal development. Together, we hope that this volume will be an essential part of many laboratory libraries. We would be pleased if the book will be found more often on the bench top than in the book shelf.

L. Hennig
C. Köhler

Contents

Contributors

ERNST AICHINGER • *Department of Biology, Swiss Federal Institute of Technology (ETH), Zurich, Switzerland*

JOSÉ AKER • *Laboratory of Biochemistry, Wageningen University, Wageningen, Netherlands*

CRISTINA ALEXANDRE • *Department of Biology, Swiss Federal Institute of Technology (ETH), Zurich, Switzerland*

GERRIT T.S. BEEMSTER • *Department Plant Systems Biology, Flanders Institute for Biotechnology & Department Plant Biotechnology and Genetics, Ghent University, Ghent, Belgium; Department of Biology, University of Antwerp, Antwerp, Belgium*

PHILIP N. BENFEY • *Department of Biology and NIH Center for Systems Biology, Duke University, Durham, NC, USA*

STEVEN BERNACKI • *Department of Plant Biology, North Carolina State University, Raleigh, NC, USA*

LORENZO BORGHI • *Department of Biology, Swiss Federal Institute of Technology (ETH), Zurich, Switzerland*

JAN WILLEM BORST • *Laboratory of Biochemistry, Wageningen University, Wageningen, Netherlands*

CHRISTOPH BÜCHERL • *Laboratory of Biochemistry, Wageningen University, Wageningen, Netherlands*

CRISTEL C. CARLES • *Plant Gene Expression Center, USDA-ARS/UC Berkeley & Department of Plant and Microbial Biology, University of California, Berkeley, CA, USA*

SARAH J. COOKSON • *Laboratoire d'Ecophysiologie des Plantes sous Stress Environnementaux UMR759, INRA-SUPAGRO, Montpellier, France*

FREDERIK COPPENS • *Department Plant Systems Biology, Flanders Institute for Biotechnology & Department Plant Biotechnology and Genetics, Ghent University, Ghent, Belgium*

TAMAS DALMAY • *School of Biological Sciences, University of East Anglia, Norwich, Norfolk, England*

ROBERT C. DAY • *Department of Biochemistry, University of Otago, Dunedin, Otago, New Zealand*

HANS J. DE JONG • *Laboratory of Genetics, Wageningen University, Wageningen, Netherlands*

SACCO DE VRIES • *Laboratory of Biochemistry, Wageningen University, Wageningen, Netherlands*

STIJN DHONDT • *Department Plant Systems Biology, Flanders Institute for Biotechnology & Department Plant Biotechnology and Genetics, Ghent University, Ghent, Belgium*

VIVIEN EXNER • *Department of Biology, Swiss Federal Institute of Technology (ETH), Zurich, Switzerland*

FABIO FIORANI • *Department Plant Systems Biology, Flanders Institute for Biotechnology & Department Plant Biotechnology and Genetics, Ghent University, Ghent, Belgium*

ELISA FIUME • *Plant Gene Expression Center, USDA-ARS/UC Berkeley & Department of Plant and Microbial Biology, University of California, Berkeley, CA, USA*

JENNIFER C. FLETCHER • *Plant Gene Expression Center, USDA-ARS/UC Berkeley & Department of Plant and Microbial Biology, University of California, Berkeley, CA, USA*

PAUL FRANSZ • *Swammerdam Institute for Life Sciences, University of Amsterdam, Amsterdam, Netherlands*
JIŘÍ FRIML • *VIB Department of Plant Systems Biology, University of Ghent, Ghent, Belgium*
CHRISTINE GRANIER • *Laboratoire d'Ecophysiologie des Plantes sous Stress Environnementaux UMR759, INRA-SUPAGRO, Montpellier, France*
RITA GROß-HARDT • *Center for Plant Molecular Biology (ZMBP), University of Tübingen, Tübingen, Germany*
CHAN MAN HA • *Plant Gene Expression Center, USDA-ARS/UC Berkeley & Department of Plant and Microbial Biology, University of California, Berkeley, CA, USA*
RENZE HEIDSTRA • *Molecular Genetics Group, Department of Biology, Utrecht University, Utrecht, Netherlands*
LARS HENNIG • *Department of Biology, Swiss Federal Institute of Technology (ETH), Zurich, Switzerland*
PIERRE HILSON • *Flanders Interuniversity Institute for Biotechnology (VIB), Department of Plant Systems Biology, Ghent University, Ghent, Belgium*
MARTIN HÜLSKAMP • *Botanical Institute III, University of Cologne, Cologne, Germany*
ERIKA ISONO • *Department of Plant Systems Biology, Technical University Munich – Weihenstephan, Munich, Germany*
ANJALI S. IYER-PASCUZZI • *Department of Biology and NIH Center for Systems Biology, Duke University, Durham, NC, USA*
MARK A. JOHNSON • *Department of Molecular Biology, Cell Biology and Biochemistry, Brown University, Providence, RI, USA*
JASON JONES • *Department of Biological Sciences, Louisiana State University, Baton Rouge, LA, USA*
JI HYUNG JUN • *Plant Gene Expression Center, USDA-ARS/UC Berkeley & Department of Plant and Microbial Biology, University of California, Berkeley, CA, USA*
MANSOUR KARIMI • *Flanders Interuniversity Institute for Biotechnology (VIB), Department of Plant Systems Biology, Ghent University, Ghent, Belgium*
NAOHIRO KATO • *Department of Biological Sciences, Louisiana State University, Baton Rouge, LA, USA*
JIN SUN KIM • *Plant Gene Expression Center, USDA-ARS/UC Berkeley & Department of Plant and Microbial Biology, University of California, Berkeley, CA, USA*
STEFAN KIRCHER • *Albert-Ludwigs-University of Freiburg, Freiburg im Breisgau, Baden-Württemberg, Germany*
CLAUDIA KÖHLER • *Department of Biology, Swiss Federal Institute of Technology (ETH), Zurich, Switzerland*
BENEDIKT KOST • *Uppsala BioCenter, Plant Biology and Forest Genetics, Swedish University of Agricultural Sciences, Ultuna, Uppsala, Sweden*
THOMAS KRETSCH • *Albert-Ludwigs-University of Freiburg, Freiburg im Breisgau, Baden-Württemberg, Germany*
CATHERINE MASSONNET • *Laboratoire d'Ecophysiologie des Plantes sous Stress Environnementaux UMR759, INRA-SUPAGRO, Montpellier, France*
YVONNE MÖLLER-STEINBACH • *Department of Biology, Swiss Federal Institute of Technology (ETH), Zurich, Switzerland*
NIR OHAD • *Department of Plant Sciences, Tel-Aviv University, Tel-Aviv, Israel*
PENKA PAVLOVA • *Laboratory of Genetics, Wageningen University, Wageningen, Netherlands*

SERENA PERILLI • *Dipartimento di Genetica e Biologia Molecolare, Laboratory of Functional Genomics and Proteomics of Model Systems, Università La Sapienza, Rome, Italy*
HELENA R. PIRES • *Plant Gene Expression Center, USDA-ARS/UC Berkeley & Department of Plant and Microbial Biology, University of California, Berkeley, CA, USA*
NIKI ROBERTSON • *Department of Plant Biology, North Carolina State University, Raleigh, NC, USA*
BART RYMEN • *Department Plant Systems Biology, Flanders Institute for Biotechnology & Department Plant Biotechnology and Genetics, Ghent University, Ghent, Belgium*
SABRINA SABATINI • *Dipartimento di Genetica e Biologia Molecolare, Laboratory of Functional Genomics and Proteomics of Model Systems, Università La Sapienza, Rome, Italy*
MICHAEL SAUER • *Centro Nacional de Biotecnologia CSIC Madrid, Madrid, Spain*
CLAUS SCHWECHHEIMER • *Department of Plant Systems Biology, Technical University Munich – Weihenstephan, Weihenstephan, Germany*
RÜDIGER SIMON • *Institute of Genetics, Heinrich-Heine University Düsseldorf, Düsseldorf, Germany*
YVONNE STAHL • *Institute of Genetics, Heinrich-Heine University Düsseldorf, Düsseldorf, Germany*
FEDERICO TESSADORI • *Swammerdam Institute for Life Sciences, University of Amsterdam, Amsterdam, Netherlands*
OLIVIER TURC • *Laboratoire d'Ecophysiologie des Plantes sous Stress Environnementaux, Montpellier, France*
COLIN G. N. TURNBULL • *Division of Biology, Imperial College London, London, UK*
JOACHIM F. UHRIG • *Botanical Institute III, University of Cologne, Köln, Germany*
ANDRÁS VICZIÁN • *Institute of Plant Biology, Hungarian Academy of Science, Biological Research Center, Szeged, Hungary*
CORINA B. R. VILLAR • *Department of Biology, Swiss Federal Institute of Technology (ETH), Zurich, Switzerland*
RONNY VÖLZ • *Center for Plant Molecular Biology (ZMBP), University of Tübingen, Tübingen, Germany*
GUY WACHSMAN • *Molecular Genetics Group, Department of Biology, Utrecht University, Utrecht, Netherlands*
SHAUL YALOVSKY • *Department of Plant Sciences, Tel-Aviv University, Tel-Aviv, Israel*

Chapter 1

Growth Protocols for Model Plants in Developmental Biology

Lars Hennig

Abstract

Arabidopsis is the dominating model species for plant developmental biology, but other species serve as models for processes that cannot be studied in *Arabidopsis*, such as compound leaf or wood formation, or to test the universality of developmental mechanisms initially identified in *Arabidopsis*. Research in plant developmental biology depends critically on robust growth protocols that will support reproducible development. Here, protocols are given to grow *Antirrhinum*, *Arabidopsis*, *Brachypodium*, maize, *Medicago*, *Petunia*, rice, and tomato in the laboratory.

Key words: *Antirrhinum*, *Arabidopsis*, *Brachypodium*, maize, *Medicago*, *Petunia*, rice, tomato.

1. Introduction

Research in plant developmental biology depends critically on robust growth protocols that will support reproducible development. Although *Arabidopsis thaliana* is the dominating model species for plant developmental biology, other species serve as models for processes that cannot be studied in *Arabidopsis*, such as compound leaf or wood formation, or to test the universality of developmental mechanisms initially identified in *Arabidopsis*.

A. thaliana is a member of the mustard family (*Brassicaceae*) with a broad natural distribution throughout Europe, Asia, and North America, and many accessions (ecotypes) can be obtained from stock centers. The most commonly used accessions are Columbia and Landsberg *erecta*. *Arabidopsis* has a life cycle of only 6 weeks (for a review about *Arabidopsis* as model species, *see* (1)).

L. Hennig, C. Köhler (eds.), *Plant Developmental Biology*, Methods in Molecular Biology 655,
DOI 10.1007/978-1-60761-765-5_1, © Springer Science+Business Media, LLC 2010

Snapdragon (*Antirrhinum majus*) is a member of the speedwell family (*Plantaginaceae*) and native to the Mediterranean. *A. majus* has a life cycle of 3–4 months. It has been used as a model for biochemical and developmental genetics for about a century, and many developmental regulatory genes were identified in *A. majus* by transposon tagging. *A. majus* is used to study processes such as the specification of flower and floral organ identity and leaf and flower asymmetry (for a review about *A. majus* as model species, *see* (2)).

Tomato (*Solanum lycopersicum*) is a domesticated member of the nightshade family (*Solanaceae*) and originated in western South America. Interesting developmental features of tomato include fleshy fruits, sympodial shoots, and compound leaves. Micro-Tom is an extremely small tomato variety (10–20 cm high), and due to its low-space requirements it is widely used for molecular research (3) (for a review about tomato as model species, *see* (4)).

Petunia is another member of the nightshade family originating from South America. *P. hybrida*, a hybrid of *P. axillaris* and *P. integrifolia*, is most commonly used in research. *P. hybrida* has a life cycle of only 8–12 weeks. A major attraction of *P. hybrida* is the presence of the extremely active endogenous dTph1 transposon system, which allows for efficient forward and reverse genetics (for reviews about *Petunia* as model species, *see* (5, 6)).

Barrel medic (*Medicago truncatula*) is a member of the pea family (*Fabaceae*) and native to the Mediterranean. It has a small diploid genome, is self-fertile, has a rapid generation time and prolific seed production and is also amenable to genetic transformation. Plants are 10- to 60-cm high. *M. truncatula* has a life cycle of 3–4 months. It serves mainly as model for nodulation and symbioses with nitrogen-fixing rhizobia, *Sinorhizobium meliloti*, and arbuscular mycorrhizal fungi (for a review about *M. truncatula* as model species, *see* (7)).

Rice (*Oryza sativa*) is a member of the grass family (*Poaceae*) and native to tropical and subtropical southern Asia. Plants grow 1–1.8 m tall. Rice has a life cycle of 4–6 months. It has a small diploid genome and is the most widely used model species for monocotyledonous plants (for a review about rice as model species, *see* (8)).

Maize (*Zea mays*) is another member of the grass family and was domesticated in Mesoamerica. Plants grow 2–3 m tall. Maize has a rich tradition in developmental genetics and large mutant collections exist (for a review about maize as model species, *see* (9)).

Purple false brome (*Brachypodium distachyon*), a third member of the grass family that is a model species for plant developmental biology, is native to southern Europe, northern Africa, and southwestern Asia. *B. distachyon* is a small (20-cm high), self-fertile, inbreeding annual weed. It has a life cycle of less than

4 months. *B. distachyon* has the simplest genome described in grasses to date, comparable to the *Arabidopsis* genome and several times smaller than the rice genome. *B. distachyon* is an emerging model for temperate grasses such as wheat (for a review about *B. distachyon* as model species, *see* (10)).

Detailed growth protocols for *Arabidopsis, Medicago, Antirrhinum,* and tomato can also be found elsewhere (11–13).

Poplar (*Populus trichocarpa*) is a member of the willow family (*Salicaceae*) and native to western North America. It is the preferred model to study perennial life cycles, bud dormancy, and wood formation (14). Because poplar has a long life cycle of several years, work with poplar differs substantially from raising annual or biannual herbaceous plants and will not be covered here. Models for nonflowering plants include the moss *Physcomitrella patents,* the green algae *Chlamydomonas reinhardtii,* and the fern *Ceratopteris richardii* and will also not be covered here.

2. Materials

2.1. Growing Arabidopsis

1. Gardening soil (*see* **Note 1**).
2. Pots (5–10 cm diameter).
3. Greenhouse or growth cabinet (22°C, 120–150 μmol m^{-2} s^{-1}) (*see* **Note 2**).

2.2. Growing Petunia

1. Gardening soil (*see* **Note 3**).
2. Pots (10–25 cm diameter).
3. Greenhouse (20–25° day, 18–20° night; >250 μmol m^{-2} s^{-1}, 16 h light).

2.3. Growing Tomato

1. Gardening soil (pH 6.0–6.8).
2. Pots (15 cm diameter).
3. Greenhouse or growth cabinet (18°C–24°C, 600–700 μmol m^{-2} s^{-1}).

2.4. Growing Antirrhinum

1. Gardening soil.
2. Pots (5–10 cm diameter).
3. Greenhouse (17°C–23°C; <70% humidity; >150 μmol m^{-2} s^{-1}; 16 h light).

2.5. Growing Brachypodium

1. Mix of gardening soil with vermiculite (2:1, v/v).
2. Pots (5–10 cm diameter).
3. Greenhouse or growth cabinet (24°C day, 18°C night; >150 μmol m^{-2} s^{-1}; 16 h light).

2.6. Growing Medicago

1. Mix of sand and gardening soil (1:2–1:3, v/v) (*see* **Note 4**).
2. Pots (8 and 20 cm diameter).
3. Greenhouse or growth cabinet (20–25°C day, 15–21°C night; 200–600 μmol m^{-2} s^{-1}; 16 h light).
4. Sand paper.

2.7. Growing Maize

1. Mix of sand and gardening soil (1:2–1:3, v/v).
2. Pots (10 and 25 cm diameter).
3. Greenhouse or growth cabinet (28°C day, 18°C night; >600 μmol m^{-2} s^{-1}; 16 h light).

2.8. Growing Rice

1. Growth medium: 1× MS salts (Sigma), 10% sucrose, 0.8% bacto agar, 100 mg/L inositol, 0.05 mg/L biotin, 0.5 mg/L pyridoxine HCl, 0.5 mg/L thiamin HCl, 5 mg/L nicotine acid, 0.5 mg/L folic acid, and 2 mg/L glycine (*see* **Note 5**).
2. Nutrient solution: 0.70 mM K_2SO_4, 0.10 mM KCl, 0.10 mM KH_2PO_4, 2.0 mM $Ca(NO_3)_2$, 0.50 mM $MgSO_4$, 10 μM H_3BO_3, 0.50 μM $MnSO_4$, 0.50 μM $ZnSO_4$, 0.20 μM $CuSO_4$, 0.01 μM $(NH_4)_6Mo_7O_{24}$, and 100 μM Fe(III)-EDTA (pH 5.5) (*see* **Note 6**).
3. Mix of vermiculite and gardening soil (1:1, v/v).
4. Pots (18 cm diameter, 3 L).
5. Sterile transparent containers (5 cm diameter, 10 cm high).
6. Greenhouse or growth cabinet (28–32°C day, 20–25°C night; 80% humidity; >500 μmol m^{-2} s^{-1}; 12 h light).

3. Methods

3.1. Growing Arabidopsis

1. Fill pots with soil and compress very lightly to give a firm bed.
2. Sow the seeds onto the surface of the moist soil by scattering them carefully from a piece of folded cardboard; cover pots with a transparent lid to keep humidity high (*see* **Notes 7** and **8**).
3. Stratification at 4°C for 2–5 days improves germination rate and synchrony (*see* **Note 9**).
4. Transfer pots to growth cabinets or greenhouse, remove covers after 5–7 days, and then water pots from below so that pots can soak up water (*see* **Notes 10** and **11**).

5. Water plants until at least 90% of the seed pods have dried completely. Allow plants to dry slowly for maximum viable seed production.
6. After harvest, separate seeds from dry vegetative material and chaff using nylon mesh.
7. Seeds can be stored in paper bags in a dry atmosphere at room temperature for at least 3 years.

3.2. Growing *Petunia*

1. Fill pots with soil and compress very lightly to give a firm bed.
2. If desired, surface-sterilize seeds (1 min in 70% ethanol, 5 min in 2% bleach (sodium hypochlorite); rinse 5 times with sterile water).
3. Sow the seeds onto the surface of the moist soil.
4. Germination will take 3–7 days, flowering will occur after 10–12 weeks; seeds can be harvested after 13 weeks.
5. Harvest capsules and dry.
6. Store seeds at 4°C at low humidity.

3.3. Growing Tomato

1. Fill pots with soil and compress very lightly to give a firm bed.
2. Sow the seeds 3 cm apart in flats containing soil (0.5–2 cm from the surface of the soil) or wet filter paper. Cover with a plastic cover until seedlings emerge (7–14 days) to keep the substrate moist (*see* **Note 12**).
3. Place flats containing soil in a greenhouse or growth cabinet. Place flats with filter paper, until germination, in the dark at room temperature.
4. Transplant seedlings into pots.
5. Once flowering begins, shake the tomato plants gently once or twice each week to promote pollination.
6. Harvest ripe fruits (90–120 days after germination), cut in half with a knife, and squeeze the seeds into a container.
7. Mix seeds with excess tap water and incubate for ~3 d at room temperature to allow removal of the gelatinous seed coating.
8. Wash seeds extensively with tap water to remove the coating.
9. Dry seeds on a paper towel overnight at room temperature.

3.4. Growing *Antirrhinum*

1. Fill pots with soil and compress very lightly to give a firm bed.
2. Sow seeds on the surface of the soil and cover pots with a clear plastic cover to keep the soil moist (*see* **Note 13**).
3. Keep the pots at ~17°C.

4. After germination (7–10 days) (*see* **Note 14**), keep plants under high light at 17–23°C (*see* **Note 15**).
5. Water the plants as necessary (*see* **Note 16**).
6. Expect plants to flower 3–4 months after sowing.

3.5. Growing Brachypodium

1. Sow seeds into the soil.
2. Stratify seeds at 4°C for 1 week after sowing. To promote flowering by vernalization, extend incubation at 4°C to 3 weeks.
3. Transfer to growth cabinet or green house.

3.6. Growing Medicago

1. For mechanical scarification, place seeds on a fine-grade sand-paper sheet and rub them gently with another piece of sand paper until there are visible signs of abrasion (*see* **Note 17**).
2. Sow seeds on soil and cover with a fine dry sand/soil mixture (1:1, sieved to eliminate larger particles; 10–15 mm thick) for efficient rooting. Evenly spread and gently pack down this top layer using a flat tool (e.g., a Petri dish) and humidify it with a fine water spray. Cover with a clear plastic cover to maintain high humidity.
3. For dormant seeds, incubate for 48–72 h at 4°C.
4. If reduced growth and a shorter life cycle are desired, vernalize seedlings by incubating at 4°C for 10–14 days. Vernalized plants can flower as early as after 30 days (cultivar Jemalong), usually produce at least 15 pods (equivalent to 150 seeds), and can be grown at a density of up to 60 plants/m^2. Nonvernalized plants usually flower after 60–70 days, can yield several thousands of seeds per plant but need up to 1 m^2 per plant.
5. Incubate at 20°C in the dark until all viable seedlings have sprouted from the substrate (after 2–3 days).
6. Transfer into growth room, gradually remove (or puncture) the cover, and maintain watering with a fine spray for the first week.
7. Plant seedlings into small pots (8 cm). Once plants have at least 5 leaves (after ~2 weeks), transplant into large pots (20 cm) and transfer to glass house (*see* **Note 18**).
8. After pod harvest, allow pods to dry out at room temperature for ≥1 week at low humidity. Store dry, intact pods, at room temperature and low humidity, in strong paper envelopes or screw-cap plastic vials with punched holes in the cap to allow air circulation. Seeds normally retain good viability for at least 3–5 years under such conditions. Once removed from the fruits, seeds should be used within 1 month.

3.7. Growing Maize

1. If desired, surface-sterilize seeds with 6% hypochlorite for 10 min and wash 5 times with sterile water.
2. Plant seeds 1–3 cm deep into the soil in small pots (10 cm).
3. Place pots in growth room or greenhouse and water every 2–3 days with nutrient solution. Expect germination approximately 2–4 days after planting.
4. Transfer to large pots (25 cm) as required (usually after 15 days). Expect flowering after 6–12 weeks, depending on genotype.
5. Harvest seeds after 3–5 month.
6. Store seeds at 8°C at very low humidity.

3.8. Growing Rice

1. Surface-sterilize seeds: Shake in 70% ethanol for 3 min, shake in 6% sodium hypochlorite for 10 min, and wash four times with sterile water.
2. Place seeds on growth medium in sterile transparent containers and transfer to greenhouse.
3. Grow plantlets in the containers for about 2 weeks (until the leaves reach the lid of the container) and then transfer to soil or a hydroponics system. For transfer to soil, plantlets may be first grown in small pots (10 cm) and after another 2–3 weeks in large pots (18 cm) (*see* **Note 19**).
4. Plants will flower after about 4–5 months and first seeds can be harvested 4 weeks later (*see* **Note 20**).
5. Store seeds under dry conditions at 4°C.

4. Notes

1. Many commercially available soils work well. However, compost composition changes with the season and batch-to-batch variability can be high. Substrates based on peat are usually more consistent. Because water control is important, include a layer of perlite or vermiculite at the bottom of the pots to aid drainage. Alternatively, premixed soils that include vermiculite can be used.
2. Use continuous light photoperiods for fastest progression to flowering (after ~25 d). Long-day photoperiods (16 h of light) will result in slightly more vigorous plants (flowering after ~30 d). Use short-day photoperiods for prolonged vegetative development (flowering after ~60 d). Note that time estimates are for accession Columbia and will differ for other accessions.

3. Many commercially available soils work well. For mycorrhiza experiments, use a sand:soil mixture (2:1).
4. *M. truncatula* grows best on well-drained, fairly dense substrates such as mixtures of sand and soil. Alternatively, use mix of perlite and soil (1:3, v/v) or 100 % perlite, sand, or vermiculite. Perlite-grown plants are also ideal for nodulation assays. Wash perlite and vermiculite until pH is ~7.0 before use.
5. Less rich medium works well for germination as well. On the extreme side, seeds can be germinated on water-saturated Whatman paper.
6. Adjust the pH of the nutrient solution daily to 5.5 with 1 M HCl. Renew weekly.
7. Do not bury seeds in soil; *Arabidopsis* seeds require light for germination.
8. Seeds can be mixed with clean sand or glass beads for even distribution.
9. Vernalization is not required for common laboratory accessions, but exposure to 4°C for 6 weeks will accelerate flowering of many accessions from the wild.
10. One of the most common beginner's mistake is excess watering. Never allow excess water to remain in the tray.
11. *Arabidopsis* is predominantly self-fertilizing, but it is still advised to prevent direct contact between flowers from different lines. This can be easily achieved using plastic sleeves. Alternatively, plants can be fixed to stakes of metal wire or plastic rods.
12. Seeds of some accessions germinate poorly. To improve the germination rate, treat seeds in 2.7% bleach (sodium hypochlorite) for 30 min at room temperature and wash off the bleach completely by rinsing the seeds in water before sowing.
13. To facilitate sowing, seeds can be suspended in 0.1% agar and pipetted onto the soil.
14. Germination rate can be increased by imbibing seeds in 10 μM gibberellin solution for 3–5 d at 4°C before sowing.
15. Flowering is promoted by long days; a nighttime drop in temperature to 15–17°C increases apical meristem size and encourages robust stem growth.
16. To prevent wilting, it may be necessary to water plants twice daily. Increasing the size of pots and placing pots on capillary matting can reduce the need for watering. Avoid wetting the foliage to prevent fungal infections. All *Antirrhinum* species are intolerant to waterlogged soil. Avoid leaving pots in standing water.

17. The hydrophobic, waxy *M. truncatula* seed coat must be scarified in order to allow the penetration of water and oxygen that trigger germination.
18. Because *M. truncatula* is salt sensitive, water plants with deionised water instead of tap water and include complete fertilizer once every 1–2 weeks. Excessive watering will cause roots to rot with leaf wilting as the most common over-watering symptom. Therefore, do not over-water plants and allow the soil to partially dry out between watering.
19. Soil should have a low content of organic matter such as peat. Add 1 g of fertilizer (15% nitrogen, 10% phosphorus, 15% potassium, 2% magnesium, 0.05% boron, 0.1% copper, 0.05% iron, 0.1% manganese, 0.0001% cobalt, 0.0083% molybdenum, and 0.025% zinc) per 1 L of soil. Fill pots only 2/3 of their height with soil. Soil should be kept well watered at all times. Add water to the soil surface one to three times per day or place pots in tanks containing about 5–10 cm of room temperature water. Never use cold water for irrigation because roots are sensitive to low water temperature.
20. Most rice varieties are short-day plants that will flower sooner under 10-h than under 12-h photoperiods.

Acknowledgments

The author would like to thank Didier Reinhart, Frank Hochholdinger, and Judith Wirth for sharing information. Research in the author's laboratory is supported by grants from the Swiss National Science Foundation [3100AO-116060] and ETH [TH-16/05-2].

References

1. Meinke, D. W., Cherry, J. M., Dean, C., Rounsley, S. D., and Koornneef, M. (1998) *Arabidopsis thaliana*: A model plant for genome analysis. *Science* 282, 679–682.
2. Schwarz-Sommer, Z., Davies, B., and Hudson, A. (2003) An everlasting pioneer: the story of *Antirrhinum* research. *Nat Rev Genet* 4, 657–666.
3. Meissner, R., Jacobson, Y., Melamed, S., Levyatuv, S., Shalev, G., Ashri, A., Elkind, Y., and Levy, A. (1997) A new model system for tomato genetics. *Plant J* 12, 1465–1472.
4. Kimura, S. and Sinha, N. (2008) Tomato (*Solanum lycopersicum*): A model fruit-bearing crop. *Cold Spring Harb Protoc*, doi:10.1101/pdb.emo105.

5. Angenent, G. C., Stuurman, J., Snowden, K. C., and Koes, R. (2005) Use of *Petunia* to unravel plant meristem functioning. *Trends Plant Sci* 10, 243–250.
6. Gerats, T. and Vandenbussche, M. (2005) A model system for comparative research. *Petunia Trends Plant Sci* 10, 251–256.
7. Cook, D. R. (1999) *Medicago truncatula*-a model in the making! *Curr Opin Plant Biol* 2, 301–304.
8. Shimamoto, K. and Kyozuka, J. (2002) Rice as a model for comparative genomics of plants. *Annu Rev Plant Biol* 53, 399–419.
9. Candela, H. and Hake, S. (2008) The art and design of genetic screens: Maize. *Nat Rev Genet* 9, 192–203.
10. Draper, J., Mur, L. A., Jenkins, G., Ghosh-Biswas, G. C., Bablak, P., Hasterok, R., and Routledge, A. P. (2001) *Brachypodium distachyon*. A new model system for functional genomics in grasses. *Plant Physiol* 127, 1539–1555.
11. Barker, D. G., Pfaff, T., Moreau, D., Groves, E., Ruffel, S., Lepetit, M., Whitehand, S., Maillet, F., Nair, R. M., Journet, E.-P. (2006) Growing *M. truncatula*: Choice of substrates and growth conditions. In: The Medicago truncatula handbook. Mathesius, U., Journet, E.-P., Sumner, L. W. (eds.), http://www.noble.org/MedicagoHandbook/
12. Hudson, A., Critchley, J., and Erasmus, Y. (2008) Cultivating *Antirrhinum*. *Cold Spring Harb Protoc* doi:10.1101/pdb.prot5051.
13. Kimura, S. and Sinha, N. (2008) How to grow tomatoes. *Cold Spring Harb. Protoc*, doi:10.1101/pdb.prot5051.
14. Jansson, S. and Douglas, C. J. (2007) *Populus*: A model system for plant biology. *Ann Rev Plant Biol* 58, 435–458.

Chapter 2

Grafting as a Research Tool

Colin G.N. Turnbull

Abstract

Grafting as a means to connect different plant tissues has been enormously useful in many studies of long-distance signalling and transport in relation to regulation of development and physiology. There is an almost infinite number of pairwise graft combinations that can be tested, typically between two different genotypes and/or between plants previously exposed to different environmental treatments. Grafting experiments are especially powerful for unambiguous demonstration of spatial separation of source and target, including genetic complementation of mutant phenotypes across a graft union, direct detection of transmitted molecules in receiving tissue or vascular sap, and activation or suppression of molecular targets due to signal transmission. Although grafting has a long history in research, only in the past decade has it been applied extensively to the Arabidopsis model. This chapter compares the main Arabidopsis grafting methods now available and describes seedling grafting in detail. Information is also provided on grafting of other common research model species, together with outlines of some successful applications.

Key words: Grafting, long-distance signalling.

1. Introduction

1.1. Context and History

Grafting as a biological concept is the deliberate connection of tissues from two different organisms or organs. Today, grafting techniques have widespread practical applications in both animal (skin grafts, organ transplants) and plant (horticultural manipulation) contexts. The principles are somewhat similar in both kingdoms: there is a requirement for some tissue regeneration to provide a stable graft union between the two parts, and fully functional grafts also generally need vascular connections across the union. In plants, the vascular connections are achieved by

L. Hennig, C. Köhler (eds.), *Plant Developmental Biology*, Methods in Molecular Biology 655,
DOI 10.1007/978-1-60761-765-5_2, © Springer Science+Business Media, LLC 2010

regeneration and are facilitated by alignment of existing vascular tissues when making the graft. In some types of animal grafts, especially of major organs, major vascular connections can be made surgically, but there are usually also substantial regenerative processes.

The history of plant grafting dates back many centuries, with evidence for a range of practices in ancient Middle Eastern, Roman, and Chinese cultures. These were exclusively horticultural applications, enabling improvement, for example, in grapevine, olive, or vegetable management and productivity. Grafting applications in woody plants often relate to modified development. Examples include dwarfing rootstocks that restrict scion growth and/or enhance flowering and cropping efficiency, universal rootstocks such as salt-tolerant grapevines (1), compatible with many different scion genotypes, and maintenance and propagation of clonal genotypes as scions where ability to form their own roots is limited (2).

1.2. Grafting Concepts

There is a wealth of literature on the principles and practices of grafting of horticultural species (e.g. 2), extending well beyond the scope of this research-focused chapter. However, a brief coverage of some of the concepts and strategies is merited.

Most commonly two plants are connected, with the shoot piece known as a *scion* and the root piece called a *rootstock*. Precise description of the source tissues and the point of connection is important because in many cases the rootstock comprises not just true root tissues and may include hypocotyl (e.g. Arabidopsis seedling grafts), cotyledons and epicotyls (e.g. pea seedling grafts), or substantial portions of leafy stem (e.g. mature tomato or cucurbit grafts). Leaf- or shoot-derived signals and other molecules may therefore emanate from rootstocks as well as scions. The net vascular flow will impact on direction and velocity of transmission of signals. The xylem stream is almost always from root to shoot, following the transpirational path. However, phloem flows from photoassimilate source to sink. In the case of a young (sink) shoot grafted to a mature leafy (source) rootstock, the direction is likely to be rootstock to scion. If instead the scion possesses mature (source) leaves and/or the rootstock is partially or wholly defoliated, then phloem flow across the graft may be in the opposite direction (3). Such manipulations have enabled substantial progress on elucidating the sources, conduits, and targets for many long-distance signals.

In some instances, both in commercial horticulture and in developmental research, more complex graft configurations are constructed. In particular, grafts with two different shoot genotypes on a single root system, sometimes called Y-grafts, allow tests of communication between the two shoots. The converse arrangement, two roots under a single shoot system (inverted Y), enables, for example, diagnosis of relative contribution of different root genotypes to providing materials to the shoot, including positive or negative regulatory signals.

1.3. Research Model Species for Grafting

Although Arabidopsis is the species of choice for much of current plant developmental research, and grafting tools are now readily applicable (*see* below), much past and continuing effort has been invested in other model and crop species. Particular focus has been on herbaceous members of Solanaceae, Cucurbitaceae, and Fabaceae. Most research has been intra-specific, but in many cases inter-specific or inter-generic grafts have been successful. Not all species can be inter-grafted, especially those that are genetically distant. This is sometimes referred to as incongruity. However, there are a few examples of inter-familial grafts (4). Some plant species are natural parasites that form connections to their (phylogenetically unrelated) hosts. These junctions closely resemble grafts, especially in the development of vascular continuity, and allow insights into translocation across the host–parasite interface (5, 6). Ordinarily, grafts attempted between such highly unrelated species would be unsuccessful due to genetic incongruity, suggesting that host–parasite interactions may involve positive recognition and/or suppression of non-self rejection.

Some taxonomic groups are extremely recalcitrant to grafting. The most significant of these are the monocots, which include all grasses and cereals, and therefore many major crop and research model species. The fundamental problem in monocots appears to be their vascular anatomy which is unsuited to regeneration of successful vascular connections. In particular, the vascular cambium has very limited regenerative potential and monocots do not undergo secondary growth. Second, it is difficult to align the vascular bundles across the graft union because they are typically scattered through the stem, often in multiple rings rather than being positioned in a single ring. A third complication is that many cereal seedlings have a high propensity for generation of adventitious roots. As with dicot grafts, adventitious roots would most likely hinder successful graft formation even where a scion and rootstock junction could be constructed.

1.3.1. Solanaceae

Solanaceous species, such as tomato, potato, tobacco, and petunia, appear particularly amenable to grafting, related to their regenerative abilities as previously demonstrated in some of the earliest successful plant tissue culture experiments. There are many reports of inter-specific grafts and some of inter-generic grafts, e.g. *Solanum* to *Nicotiana* (7). In most instances, simple shoot–root grafts are constructed using either seedlings or leafy shoots connected by side (*see* **Fig. 2.1A**), cleft, or whip grafting methods.

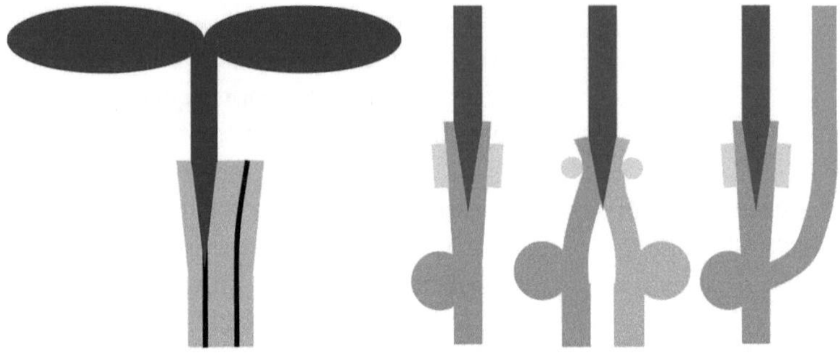

Fig. 2.1. **Diagrams of example graft combinations for species other than Arabidopsis. A**. Side graft, e.g. tomato or cucurbit, where scion and rootstock differ in diameter perhaps due to age. Note that scion is aligned with rootstock vasculature; **B**. epicotyl graft in pea; **C**. 'inverted Y' epicotyl graft in pea connecting two different rootstocks to a single scion. In many cases, only two genotypes would be combined; **D**. two-shoot epicotyl graft in pea, essentially a single graft where an axillary shoot is also allowed to grow from the rootstock. For pea grafts, *see* also (15, 37, 38).

1.3.2. Cucurbits

Several cucurbit species, especially pumpkin and cucumber, have been used widely in grafting experiments (8). Much of the emphasis from pioneering laboratories (9) has been on detection of long-distance translocation of phloem-borne molecules, especially proteins and nucleic acids. Some of these studies have radically changed our understanding of mechanisms of communication and coordination of development and stress responses in plants. There are also many commercial applications of cucurbit grafting, and many different ways of assembling grafts in these species, recently reviewed in depth (10). As with the Solanaceae, intra- and inter-specific grafts are possible. Most techniques are variations on connecting a single scion to a single rootstock, and most are done at young seedling stages. One version is illustrated in **Fig. 2.1A**. It is worth noting the possibility of high throughputs, perhaps for screening purposes, using automation either with operator assistance or with complete robotic control, enabling up to 750 grafts per hour (10, 11).

1.3.3. Legumes

Legumes (i.e. members of the Fabaceae) such as soybean and pea have had a unique place in grafting because of the complex long-distance signalling systems that enable development of rhizobial symbiotic nitrogen fixation (12–14). Legumes have also been a preferred genetic and/or physiological model for studies of many other long-distance signalling processes.

In pea and some other legume and non-legume species, germination is hypogeal, meaning that limited hypocotyl extension occurs and most seedling shoot elongation is from the epicotyl. For the purposes of grafting such species, various forms of epicotyl connections have been very successful (*see* **Fig. 2.1**).

In epicotyls grafts, the rootstock includes a significant component of shoot-derived tissue including cotyledons, cotyledonary node with axillary buds, and hypocotyl. It is also possible to construct grafts with the union in an internode further up the stem, so that the rootstock possesses true leaves. Epicotyl single grafts (*see* **Fig. 2.1B**) also readily enable development of two-shooted plants. If one axillary bud at the cotyledonary node is allowed to grow out, that will have the genotype of the rootstock (*see* **Fig. 2.1C**). The opposite configuration, a two-rootstock graft, requires slightly more ingenuity, first in growing two plants in close proximity and then in successfully connecting three tissue pieces (two rootstocks, one shoot) together (*see* **Fig. 2.1D**). More complex combinations and variations on these themes have also been generated for particular applications (15). Because pea epicotyl grafting is slightly different to hypocotyl and stem procedures for other species, a basic protocol is outlined in **Note 1**.

1.3.4. Arabidopsis

As discussed above, classic investigative tools such as grafting have long been applied to many model species, but the rosette habit and diminutive stature of Arabidopsis hampered research into long-distance transport and signalling. With the relatively recent availability of reliable methods for Arabidopsis, there is now enormous potential to exploit the wealth of molecular and genetic resources through the investigative and diagnostic leverage of grafting. Grafting has certain advantages over other means to generating spatially resolved data on long-distance communication. First, almost any pair of genotypes can be tested. Second, native genes expressed under normal regulatory mechanisms can be used, typically present in one graft partner while the other part carries a corresponding mutation, and thus signal source and site of action (e.g. shoot or root) can readily be deduced. Comparable transgenic routes to generating similar data typically employ inducible or tissue-specific promoters, but these are rarely regulated in exactly the same manner or to the same strength as the native gene.

Although this chapter focuses extensively on Arabidopsis seedling grafts (16), there are several other reported grafting methods for this species. These have employed plants at much later developmental stages with large rosettes and/or with developing inflorescences. One consequence of increased size at grafting time is probably relative ease of construction of the grafts, although published method descriptions indicate considerable care is still required to achieve reasonable success rates. The earliest report of Arabidopsis grafting (17) employed transverse cut connections, akin to **Fig. 2.2A**, on inflorescence stems. This approach was recently further refined (4), in particular, demonstrating that wedge–cleft grafting was more

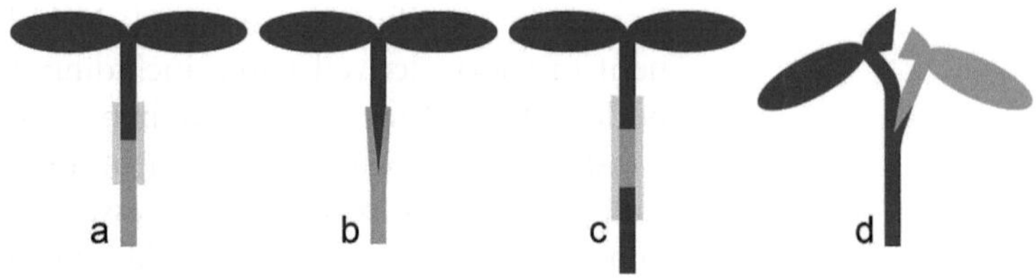

Fig. 2.2. **Diagrams of Arabidopsis seedling hypocotyls grafts. A**. 90-degree butt graft, with silicone tubing support (*light grey*); **B**. wedge-cleft graft; **C**. interstock butt graft; **D**. two-shoot wedge Y-graft. Different grey shades denote tissues of different genotypes. Roots are not shown here. Photographs of similar grafts can be found in (16).

effective. Other groups instead grafted tissues at the rosette base (18, 19). Although grafts on mature plants are amenable to many types of experimental enquiry, in cases where the signalling process or developmental event is initiated or committed relatively early in the plant's life (e.g. long-day flowering, shoot branching), it is more likely that seedling grafts will provide meaningful data.

The main protocols detailed in this chapter are for Arabidopsis seedling grafts and are adapted from (16, 20). Examples of applications of these techniques are given in **Table 2.1**.

Table 2.1
Examples of research exploiting the Arabidopsis seedling micro-grafting platform. Many other subjects are not listed here either because they have not yet reached publication stage or because results have confirmed absence of graft-transmissible action

Research topic	Example references
Flowering time	21–23
Shoot branching	16
Shoot development	24
Fe uptake/transport	25
Na transport	26
P uptake/transport	27
Systemic resistance	28
Systemic silencing	29
MicroRNA signalling	30
Auxin signalling	31
Cytokinin feedback and regulation	32, 33
ABA and hydraulic root signals	34

2. Materials

1. Arabidopsis seed. Almost any genotype with reasonably normal seedling development is suitable. Use seed with good viability, otherwise uneven plant size causes problems. Most methods were developed with Columbia and Landsberg *erecta* lines, but success has been obtained in Wassilewskija and C24 backgrounds, and with grafts between ecotypes. *See* **Note 2**.
2. Square Petri dishes, 100–150 mm × 15 mm deep.
3. Sterile filters, 47 mm diameter, cellulose-based, Millipore or similar.
4. ATS (*Arabidopsis thaliana* salts) (35) or $^1/_2$ strength MS (Murashige & Skoog) salts.
5. Phytagel, Gelrite or agar.
6. Sucrose.
7. Micropore tape, or Nescofilm or Parafilm.
8. Fine forceps, straight and curved, with ultra-fine points (e.g. Dumont styles 5 and 7).
9. Standard #11 and #15 scalpel blades.
10. Microsurgery knives, ultra-thin blade, 15° point (e.g. Fine Science Tools catalog no. 10315-12).
11. Micro scissors (e.g. FST Catalog No. 15000-04).
12. Fine silicone tubing, 0.3 mm i.d. (e.g. VWR catalog no. 228-0228, manufactured by SF Medical).
13. Suitable Arabidopsis potting mix such as Levington's F2S mixed 4:1 with vermiculite.
14. Covered growing trays – 24 or 40 cell inserts in standard 35 × 23 cm trays.

3. Methods: *Arabidopsis* Grafting on Gel Plates

3.1. Plate and Seedling Preparation

1. Pour sterile Petri plates of strong gel (Phytagel, 0.6% or Agar, 2%) + 1/2 MS salts or ATS salts, with or without 1% sucrose (*see* **Note 3**).
2. Cut sterile Millipore filters in two and place 2 × $^1/_2$ circles side by side on top of gel. Filters act as support raft to prevent grafts sinking into gel when cutting.

3. Surface sterilise seed: 70% ethanol for 30 s, rinse in sterile water, incubate in 1% hypochlorite (1/10 dilution household grade bleach) for 10 min with occasional agitation, and then five rinses with sterile water.
4. Plate seeds as suspension in water onto filters, using a cut-off micropipette tip to prevent blocking with seed. Spread with forceps or needle. Normal spacing is one row of 12 seeds near top of straight edge of each $^1/_2$ circle. Include approximately 10 extra seeds on lower part of each $^1/_2$ circle to allow for casualties or poor germination. Seal plates with porous tape or with Nescofilm/Parafilm with two small holes pierced to promote ventilation.
5. Stratify at 4°C for 2–3 days.
6. Move to growth room at 21–23°C day and 18–23°C night for 3 days. Select photoperiod according to experimental needs, typically 8–16 h photoperiod and approximately 100–120 μmol/m^2/s light. Keep plates vertical.
7. If a second growth cabinet is available, then move to 27°C day for 2–3 days before and after grafting. This reduces adventitious rooting problems.
8. Plants are best for grafting at 4–9 days old. Two-shoot grafts need to be a minimum of 5 days old. *See* **Note 4**.

3.2. Making the Grafts

There is a choice of four main methods (*see* **Fig. 2.2**), although variations can be explored. For all versions, make all cuts no lower than midway down the hypocotyls. When plate is complete, add 200 μL of water, reseal, and return to growth room. *See* **Note 5** for additional tips.

1. *Single shoot–root graft by cut-and-butt union* (*see* **Fig. 2.2A**). Both scion and rootstock are cut transversely. In essence, this is similar to a method developed for somewhat larger petunia seedlings (36), except that the graft is held in place with a silicone tubing splint (*see* **Note 6**). Slide tubing over rootstock first, then push in scion. Make sure root and shoot align precisely. Growth conditions and genetic background can affect hypocotyl diameter, creating a slack or tight fit inside the tubing.
2. *Single shoot–root graft by 'V' wedge–cleft connection* (*see* **Fig. 2.2B**). Make rootstock by cutting hypocotyl at 90°; then make a ~0.5 mm long slit down the middle of hypocotyl. Create scion by cutting very shallow-angled V shape. Make both the scion cuts from top to bottom of hypocotyl, starting from epidermis and cutting towards centre. With first scion cut, do not sever hypocotyl completely, because if root is detached, the plant moves around a lot when making the second cut. Push scion gently into slit

(cleft) in rootstock. Ideally, scions are symmetrical wedges of same length as slit in rootstock, as this will ensure maximum tissue contact.

3. *Interstock graft by double butt connection* (*see* **Fig. 2.2C**). This is almost identical in construction to method 1, except that a short interstock hypocotyl piece (0.5–1 mm long) is inserted between scion and rootstock, and longer tubing is used. There is no reason why the pieces could not be from three different genotypes.
4. *Two-shoot Y-graft by wedge scion into side of intact plant rootstock* (*see* **Fig. 2.2D**). The rootstock plant keeps all of its root and shoot, but a shallow-angled slit is made into the side of the hypocotyl, no more than half way across diameter. Note that vascular strands are central in Arabidopsis hypocotyls. Too deep a cut in the rootstock piece will completely sever the vasculature and greatly reduce grafting success. Make V-shaped scion as in method 2 and push wedge into rootstock slit. Removal of the majority of one cotyledon from each shoot is normally required to enable correct graft alignment, as shown in **Fig. 2.2D**.

3.3. Post-grafting Maintenance and Transplanting

1. From 3–4 days after grafting, inspect plates for adventitious roots forming on scion. Excise all such roots with micro-scissors or crush with fine forceps. Note that roots may emerge from within silicone tubing support, if used. Although grafts can recover, vigorous adventitious rooting normally indicates poor graft connection and probably a weak rootstock. Continue inspection every 2–3 days until after transfer to soil. Once shoot growth is re-established, root problems usually diminish. Incidence of adventitious rooting is reduced by increasing growth temperature around time of grafting (*see* **Section 3.1** and step 7).
2. Transfer plants to soil as soon as graft union is functional, usually 7–12 days after grafting (*see* **Note 7**). At this stage, axenic conditions can be dispensed with. Test strength of the union by very gently pushing against shoot and observe whether shoot and/or root growth is re-established. Keep everything wet during transfer – add extra water to plates to reduce surface tension, saturate potting mix, spray transferred plants frequently with fine mister (laundry sprayer), and spray inside of incubator lid.
3. Pick plants off plates really carefully – 'hook up' with fine curved forceps or grab edge of cotyledon. Do not crush hypocotyls or roots. With Y-grafts, be careful not to bend graft union – it will probably break! One way is to put one point of forceps under cotyledon of each shoot, then

hold *very* still while moving to planting tray. Drop roots into pre-bored hole (seeker handle is about the right diameter), gently push potting mix across to hold roots in place, but without crushing or kinking roots and especially without burying graft union.

4. Put transparent covers on tray. Keep vents closed for first 3 days or so; then open vents for another 3 days. Remove lid after a week or so. Keep growth room humidity initially high (70%+) if possible. Often there are a few casualties after lid is removed – these have weak root systems (poor grafts or adventitious root removal was too much for them) and would not be usable for experiments. Good grafts will rapidly resume normal development and growth rate and will be only marginally retarded by the few days after grafting when growth was stopped.

3.4. Data Collection and Analysis Methods

Grafting experiments invariably aim to test, directly or indirectly, for some influence acting across the graft union. Three principal types of data can be collected: phenotypic, molecular movement, and molecular target. Because the appropriate approaches are very much specific to the nature of the experiment, only an outline of possible techniques is given here. An indicative list of research topics successfully exploiting Arabidopsis seedling grafts is given in **Table 2.1**. Regardless of aims and details of design, all experiments should include self-grafted and non-grafted plants of each genotype, to provide reference phenotypes and to assess whether the grafting process itself has an influence.

For *phenotyping*, the type of data collected depends on the developmental process under study.

For *molecule movement,* especially transmitted proteins and RNA, detection and quantification can be achieved through many different techniques including imaging, mass spectrometry, pull downs, immunolocalisation, in situ hybridisation, northern blots, and RT-PCR. In all cases, unambiguous data are most easily generated where the transmitted molecule is not expressed in the graft receiver tissue. This is readily achieved through use of mutants, transgenes, and inter-specific grafts.

For *molecular targets*, it is helpful to have some prior knowledge of likely molecular consequences in the receiving tissue following successful transmission of signals or other compounds across the graft union. Many standard approaches are suitable including comparative qRT-PCR, GFP silencing, and quantification of biochemical targets such as enzyme activities.

4. Notes

1. Pea epicotyl grafting. *See* also (37, 38).
   - Sow pea seeds in compost approximately 2 cm below surface. This forces extension of the epicotyls, providing sufficient tissue length for grafting.
   - Grow under standard controlled environment or glasshouse conditions, ideally 20–25°C days, maximum 18°C nights, until shoots have emerged but before much leaf expansion occurs. This typically takes 6–8 days.
   - Remove compost to expose epicotyl and cotyledons. Wash off compost with squeeze bottle.
   - Cut off shoot of rootstock plant at top of epicotyl, i.e. just below first node with scale leaf. Cut vertical slit, ~10 mm long, down the middle of epicotyl using sharp razor blade. Place short piece of silicone tubing (~3 mm i.d., 2–3 mm long) over rootstock and slide down.
   - Cut off shoot of scion plant at base of epicotyl, i.e. just above cotyledonary node. With same blade, cut base of scion to wedge ('V') shape, same length as rootstock slit.
   - Push scion into rootstock slit. Slide up tubing to hold graft union firmly together. The assembled graft should look like **Fig. 2.1B**. Refer to **Fig. 2.1C**, D and (15) for other configurations.
   - Ensure saturating humidity by placing 2 L plastic soft drink (soda) bottles, with base cut off, over plants and push gently into compost. Alternatively, cover pot with large clear plastic bag held over pot with elastic band to avoid bag touching graft.
   - Grafts take 5–7 days to form functional unions. During this period, keep plants out of direct sun if in glasshouse. Maintain compost moisture, but do not over-water. Gradually reduce humidity by unscrewing, then removing bottle cap (or by cutting off corners of plastic bag), and finally removing bottle or bag completely.
   - During graft healing, inspect plants for formation of adventitious roots on scion and remove these. Also, except where shoot from rootstock is desired, cut or scrape out all cotyledonary axillary bud growth, as these may otherwise adversely affect scion vigour.
2. A choice of suitable marker lines is available for visual verification of graft integrity. Col-5 (*gl1-1*) and *ttg* in Ler are

glabrous and therefore useful shoot markers. Apart from effects on trichomes and flavonoids, both of these can generally be considered as WT. Constitutive reporter genes likewise allow confirmation of root and shoot origins during experiment or at end. *35S::GUS* works well but *LUC*, *GFP* equivalents, and other strong promoters are also suitable. Multiple marker lines such as *35S::GUS* Col-5 can be used. Some mutants have useful diagnostic early seedling phenotypes, such as long hypocotyls. These can still be grafted but sometimes need to offset planting times to equalise size.

3. Plants on sucrose media will be much bigger in all respects and therefore easier to handle. However, sucrose has a big influence on flowering time of many mutants and may affect other developmental process.
4. Two-shoot Y and single V (Wedge) grafts can be easier to cut and construct with slightly curved hypocotyls: rotate pairs of scion and rootstock plates 45° left and right 1 day before grafting, to induce phototropic bending.
5. Additional tips
   - Method 1 in **Section 3.2** (*see* **Fig. 2.2A**) is the most commonly adopted protocol, suitable for the majority of applications.
   - All methods require practice, patience, and a steady hand (low caffeine in bloodstream makes a difference!)
   - Ideally work under a good stereo-dissecting microscope with a uniform cold light source. All the work is at a very fine scale: hypocotyl diameters are 0.2–0.25 mm and all tissues are extremely delicate.
   - Wear gloves, and sterilise with 70% ethanol whenever hands are removed from cabinet or are in contact with non-sterile materials. New grafts are very susceptible to disease. Because of the lengthy and precise manipulations, plates are open for extended periods and hands are very close to the plant material, increasing chances of contamination.
   - Label all plates top and bottom, and sow only one genotype per plate.
   - Leave rootstocks in place on original plate, and bring scions from another plate.
   - Make sure that each seedling can be identified by position on plate, stage of cutting, etc. All genotypes look identical once moved around.
   - A standard #15 scalpel blade is sufficient for all coarse cuts – cotyledon trimming, etc; use a 15° microscalpel (Fine Science Tools) for all fine cuts – 90° cuts, slits,

V-cuts, wedges, and angled cuts. Practice is needed in precise cutting: in particular, it is important to apply a very fine sawing action with the knife, rather than pushing straight down or using large strokes. Also, microscalpels are delicate and easily bent by unwanted contact with Petri dish or other hard surfaces.

- Reduce evaporation rates by slowing laminar flow speed to minimum safe flow rate. A baffle can also be placed behind microscope to deflect airflow away from plate.
- Watch for liquid disappearing from plate surface. Top up with 100–200 μL of sterile water as often as needed. Grafts are easiest to see and cut when plates are not swimming wet, but too dry and they die. Drying also increases surface tension, which makes shifting plants around and graft alignment a bit harder. Therefore re-wet plates after doing all the cutting.
- Align shoots and root for maximum possible cut contact area. Especially where tubing supports are not used, it sometimes helps to prop up shoot slightly by putting a small piece of filter or spare cotyledon as support under the graft union. Alternatively, trim cotyledons on one side of shoot so hypocotyl lies completely flat and hence lines up with root. Note that excessive cotyledon removal will reduce vigour.
- With practice, it is possible to assemble 20–25 single grafts or 12–15 Y-grafts per hour. It is not recommended to exceed 4–5 h per day because fine coordination and concentration decline with fatigue. A final success rate of >80% is achievable for simple grafts, but Y-grafts are more challenging, with 25–30% success being typical.
- Number of grafts required per treatment or combination depends on nature of the experiment. For phenotyping, 10 successful grafts of each combination may suffice, but more may be required for subtle phenotypes or where there is inherent biological variability. A smaller number may be adequate for imaging experiments, e.g. GFP movement across a graft, and for molecular analysis, e.g. three biological replicates for a qRT-PCR experiment. Total number of grafts to be assembled for each graft combination is therefore related to success rate and replication required.

6. Use 0.3 mm inner diameter silicone tubing, ethanol sterilised and rinsed with sterile water. Cut into pieces of approximately 30 mm lengths. Although not essential because silicone tubing can stretch, it is helpful to slit it lengthways to prevent later constriction around the graft union as the

plant grows in diameter. This is most easily done with a #11 scalpel blade point pushed inside the tubing with sharp side up. Hold scalpel still and pull tubing over blade surface with fine forceps. Finally chop into approximately 1 mm length to make the graft support splints.

7. In most cases, transfer to soil is the preferred option. However, plants can be maintained on plates, for example for root sampling or imaging. In this case, it is recommended to move the successful grafts to provide a wider spacing to allow for growth. Note that phenotypes of shoots and roots on plates may both differ substantially from those in more normal environments.

Acknowledgements

I am grateful to Jon Booker and Ottoline Leyser for very substantial contributions to development and refinement of Arabidopsis grafting methods and to Christine Beveridge for expert instruction in pea grafting. Financial support from the The Royal Society and the Gatsby Charitable Foundation enabled the original development of several of the techniques described here.

References

1. Walker, R. R., Blackmore, D. H., Clingeleffer, P. R., and Iacono, F. (1997) Effect of salinity and Ramsey rootstock on ion concentrations and carbon dioxide assimilation in leaves of drip-irrigated, field-grown grapevines (*Vitis vinifera* L. cv. Sultana). *Aust J Grape Wine Res* **3**, 66–74.
2. Hartmann, T. H., Kester, E. D., Davies, T. F., and Geneve, L. R. (1997). Plant Propagation: Principles and Practices. Prentice Hall, Englewood Cliffs, NJ.
3. Tournier, B., Tabler, M., and Kalantidis, K. (2006) Phloem flow strongly influences the systemic spread of silencing in GFP *Nicotiana benthamiana* plants. *Plant J* **47**, 383–394.
4. Flaishman, M. A., Loginovsky, K., Golobowich, S., and Lev-Yadun S. (2008) *Arabidopsis thaliana* as a model system for graft union development in homografts and heterografts. *J Plant Growth Regul* **27**, 231–239.
5. Roney, J. K., Khatibi, P. A., and Westwood, J. H. (2007) Cross-species translocation of mRNA from host plants into the parasitic plant dodder. *Plant Physiol* **143**, 1037–1043.
6. David-Schwartz, R., Runo, S., Townsley, B., Machuka, J., and Sinha, N. (2008). Long-distance transport of mRNA via parenchyma cells and phloem across the host-parasite junction in *Cuscuta*. *New Phytol* **179**, 1133–1141.
7. Kaddoura, R. L. and Mantell, S. H. (1991) Synthesis and characterization of *Nicotiana-Solanum* graft chimeras. *Ann Bot* **68**, 547–556.
8. Tiedemann, R. (1989). Graft union development and symplastic phloem contact in the heterograft *Cucumis sativus* on *Cucurbita ficifolia*. *J Plant Physiol* **134,** 427–440.
9. Ruiz-Medrano, R., Xoconostle-Cazares, B., and Lucas, W. J. (1999) Phloem long-distance transport of *CmNACP* mRNA: Implications for supracellular regulation in plants. *Development* **126**, 4405–4419.
10. Davis, A. R., Perkins-Veazie, P., Sakata, Y., López-Galarza, S., Maroto, J. V., Lee, S. G, Huh, Y. C., Sun, Z., Miguel, A., King, S. R.,

Cohen, R., and Lee, J. M. (2008). Cucurbit Grafting. *Crit Rev Plant Sci* **27**, 50–74.
11. Kubota, C., McClure, M. A., Kokalis-Burelle, N., Bausher,M. G., and Rosskopf, E. N. (2008) Vegetable grafting: History, use and current technology status in North America. *HortScience* **43**, 1664–1669.
12. Delves, A. C., Mathews, A., Day, D. A., Carter, A. S., Carroll, B. J., and Gresshoff, P. M. (1985) Regulation of the soybean–rhizobium nodule symbiosis by shoot and root factors. *Plant Physiol* **82**, 588–590.
13. Searle, I. R., Men, A. E., Laniya, T. S., Buzas, D. M., Iturbe-Ormaetxe, I., Carroll, B. J., and Gresshoff, P. M. (2003) Long-distance signaling in nodulation directed by a CLAVATA1-like receptor kinase. *Science* **299**, 109–112.
14. Oka-Kira, E. and Kawaguchi, M. (2006) Long-distance signaling to control root nodule number. *Curr Opin Plant Biol* **9**, 496–502.
15. Foo, E., Turnbull, C. G. N., and Beveridge, C. A. (2001) Long-distance signaling and the control of branching in the *rms1* mutant of pea. *Plant Physiol* **126**, 203–209.
16. Turnbull, C. G. N., Booker, J. P., and Leyser, H. M. O. (2002) Micrografting techniques for testing long-distance signalling in Arabidopsis. *Plant J* **32**, 255–262.
17. Rhee, S. Y. and Somerville, C. R. (1995) Flat-surface grafting in *Arabidopsis thaliana*. *Plant Mol Biol Rep.* **13**, 118–123.
18. Ayre, B. G. and Turgeon, R. (2004) Graft transmission of a floral stimulant derived from *CONSTANS*. *Plant Physiol* **135**, 2271–2278.
19. Chen A., Komives, E. A., and Schroeder, J. I. (2006) An improved grafting technique for mature Arabidopsis plants demonstrates long-distance shoot-to-root transport of phytochelatins in Arabidopsis. *Plant Physiol* **141**, 108–120.
20. Bainbridge, K., Bennett, T., Turnbull, C., and Leyser, O. (2006) Grafting. In: Arabidopsis Protocols, 2nd edition, Methods in Molecular Biology Volume 323, pp. 39–44. Salinas, J. and Sanchez-Serrano, J.J., eds., Humana Press, Totowa, NJ, ISBN: 1-58829-395-5.
21. An, H.L., Roussot, C., Suarez-Lopez, P., Corbesier, L., Vincent, C., Pineiro, M., Hepworth, S., Mouradov, A., Justin, S., Turnbull, C., and Coupland, G. (2004) CONSTANS acts in the phloem to regulate a systemic signal that induces photoperiodic flowering of Arabidopsis. *Development* **131**, 3615–3626.
22. Corbesier, L., Vincent, C., Jang, S., Fornara, F., Fan, Q., Searle, I., Giakountis, A., Farrona, S., Gissot, L., Turnbull, C., and Coupland, G. (2007) FT protein movement contributes to long-distance signaling in floral induction of Arabidopsis. *Science* **316**, 1030–1033.
23. Notaguchi, M., Abe, M., Kimura, T., Daimon, Y., Kobayashi,T., Yamaguchi, A., Tomita, Y., Dohi, K., Mori, M., and Araki, T. (2008) Long-distance, graft-transmissible action of Arabidopsis FLOWERING LOCUS T protein to promote flowering. *Plant Cell Physiol* **49**, 1645–1658.
24. Green, L. S. and Rogers, E. E. (2004) *FRD3* controls iron localization in Arabidopsis. *Plant Physiol* **136**, 2523–2531.
24. Van Norman, J. M., Frederick, R. L., and Sieburth, L. E. (2004) *BYPASS1* negatively regulates a root-derived signal that controls plant architecture. *Curr Biol* **14**, 1739–1746.
26. Rus, A., Baxter, I., Muthukumar, B., Gustin, J., Lahner, B., Yakubova, E., and Salt, D. E. (2006) Natural variants of *AtHKT1* enhance Na^+ accumulation in two wild populations of Arabidopsis. *PLOS Genet* **2**, 1964–1973.
27. Bari, R., Pant, B. D., Stitt, M., and Scheible, W. R. (2006) *PHO2*, *microRNA399*, and *PHR1* define a phosphate-signaling pathway in plants. *Plant Physiol* **141**, 988–999.
28. Xia, Y. J., Suzuki, H., Borevitz, J., Blount, J., Guo, Z. J., Patel, K., Dixon, R. A., and Lamb, C. (2004) An extracellular aspartic protease functions in Arabidopsis disease resistance signaling. *EMBO J* **23**, 980–988.
29. Brosnan, C. A., Mitter, N., Christie, M., Smith, N. A., Waterhouse, P. M., and Carroll, B. J. (2007) Nuclear gene silencing directs reception of long-distance mRNA silencing in Arabidopsis. *Proc Natl Acad Sci USA* **104**, 14741–14746.
30. Pant, B. D., Buhtz, A., Kehr, J., and Scheible, W .R. (2008) MicroRNA399 is a long-distance signal for the regulation of plant phosphate homeostasis. *Plant J* **53**, 731–738.
31. Wilmoth, J. C., Wang, S. C., Tiwari, S. B., Joshi, A. D., Hagen, G., Guilfoyle, T. J., Alonso, J. M., Ecker, J. R., and Reed, J. W. (2005) *NPH4/ARF7* and *ARF19* promote leaf expansion and auxin-induced lateral root formation. *Plant J* **43**, 118–130.
32. Foo, E., Morris, S. E., Parmenter, K., Young, N., Wang, H., Jones, A., Rameau, C., Turnbull, C. G. N., and Beveridge, C.A. (2007) Feedback Regulation of xylem cytokinin content is conserved in pea and Arabidopsis. *Plant Physiol* **143**, 1418–1428.

33. Matsumoto-Kitano, M., Kusumoto, T., Tarkowski,P., Kinoshita-Tsujimura, K., Vaclavikova, K., Miyawaki, K., and Kakimoto, T. (2008) Cytokinins are central regulators of cambial activity. *Proc Natl Acad Sci USA* **105**, 20027–20031.
34. Christmann, A., Weiler, E. W., Steudle, E., and Grill, E. (2007) A hydraulic signal in root-to-shoot signalling of water shortage. *Plant J* **52**, 167–174.
35. Wilson, A. K., Pickett, F. B., Turner, J. C., and Estelle, M. (1990) A dominant mutation in Arabidopsis confers resistance to auxin, ethylene and abscisic acid. *Mol Gen Genet* **222**, 377–383.
36. Napoli, C. (1996) Highly branched phenotype of the petunia *dad1*-1 mutant is reversed by grafting. *Plant Physiol* **111**, 27–37.
37. Murfet, I. C. (1971) Flowering in *Pisum:* Reciprocal grafts between known genotypes. *Aust J Biol Sci* **24**, 1089–1101.
38. Beveridge, C. A., Ross, J. J., and Murfet, I. C. (1994) Branching mutant *rms-2* in *Pisum sativum* – grafting studies and endogenous indole-3-acetic-acid levels. *Plant Physiol* **104**, 953–959.

Chapter 3

Virus-Induced Gene Silencing as a Reverse Genetics Tool to Study Gene Function

Steven Bernacki, Mansour Karimi, Pierre Hilson, and Niki Robertson

Abstract

Reverse genetics has proven to be a powerful approach to elucidating gene function in plants, particularly in *Arabidopsis*. Virus-induced gene silencing (VIGS) is one such method and achieves reductions in target gene expression as the vector moves into newly formed tissues of inoculated plants. VIGS is especially useful for plants that are recalcitrant for transformation and for genes that cause embryo lethality. VIGS provides rapid, transient knockdowns as a complement to other reverse genetics tools and can be used to screen sequences for RNAi prior to stable transformation. High-throughput, forward genetic screening is also possible by cloning libraries of short gene fragments directly into a VIGS plasmid DNA vector, inoculating, and then looking for a phenotype of interest. VIGS is especially useful for studying genes in crop species, which currently have few genetic resources. VIGS facilitates a rapid comparison of knockdown phenotypes of the same gene in different breeding lines or mutant backgrounds, as the same vector is easily inoculated into different plants. In this chapter, we briefly discuss how to choose or construct a VIGS vector and then how to design and carry out effective experiments using VIGS.

Key words: Virus-induced gene silencing (VIGS), RNAi, geminivirus, *Cabbage Leaf Curl Virus* (CaLCuV), TRV, *Arabidopsis*, *Nicotiana benthamiana*, functional genomics.

1. Introduction

Virus-induced gene silencing (VIGS) derives from a mechanism of posttranscriptional gene regulation for defense against infecting viruses (1). Related pathways are also used to regulate endogenous small RNAs and maintain transcriptionally silenced regions of the genome (1–3). The mechanism of VIGS involves small interfering RNAs (siRNAs) between 21 and 24 nucleotides in length that are generated from viral sequences. Individual

L. Hennig, C. Köhler (eds.), *Plant Developmental Biology*, Methods in Molecular Biology 655,
DOI 10.1007/978-1-60761-765-5_3, © Springer Science+Business Media, LLC 2010

siRNAs, in a complex with ARGONAUTE 1, are targeted to viral RNAs that, if exact homology occurs, are then degraded (4). By engineering the VIGS vector to transcribe sequences with gene-specific homology, endogenous genes are also silenced.

VIGS as a functional genomics tool has both benefits and drawbacks. Experiments using VIGS have the complication of introducing a virus into the plant. Even though symptoms can be minimal, viral genes are expressed that interact with host metabolism, transcription, and other functions. Most of these interactions are cell-autonomous, while gene silencing is pervasive throughout infected tissue (5). However, VIGS-related phenotypes must always be studied by comparing 'empty' VIGS vector-inoculated plants with the experimental VIGS vector. Once the expression of the target gene in the controls and experimental plants has been determined, phenotypic changes can more easily be interpreted. Phenotypic changes should then be analyzed and documented by other methods, such as RT-PCR of related and/or predicted interacting genes, in situ or immunolocalization studies, chlorophyll measurements, stress response assays, developmental timing, and morphological changes in cells, tissues, and organs (6–10). In each case, the impact of the VIGS vector on the trait should be measured and considered. Another drawback is that plants do not inherit the VIGS vector or endogenous gene-silencing signals, so it is currently not possible to study genes required for gametogenesis or ovule development. Although VIGS vectors can be inoculated into plants at the cotyledon stage (in cotton) or 4–6 leaf stage (*Arabidopsis*), it will still take up to 3 weeks for silencing to occur. Nevertheless, VIGS vectors can be used to silence genes throughout both vegetative and floral meristems and can likely silence genes in the maternally derived seed coat (11).

These drawbacks are offset by a number of benefits to working with VIGS. Engineering VIGS vectors with different genes is easier than engineering RNAi cassettes for transformation because inverted repeats are not required. Analyzing VIGS phenotypes is faster than using methods that require plant transformation. This is a particular advantage for species recalcitrant to transformation, which include many crop plants. VIGS is useful in model systems when the target genes are embryo-lethal, partly because VIGS reduces the amount of target mRNA but does not eliminate it from the cytoplasm (12). By varying the time of inoculation, the gene of interest can be down-regulated after plants reach different developmental stages, and in most cases VIGS can be maintained well past maturity (11, 12). Moreover, VIGS is very useful for down-regulating entire gene families using one or more conserved regions of the gene to initiate silencing. VIGS can be used to silence two or more unrelated genes, as well (8). For these reasons, VIGS is an extremely useful functional genomics tool.

There are many different VIGS systems available for use with different host plants. Our laboratory focuses on the single-stranded DNA viruses known as geminiviruses. We have developed three different VIGS vectors from bipartite members of the *Geminiviridae* in the genus *Begomovirus.* These are *Cabbage Leaf Curl Virus* (CaLCuV) for use with *Arabidopsis thaliana*, *Tomato Golden Mosaic Virus* (TGMV) for use with *Nicotiana benthamiana*, and *Cotton Leaf Crumple Virus* (CLCrV) for use in cotton, *Gossypium hirsutum* (11–14). Recently, VIGS vectors have been developed for several different recalcitrant plant systems including barley, cassava, and soybean (5, 15–17). However, the most widely used VIGS vector is *Tobacco Rattle Virus* (TRV), due to its wide host range and minimal symptoms (18, 19).

In this chapter, we describe the methods for conducting a VIGS experiment using a specific virus, CaLCuV. Before we describe the geminivirus system in detail, it should be noted that there are some important differences between using RNA and DNA VIGS vectors. Because geminiviruses are DNA viruses, they can be inoculated either through microprojectile bombardment or *Agrobacterium* inoculation. However, RNA VIGS vectors are generally inoculated using various *Agrobacterium* inoculation methods. There are advantages and disadvantages of both methods; microprojectile bombardment causes physical damage to the plant and requires access to a gene gun, whereas *Agrobacterium* introduces a second pathogen to the inoculation system. Since phenotypes are analyzed in newly developing tissues but not in the inoculated area, these particular disadvantages are usually of little relevance. Nevertheless, they have the potential to interfere with the phenotypes obtained from silencing some types of genes.

It is also important to consider differences in the pattern of silencing produced from DNA versus RNA VIGS vectors. In our experience, VIGS from geminivirus-derived vectors is more stable than that for TRV. Assays for TRV-mediated silencing are routinely done at 21 dpi, after which silencing appears to decline, and silencing was maximal at 14–17 days for *Barley Stripe Mosaic Virus*, but no longer present at 25 dpi (6, 16, 18). In contrast, geminivirus-induced gene silencing persists throughout the life of the plant, even when cotyledons are bombarded (11). There are also similarities in RNA and DNA virus-mediated VIGS; for example, both silence gene expression in the meristem. Geminiviruses, and their beta components, have been used to silence PCNA in the meristem, and TRV has been shown to silence the meristematic gene, LEAFY (8, 18, 20). Temperature has also been shown to have a profound effect in the extent of endogenous gene silencing from both TRV and geminivirus VIGS vectors (8, 21).

Every VIGS system has variable efficiencies and temporal dynamics based on the nature of the virus–host interaction; however, there are a few key differences to keep in mind. Due to the fact that plants have complex RNAi pathways, and viruses have evolved different mechanisms for evading or inactivating these pathways, there may be differences in the extent of endogenous gene silencing from different RNA and DNA viruses. Specialized siRNA pathways differ in their role against RNA and DNA viruses; thus geminiviruses require the RDR6 and SGS3-dependent pathway for silencing while TRV does not (9, 22). Recent evidence shows that, in addition to the posttranscriptional RNA degradation used to defend RNA viruses, geminiviruses are also subject to genome methylation (23). This may prove to be an important difference in the persistence of the silencing signal through the life of the plant. Below, we will describe in detail how to study the function of a gene of interest using a geminivirus VIGS vector.

2. Materials

2.1. Germination and Plant Growth Conditions

1. Sun Gro Metro Mix® 360 (Sun Gro Horticulture, Vancouver, BC, Canada).
2. Miracle-Gro® water-soluble all-purpose plant food (Scotts Company LLC, Marysville, OH); however, note that different fertilizer and soil may be more appropriate for other target plants.
3. Kord Standard Round Pots 3″(Kord, Brampton, ON, Canada, available from Hummert International Inc).
4. Murashige-Skoog (MS) nutrient plates: 0.43% (w/v) MS salts (Caisson Laboratories Inc., North Logan, UT), 1% (w/v) sucrose, 0.05% (w/v) MES, 0.8% Phytoblend™ agar (Caisson Laboratories Inc., North Logan, UT), pH 5.7 before adding Phytoblend™. Autoclave and pour into Fisherbrand® extra-deep dishes, 100 × 25 mm (Thermo Fisher Scientific Inc., Waltham, MA), in sterile environment. Allow to cool, cover, seal, and store at 4°C.
5. 0.1% agarose: 0.1% agarose electrophoresis grade high gelling (Thermo Fisher Scientific Inc., Waltham, MA) solution in distilled water. Autoclave and store at room temperature (*see* **Note 1**).
6. 50% Bleach: 50% bleach (6% sodium hypochloride) (v/v) solution in water.
7. Micropore™ tape (3 M, St. Paul, MN).

2.2. Inoculation by Microprojectile Bombardment

1. Qiagen Plasmid Midi Kit (Qiagen, Valencia, CA).
2. 1.0-μm gold particles (Inbio Gold, Eltham, VIC Australia).
3. Mylar macrocarriers (Bio-Rad, Hercules, CA).
4. Stopping screens (Bio-Rad, Hercules, CA).
5. 1100 PSI Rupture Disks (Bio-Rad, Hercules, CA).
6. 2.5 M calcium chloride in water, stored at –20°C.
7. 0.5 M spermidine in water, stored at –20°C.

2.3. Agrobacterium-Mediated Inoculation of VIGS Vectors

1. LB Broth, Miller (Thermo Fisher Scientific Inc., Waltham, MA) 25 g/L in distilled water. Autoclave and allow to cool before adding antibiotics.
2. Inoculation medium: To sterile autoclaved LB broth (25 g/L), add filter-sterilized 10 mM MES and 20 μ M acetosyringone, plus appropriate antibiotics (*see* **Note 2**).
3. Infiltration solution: 10 mM magnesium chloride, 10 mM MES, 200 μM acetosyringone (*see* **Note 3**).
4. 1-cc syringe (BD, Franklin Lakes, NJ, available from Fisher Scientific).
5. PrecisionGlide® needles, 26 gauge (BD, Franklin Lakes, NJ, available from Fisher Scientific).

2.4. Viral DNA Detection

1. Qiagen DNeasy Plant Mini Kit (Qiagen, Valencia, CA).
2. GoTaq® PCR Core Systems (Promega, Madison, WI).

2.5. Gene Expression Analysis

1. Qiagen RNeasy Plant Mini Kit (Qiagen, Valencia, CA).
2. SuperScript® First-Strand Synthesis System for RT-PCR (Invitrogen, Carlsbad, California).
3. DyNAmo™ SYBR® Green qPCR Kit (New England Bio Labs, Ipswitch, MA).

2.6. Designing a High-Throughput Experiment

1. Promega Wizard® SV 96 Plasmid DNA Purification System (Promega; Madison, WI).
2. Blak-Ray® Lamp Long Wave UV-365 nm (UVP, Upland, CA).
3. Promega Wizard® Magnetic 96 Plant DNA System (Promega; Madison, WI).

3. Methods

Here, we describe how to conduct a VIGS experiment using CaLCuV in *Arabidopsis*; however, we will also discuss some techniques in broader terms for adaptation to different RNA and

DNA VIGS systems. For this specific VIGS experiment, we will discuss how to design a silencing fragment, *Arabidopsis* growth conditions, microprojectile bombardment and *Agrobacterium*-mediated inoculation methods, viral detection methods, and checking RNA transcript levels of the target gene. Finally, we will describe how to set up a high-throughput screen using VIGS.

3.1. Generation of a Silencing Insert for a VIGS Vector

1. Design primers to amplify a 200–400 bp region of your gene of interest (*see* **Note 4**).
2. Clone PCR product into VIGS vector. The cloning strategy for silencing inserts will vary depending on the VIGS vector and target plant species. VIGS vectors have been designed for both traditional cloning methods and Gateway® technology, which involves transferring an insert from an entry vector to the destination vector (VIGS vector). If entry vectors with appropriate inserts are available, such as the GST entry vectors for *Arabidopsis*, this is the simplest approach (*see* **Note 5**) (**24**). Otherwise, the simplest approach is usually to incorporate restriction sites from the vector multiple-cloning site into the gene-specific forward and reverse primers, with 3–6 extra nucleotides at the 5′end of each primer to allow proper enzyme cleavage. Isolate RNA from the target plant species and perform reverse transcription (*see* **Sections 3.5** and **3.6** for details on RNA isolation and reverse transcription). Then amplify the silencing insert by PCR, digest with appropriate enzymes, and proceed with standard cloning procedures. For Gateway® recombination-mediated cloning, add half of the appropriate attB sites to the forward and reverse gene-specific primers, amplify by PCR, and then reamplify with primers encoding the entire attB recombination sites. At this point proceed with normal Gateway® recombination procedures (25). For a map of the CaLCuV vector, *see* **Fig. 3.1**.

3.2. Germination and Plant Growth Conditions

Germination procedures will vary depending on the particular plant system used. The following procedure can be used for *Arabidopsis*.

1. For uniform germination of seeds, stratify *Arabidopsis* seeds at 4°C for 72 h (*see* **Note 5**).
2. For mutant lines, or to increase germination, sterilize seeds and plate on sterile media. MS nutrient plates are normally used for *Arabidopsis*.
3. Place vernalized seeds in a 1.7-mL tube, suspend in 500 μL of 50% bleach, mix vigorously, and leave in bleach solution for 12 min.
4. Collect seeds by centrifugation for a few seconds at maximum speed. Remove the supernatant and wash the seeds in 500 μL of sterile distilled water by mixing with the pipette.

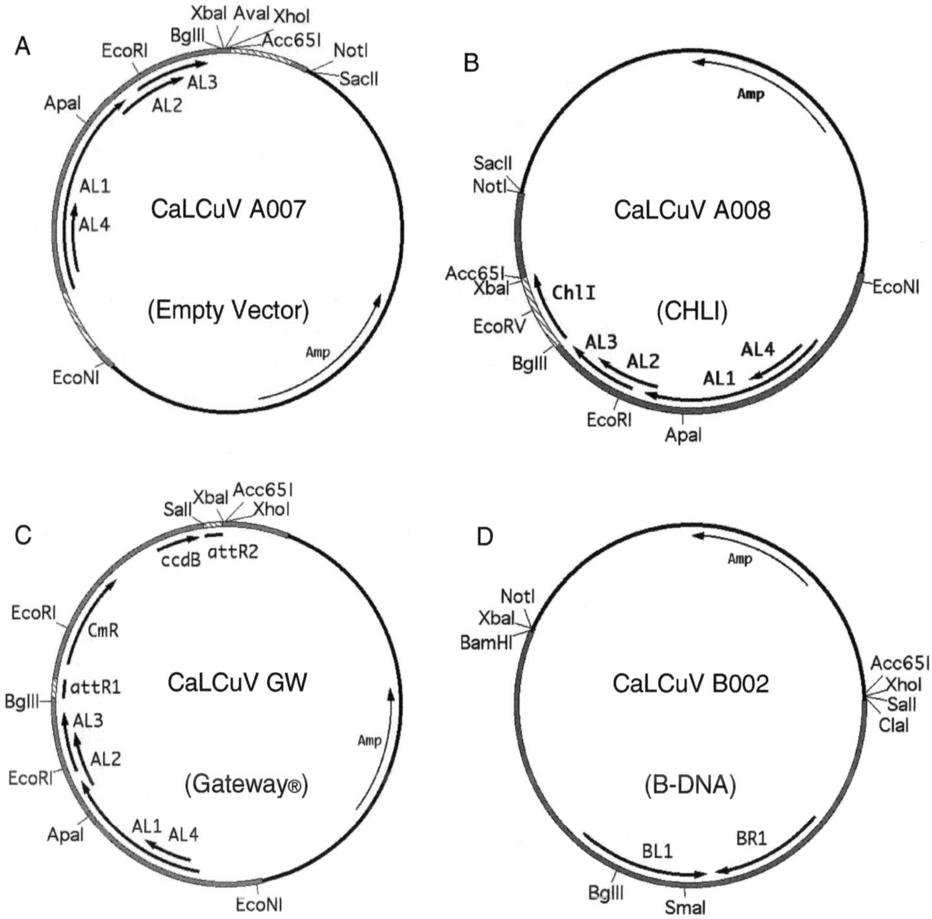

Fig. 3.1. ***Cabbage Leaf Curl Virus* (CaLCuV) VIGS cloning vectors.** During infection, the A-DNA is needed for virus replication and the B-DNA for virus movement. All of the CaLCuV vectors have an ampicillin-resistance gene. (**A**) The CaLCuV A-DNA plasmid has a multiple cloning site (MCS) at the 3′ end of the *AL3* gene for cloning up to 800 bp of silencing insert. (**B**) CaLCuV:CHLI has 380 bp of the *CHLI* gene inserted into the MCS. (**C**) The CaLCuV-GW vector has a Gateway® recombination site. (**D**) The plasmid for CaLCuV B-DNA has to be coinoculated with all A-DNA-derived plasmids for systemic infection and silencing.

5. Centrifuge, and rinse in sterile water two more times.
6. Resuspend the seeds in an appropriate amount of sterilized 0.1% agarose solution in a sterile hood. When mixed by pipette, the seeds should stay suspended in the solution, and there should be enough solution to facilitate plating the seeds one at a time.
7. Transfer seeds to Petri plates by pipette, using either a 1000-μL tip or a 200-μL tip that has been cut with a sterile razor blade. Take care to plate the seeds one at a time, approximately 5 mm apart. Be sure to do all of this in a sterile hood using aseptic technique to avoid contamination of plates (*see* **Note 6**).

8. Seal the plates with Micropore™ tape and place in a growth chamber with appropriate light and temperature conditions. Seedlings can usually be transferred to soil 14 days after plating, although some mutant lines may require additional time. It is also possible to bombard seedlings on plates as long as they are kept sterile.
9. For germination on soil, we recommend sterilizing soil before planting to avoid fungus or algae growth. This is done by putting an appropriate amount of soil in an autoclavable container, watering the soil until it has a moist, sticky consistency, covering with tin foil and then autoclaving on liquid cycle with a 45-min sterilization time. Fertilizer, such as Miracle Grow®, can be added to soil before autoclaving. Once seedlings reach the 3–5 leaf stage, transplant to individual pots.
10. VIGS is very sensitive to environmental conditions. For *Arabidopsis*, we find that 25°C is optimal for silencing. At lower temperatures (19–21°C), the extent of endogenous gene silencing is reduced, and the viral infection is more robust. We grow our plants under short-day conditions of 8 h. light/16 dark to promote vegetative growth. However, plants can be moved to long-day conditions after inoculation to promote flowering depending on the trait of interest. Once silencing is initiated using CaLCuV, it persists throughout the life of the plant.

3.3. Inoculation by Microprojectile Bombardment

For inoculation, plants should be around the 8–10 leaf stage. The older the plants, the lower the infection efficiency; however, late-stage inoculation can increase the number of plants that flower during the infection. Before inoculation, grow the *Escherichia coli* cultures with the appropriate VIGS constructs, including the experimental construct, the B-DNA plasmid, and control plasmids (*see* below). Isolate purified DNA in sufficient amounts for the number of plants to be inoculated for each construct. The bombardment protocol requires 1 μg plasmid DNA of each component, per plant inoculation. We recommend using the Qiagen® plasmid midi or maxi kit for DNA preparation but any method that produces sequencing-grade DNA is appropriate.

When designing a VIGS experiment, make sure to use all of the proper controls. We recommend including mock inoculations with the B-DNA plasmid alone to check for contaminating A-DNA plasmid, an 'empty' VIGS vector inoculation (or one containing non-homologous DNA of similar size to the insert), a vector with a visible marker for silencing (*ChlI* or *PDS*), and the VIGS vectors with experimental silencing fragments. We routinely inoculate at least five plants of each control and 10 plants with the experimental vectors. Our rate of infection is near 100%. Here, we will describe the protocol for using the Bio-Rad PDS-1000, but the preparation is easily adaptable for other instruments, such as a particle inflow gun (26).

1. Measure 60 mg of gold particles (Bio-Rad; Hercules, CA), suspend in 1 mL of 100% ethanol, and vortex on high for about 3 min.
2. To wash, centrifuge at 8,000×*g* for 1 min and remove the supernatant.
3. Add 1 mL of sterile distilled water and resuspend the gold particles by vortexing.
4. Repeat steps 2 and 3.
5. Aliquot 50 μL of gold particles in separate 1.7-mL tubes. These should be stored at −20°C. Each tube will yield 5 bombardments.
6. To each tube being prepared, add the following components and vortex, in this order:
   - VIGS Vector DNA 10 μg (add 5 μg each of the A and B components for CaLCuV)
   - 2.5 *M* calcium chloride-50 μL
   - 0.5 *M* spermidine-20 μL
7. Vortex for 3 min.
8. Centrifuge at 8,000×*g* for 10 s and remove the supernatant.
9. Add 250 μL of 100% ethanol and briefly vortex.
10. Repeat step 8 and then resuspend in 60 μL of 100% ethanol (*see* **Note** 7).
11. Pipette 10 μL of gold particles directly onto the center of a macrocarrier. Be sure to vortex the gold particles well, just before pipetting, to prevent uneven spreading of the particles. Allow the ethanol to evaporate.
12. For a Bio-Rad particle delivery system, place an 1100-PSI rupture disk into the retaining cap and screw onto the acceleration tube, tightening with the supplied wrench.
13. Unscrew the lid of the macrocarrier launch assembly and place a stopping screen in the cylinder.
14. Place the macrocarrier into a metal holder and then place the assembly so that the microprojectiles face down, towards the stopping screen.
15. Screw the cover back over the cylinder of the macrocarrier launch assembly.
16. Slide the assembly into the slot closest to the top of the chamber.

17. Position the platform in the chamber such that the shooting distance between the stopping screen and plant tissue is about 5–10 cm.
18. Close the chamber door.
19. Switch the Bio-Rad unit to the ON position, open the line to the compressed helium tank, and switch on the vacuum pump.
20. Press the VAC button to pull air from the chamber and, when the pressure gauge reaches 600 mm Hg, toggle it to hold. Press and hold the FIRE button to release helium until the rupture disk bursts. Then switch the VAC button to vent.
21. Remove the plant and repeat steps 12–20 for each inoculation.
22. Cover plants for 2–3 days with a clear plastic top to keep humidity high. Silencing should begin 2–3 weeks after inoculation (*see* **Fig. 3.2**).

Fig. 3.2. **CaLCuV-mediated silencing of subunit *I* of magnesium chelatase *(CHLI)* following microprojectile bombardment.** *Arabidopsis* Col-0 plants demonstrating silencing at 20 days postinoculation (DPI). (**A**) Mock inoculated, (**B**) CaLCuV vector, (**C**) CaLCuV:CHLI.

3.4. Agrobacterium-Mediated Inoculation

There are many different methods for *Agrobacterium*-mediated virus inoculation into plants, depending on the anatomy of the plant. Here, we will describe the steps for preparing an Agrobacterium–VIGS inoculum appropriate for any of these methods, and then describe some of the different techniques for delivery.

1. Initiate cultures of the appropriate *Agrobacterium* strains by growing them overnight by shaking at 280 rpm at 28°C in 3 mL of LB liquid medium containing the appropriate antibiotics.
2. On day 2, inoculate the 3-mL culture into 50 mL of inoculation medium. Shake overnight at 280 rpm at 28°C until an OD_{600} of 1.5 is reached (*see* **Note 8**).
3. Centrifuge the cultures for 10 min at 15,000×*g* and remove the supernatant.
4. Resuspend bacterial pellet in an equal volume of infiltration solution and mix equal amounts of bacteria containing

the A- and B-DNA plasmids together. Both viral genome components must be inoculated together for a productive infection.

5. Leave the bacteria at room temperature in the dark for 3–4 h.

6a. Leaf blade infiltration. Using a 1-cc needless syringe, fill the syringe with bacterial inoculum. Press the syringe tip firmly against the underside of the leaf with one finger placed firmly on the opposite side for maintaining pressure. Pressing too hard will tear the leaf while not enough pressure will decrease solution entry into the intercellular leaf spaces. Slowly push the plunger down a small amount; do not inoculate it all in one spot. When done correctly, the leaf blade around the syringe should turn dark green with fluid. Repeat this 4 or 5 times per plant to insure infection (*see* **Note 9**).

6b. Vascular tissue uptake. Cut the main stem of the plant near the base and immediately place a large drop of bacterial solution on the cut surface of the stem using a pipette. The bacterial solution should cover the entire surface of the cut stem. For *Arabidopsis*, this is done with the primary inflorescence stem cut near the rosette level. Secondary inflorescence stems will contain the virus and be silenced.

6c. Direct injection. Another method is injection into the stem using a syringe with a 26-gauge needle. A larger needle can be used, but will cause more damage to the plant. For *Arabidopsis*, fill the syringe and then puncture the plant with the needle about 10 times directly around the meristem of the plant. Depress the syringe plunger to place the Agrobacteria on top of the puncture wounds, allowing them to soak in. In other species, the inoculum can be directly injected into the stem or petiole of the leaf.

6d. Vacuum infiltration. Finally, one can also use vacuum infiltration of whole plants or cuttings (*see* **Note 10**). Submerge the cutting or upper portion of the plant in inoculum, place in a bell jar, and apply a vacuum using a vacuum pump. Pressures and times will vary greatly depending on the plant and tissue type being used for infiltration. Typical parameters are 600 mm Hg vacuum pressure for 1–3 min.

7. Cover plants to maintain high humidity for 2–3 days. Silencing should begin 2–3 weeks after inoculation (*see* **Fig. 3.3**).

3.5. Viral Detection

To analyze silencing, quantification of target mRNA is usually the only molecular assay that is required. However, detection of viral sequences is useful to ensure that the silencing vector is present

Fig. 3.3. ***Arabidopsis* Col-0 plants demonstrating phytoene desaturase (*PDS*) silencing from a TRV VIGS vector at 20 DPI.** Plants were injected with *Agrobacterium* using a syringe. (**A**) Mock inoculated, (**B**) TRV vector, (**C**) TRV:PDS.

in the tissues to be used for analysis. This is especially important if no visual marker is used. Viral detection can be done in a number of different ways. For DNA viruses, the options for detection include Southern blots of isolated DNA, squash blots, PCR-based detection, and quantitative-PCR (qPCR)-based detection. Here we will briefly describe the PCR and qPCR detection methods. For DNA-based blots, the protocols are the same as for genomic DNA, although not as much total DNA is necessary to see the viral DNA signal. Keep in mind that while a Southern blot can be quantitative, a squash blot can only be used to determine presence or absence.

For PCR-based applications, a suitable DNA extraction method for the plant system of interest should be used; many are available and appropriate for this application, including the Qiagen DNeasy® Plant Mini Kit. A standard PCR or qPCR protocol can be followed, such as the Promega GoTaq® PCR Core System, in order to check for the virus; however, we recommend using our method of primer design in order to avoid amplifying input DNA when geminivirus vectors are used (*see* **Fig. 3.4**). Once the geminivirus vectors replicate *in planta*, a unit length genome is released either by recombination or by rolling circle replication (27). Because this excludes the cloning vector, primers can be designed that appear to be facing in opposite directions for amplification as shown in **Fig. 3.4A**. Using this design, contaminating DNA would appear as a band greater than 2.8 kb, while *in planta*-replicated DNA would show up as a smaller fragment. This method, along with no-template controls, allows discrimination between input DNA and silencing DNA as well as identification of contaminating DNA. For *Cabbage Leaf Curl Virus* we use primer AL3F 5′ TCGCAACGGACAGATCCTAT 3′ and primer AL1R 5′ GACTGACCACGACAGGGTTT 3′. These primers also span the silencing insert and can therefore differentiate between vectors with inserts of different sizes, as shown in **Fig. 3.4B**.

3.6. Gene Expression Analysis

Whenever a VIGS experiment is performed, the extent of downregulation of the target gene should be determined. This analysis can be done by semiquantitative RT-PCR or qRT-PCR, depending on what is available (*see* **Note 11**). You should choose

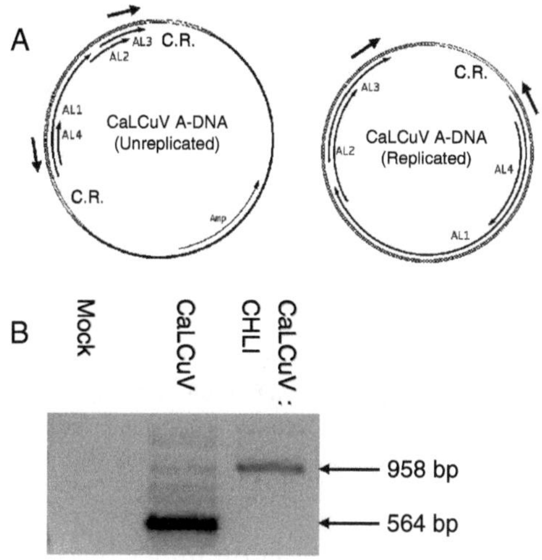

Fig. 3.4. **Primer design for detection of CaLCuV vector DNA.** (**A**) Primers for viral detection (annotated by *arrows*) are designed to amplify a region spanning the cloning vector that is larger following replication in *E. coli* than it is in plants. Once the vector has replicated out of the plasmid, *in planta*, a much smaller region containing one common region will be amplified. (**B**) Electrophoresis gel of PCR using AL3F and AL1R primers. Because the primers also amplify the silencing insert, different vectors with distinguishable insert sizes can be differentiated on an electrophoresis gel.

an appropriate RNA extraction protocol for the plant species of choice. For *Arabidopsis,* we often use the Qiagen RNeasy® Plant Mini Kit; however, other standard methods will also work. Reverse transcription can be carried out according to any standard method, such as the SuperScript® First-Strand Synthesis System for RT-PCR. Primers for the gene of interest should be designed, preferably, to span an intron in the gene to allow detection of genomic DNA contamination. The other important point is to avoid the region of the gene used for the silencing insert. For qPCR, amplicons should be between 100 and 200 bp for optimal quantification while for semiquantitative RT-PCR, amplicons should be large enough to run as a distinct band in agarose (>500 bp).

For semiquantitative PCR, a standard 50 μL PCR reaction should be conducted using 1 μL of cDNA; however, 10 μL will be removed every 5 cycles starting at cycle 10. Alternatively, several PCR reactions can be performed with one tube being removed every 5 cycles. These reactions are run to find cycle numbers where amplification is in the linear phase. Run all the samples on an agarose gel and stain with ethidium bromide to find the cycles with unsaturated PCR product. There should be clear differences of intensity in bands from different cycles (*see* **Fig. 3.5**). If most of the bands are saturated, adjust the amount of input cDNA accordingly. Once the right combination of cDNA concentration and number of ampification cycles is found, run a

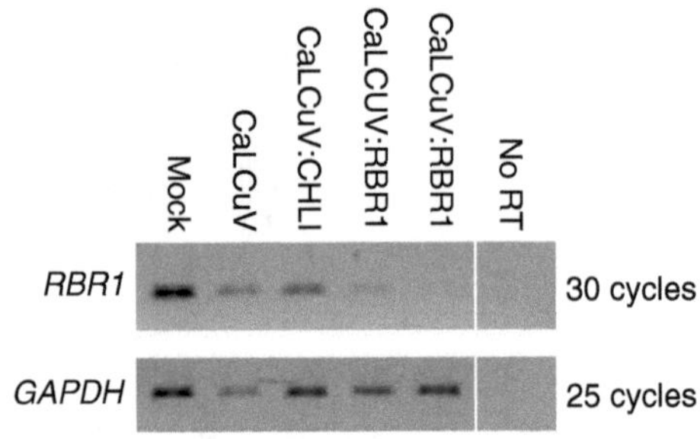

Fig. 3.5. **Silencing of the retinoblastoma-related 1 (*RBR1*) gene shown by semi-quantitative RT-PCR.** Transcript levels of *RBR1* in mock, CaLCuV vector, CaLCuV:CHLI-, and CaLCuV:RBR1-inoculated plants. The RT reaction was conducted with 100 ng RNA. Although *RBR1* is down-regulated, a second assay for vector DNA levels would be informative because geminiviruses interact with *RBR1*.

PCR reaction under these conditions and photograph the products on an agarose gel.

If possible, we recommend using qRT-PCR rather than semi-quantitative for increased accuracy and reliability. Standard qPCR kits will be fine for this application; we use the DyNAmo™ SYBR® Green qPCR Kit. After cDNA synthesis, follow any standard qPCR protocol but include a 'no-RT' enzyme control to check for contamination. For analysis, we use the comparative Ct method, also referred to as the ΔCt method (28). This requires the use of a reference gene that should show constitutive expression. For *Arabidopsis* we use *ACTIN8* or *GAPDH* as reference genes, but other genes could also be used (29, 30). For each sample, run three technical replicate reactions for the gene of interest, and three technical replicate reactions of the reference gene.

1. In each sample, average (avg) the cycle time (Ct) values, calculate the standard deviation (stdev), and subsequently the coefficient of variation (CV = stdev/avg). Any value with higher than 4% CV should be considered an outlier. However, if there is a high reoccurrence of outliers, it will be necessary to optimize methods for getting more consistent, precise results.
2. To find the ΔCt, normalize the Ct of the gene of interest (GOI) to the Ct of the reference (REF): $\Delta Ct = \text{avg } Ct_{GOI} - \text{avg } Ct_{REF}$
3. Calculate $stdev_{\Delta Ct} = ((stdev_{REF})^2 + (stdev_{GOI})^2)^{1/2}$
4. Decide what sample is the calibrator; this is the sample you will arbitrarily set at a value of 1. $\Delta\Delta Ct = \Delta Ct_{SAMPLE} - \Delta Ct_{CALIBRATOR}$
5. $stdev_{\Delta\Delta Ct} = ((stdev_{\Delta Ct})^2 + (stdev_{CALIBRATOR})^2)^{1/2}$
6. Fold-Induction $= 2^{(-\Delta\Delta Ct)}$
7. Experimental error for fold-induction is $S.D._{FOLD\text{-}INDUCTION} = (\ln 2)(stdev_{\Delta\Delta Ct})(2^{(-\Delta\Delta Ct)})$

8. Plot fold-induction on a graph using S.D.$_{FOLD\text{-}INDUCTION}$ to set error bars. Keep in mind that a logarithmic *y*-axis is often a better representation of fold-induction values. **Fig. 3.6** shows an example of this method.

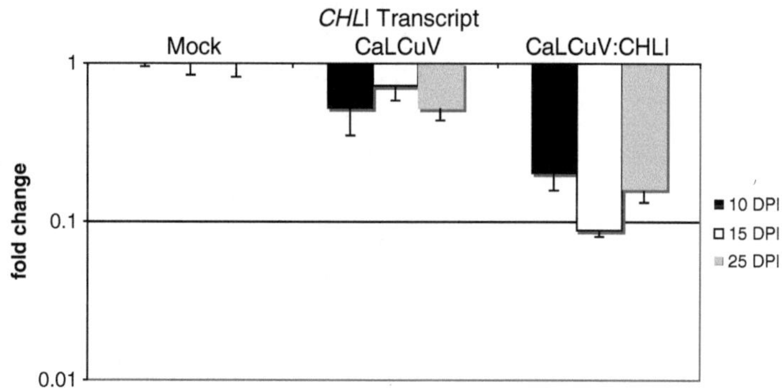

Fig. 3.6. **Down-regulation of *CHLI* using CaLCuV.** Transcript levels of *CHLI* in mock inoculated, CaLCuV vector, and CaLCuV:CHLI-inoculated plants at 10, 15, and 25 DPI.

3.7. Other Phenotypic Analyses

The experiments to assess gene function depend on the nature of the gene investigated and the prior hypotheses, but always require appropriate controls. We recommend at minimum using mock-inoculated and empty VIGS vector-inoculated plants as controls for determining function in any kind of phenotypic analysis.

3.8. Designing a High-Throughput Experiment

One of the very useful aspects of VIGS is the ability to do high-throughput experiments. In particular, this has been facilitated by the development of recombinant cloning technologies such as Gateway®. Using these technologies, it becomes feasible to build libraries of VIGS vectors with inserts for genes across an entire genome and set up genetic screens to find genes of interest. We describe the use of a silencing comarker in the following experiment, which depends on using a GFP-transgenic target plant. Examination of areas that show transgene silencing can provide more reliable information about the gene of interest.

1. An appropriate Gateway® entry vector library will be helpful for creating a VIGS library. Entry vectors should ideally have gene fragments between 200 and 600 bp in length, although Geminivirus-based VIGS vectors can use between 100 and 800 bp (*see* **Note 12**). For this experiment, it would be beneficial to put a 100-bp *GFP* fragment into each vector in addition to the endogenous gene fragment. We recommend doing this by cloning the *GFP* fragment into the CaLCuVA-GW vector by traditional cloning methods to position it just outside the attR sites (*see* **Fig. 3.1**). If fragments from a PCR-based subtractive SSH library are used, a non-Gateway VIGS vector would be more appropriate. Clone the *GFP* fragment into the VIGS vector

(pCaLCuVA.007) before creating the libraries so that it is cotranscribed with the GOI insert (*see* **Fig. 3.1**).

2. Prepare vector DNA from all plasmids in the library to be screened, for automated high-throughput plasmid preparation, preferably by a method developed for high-throughput such as the Promega Wizard® SV 96 Plasmid DNA Purification System.
3. Depending on the number of experimental vectors, mix up to 10 different vectors in equal quantity for bombardment.
4. Germinate GFP-expressing seedlings on MS agar plates as previously described, but increase the number of seedlings by a factor of 10.
5. Bombard the pools of mixed VIGS vectors onto the plate. Move the plate after each shot to cover the whole area. Fire up to 5 shots per plate.
6. Using a handheld Blak-Ray® UV lamp, monitor the GFP expression in all plants. Silencing of GFP will occur faster than endogenous genes, usually within 7–10 days. Once there is silencing in a large percentage of plants, remove the ones that are still expressing GFP, as these are also not actively silencing the GOI.
7. Depending on the pathways of interest, apply a stress or alteration to the plants in order to screen for plants that have either increased or decreased tolerance to this particular condition. For plants that have decreased tolerance, make sure to rescue them or get tissue before they die.
8. Perform DNA extractions on plants of interest, preferably with a high-throughput system such as the Promega Wizard® Magnetic 96 Plant DNA System and amplify using the previously mentioned primers AL3F and AL1R (*see* **Note 13**). Keep in mind that there could be a mixed infection in the plant of interest. As long as the inserts are not exactly of the same size, this will be seen after amplification. In the case of multiple bands, isolate all bands separately by gel extraction.
9. Sequence the amplicons in order to identify the genes of interest and repeat, using individual clones instead of pools.
10. Repeat this screen until there are no more new genes of interest emerging from the screen. These kinds of experiments can be used to help identify novel genes in a particular pathway or response that could be used for further study. Independent methods for verification of gene function are likely to be necessary for each of the genes. Although labor-intensive, using VIGS libraries for screening genes in crop plants (especially in elite lines) may be extremely useful.

4. Notes

1. A 0.1% agarose solution will easily become contaminated with bacteria or fungi. For this reason we recommend that after autoclaving only open the bottle while in the hood, and always check the bottle for contamination before using. Also, once seeds are suspended in agarose, they can be kept at 4°C for 2–3 weeks and then plated.
2. Acetosyringone should always be dissolved in DMSO and kept at –20°C.
3. Do not autoclave the acetosyringone. This should be added last.
4. Different vectors may have different capabilities for accepting insert DNA. In the case of the CaLCuV vector, the removal of the coat protein allows for up to 800 bp to be inserted into the vector. Maximum silencing efficiency in the geminivirus vectors is achieved with fragments optimally between 400 and 800 bp in length. The tandem insertion of at least two separate silencing fragments into the vector can be used for silencing more than one gene at a time. Only 200 bp of endogenous or 100 of transgene sequence homology is needed if the total length of the mRNA is at least 400 bp.
5. A Complete *Arabidopsis* Transcriptome MicroArray (CATMA) project created gene specific tags (GSTs) to most of the genes in the *Arabidopsis* genome. These vectors have been used in various applications including an RNAi project, AGRIKOLA. More information on the CATMA project can be found at http://www.catma.org.
5. For wild-type *Arabidopsis*, the seeds can be germinated directly on soil, however for mutant lines it is often better to germinate seeds on MS medium plates before transferring to soil. It is also possible to keep plants on medium for the duration of the experiment as long as measures are taken to keep the plates sterile.
6. If contamination of plates is a persistent problem, the sucrose concentration can be reduced to 0.1%.
7. There are a number of particle inflow systems that have been developed. Many of these systems work by pushing a pressurized gas through a filter containing gold particles resuspended in ethanol. This would be the appropriate stopping point for such a procedure. Most of these protocols involve dispensing 10 μL of DNA-coated gold particles onto a 13 mm Swinnex® filter (Millipore, Billerica, MA), closing the filter and attaching it to a high-pressure helium nozzle.

8. This is the OD_{600} that we use for TRV in *Arabidopsis*, but different OD_{600} values may be needed for different VIGS systems. Species that are more difficult to transform may need a higher OD_{600}.
9. Syringe inoculation works well with some species and not with others. In particular it works very well for tobacco species, but not as well for *Arabidopsis*.
10. Note that if preparing solution for vacuum infiltration, add 0.05% Silwet® to the solution to increase efficiency.
11. It should be noted that semiquantitative RT-PCR techniques for looking at gene expression changes are no longer accepted by many journals, although they can be acceptable if they are only being used to check down-regulation by RNAi. It is worth checking journal requirements in the instructions to authors for information about their policies.
12. Any protocol for making a cDNA library can be used, but the cloning vector in this case would be pCaLCuVA.007 for traditional cloning, or pCaLCuVA-GW for Gateway® recombination cloning. In particular, libraries made by subtractive, suppressive PCR are especially well suited for silencing because most of the fragments are between 100 and 800 bp in length. Make sure to transform using library-efficiency grade competent cells.
13. There are other 96-well format DNA extraction methods for *Arabidopsis* that would be efficient for such an approach as well (31).

Acknowledgments

The authors would like to thank Savithramma Dinesh-Kumar for providing the Tobacco Rattle Virus VIGS vectors.

References

1. Baulcombe, D. (2004) RNA silencing in plants. *Nature* **431**, 356–363.
2. Vazquez, F., Vaucheret, H., Rajagopalan, R., et al. (2004) Endogenous trans-acting siRNAs regulate the accumulation of Arabidopsis mRNAs. *Mol Cell* **16**, 69–79.
3. Chapman, E. J. and Carrington, J. C. (2007) Specialization and evolution of endogenous small RNA pathways. *Nat Rev Genet* **8**, 884–896.
4. Vaucheret, H. (2006) Post-transcriptional small RNA pathways in plants: mechanisms and regulations. *Genes Dev* **20**, 759–771.
5. Robertson, D. (2004) VIGS vectors for gene silencing: many targets, many tools. *Annu Rev Plant Biol* **55**, 495–519.
6. Park, J. A., Ahn, J. W., Kim Y. K., et al. (2005) Retinoblastoma protein regulates cell proliferation, differentiation, and endoreduplication in plants. *Plant J* **42**, 153–163.

7. Jin, H., Axtell, M. J., Dahlbeck, D., et al. (2002) NPK1, an MEKK1-like mitogen-activated protein kinase kinase kinase, regulates innate immunity and development in plants. *Dev Cell* **3**, 291–297.
8. Peele, C., Jordan, C. V., Muangsan, N., et al. (2001) Silencing of a meristematic gene using geminivirus-derived vectors. *Plant J* **27**, 357–366.
9. Muangsan, N., Beclin, C., Vaucheret, H., and Robertson, D. (2004) Geminivirus VIGS of endogenous genes requires SGS2/SDE1 and SGS3 and defines a new branch in the genetic pathway for silencing in plants. *Plant J* **38**, 1004–1014.
10. Senthil-Kumar, M., Rame Gowda, H. V., Hema, R., Mysore, K. S., Udayakumar, M. (2008) Virus-induced gene silencing and its application in characterizing genes involved in water-deficit-stress tolerance. *J Plant Physiol* **165**, 1404–1421.
11. Tuttle, J. R., Idris, A. M., Brown, J. K., Haigler, C. H., and Robertson, D. (2008) Geminivirus-mediated gene silencing from Cotton leaf crumple virus is enhanced by low temperature in cotton. *Plant Physiol* **148**, 41–50.
12. Jordan, C. V., Shen, W., Hanley-Bowdoin, L. K., and Robertson, D. N. (2007) Geminivirus-induced gene silencing of the tobacco retinoblastoma-related gene results in cell death and altered development. *Plant Mol Biol* **65**, 163–175.
13. Turnage, M. A., Muangsan, N., Peele, C. G., and Robertson, D. (2002) Geminivirus-based vectors for gene silencing in Arabidopsis. *Plant J* **30**, 107–114.
14. Kjemtrup, S., Sampson, K. S., Peele, C. G., et al. (1998) Gene silencing from plant DNA carried by a Geminivirus. *Plant J* **14**, 91–100.
15. Carrillo-Tripp, J., Shimada-Beltran, H., and Rivera-Bustamante, R. (2006) Use of geminiviral vectors for functional genomics. *Curr Opin Plant Biol* **9**, 209–215.
16. Holzberg, S., Brosio, P., Gross, C., and Pogue, G. P. (2002) Barley stripe mosaic virus-induced gene silencing in a monocot plant. *Plant J* **30**, 315–327.
17. Zhang, C. and Ghabrial, S. A. (2006) Development of Bean pod mottle virus-based vectors for stable protein expression and sequence-specific virus-induced gene silencing in soybean. *Virology* **344**, 401–411.
18. Ratcliff, F., Martin-Hernandez, A. M., and Baulcombe, D. C. (2001) Technical Advance. Tobacco rattle virus as a vector for analysis of gene function by silencing. *Plant J* **25**, 237–245.
19. Liu, Y., Schiff, M., Dinesh-Kumar, S. P. (2002) Virus-induced gene silencing in tomato. Plant J **31**, 777–786.
20. Cai, X., Wang, C., Xu, Y., Xu, Q., Zheng, Z., and Zhou, X. (2007) Efficient gene silencing induction in tomato by a viral satellite DNA vector. *Virus Res.* **125**, 169–175.
21. Fu, D. Q., Zhu, B. Z., Zhu, H. L., et al. (2006) Enhancement of virus-induced gene silencing in tomato by low temperature and low humidity. *Mol Cells* **21**, 153–160.
22. Akbergenov, R., Si-Ammour, A., Blevins, T., et al. (2006) Molecular characterization of geminivirus-derived small RNAs in different plant species. *Nucleic Acids Res* **34**, 462–471.
23. Raja, P., Sanville, B. C., Buchmann, R. C., and Bisaro, D. M. (2008) Viral genome methylation as an epigenetic defense against geminiviruses. *J Virol* **82**, 8997–9007.
24. Hilson, P., Allemeersch, J., Altmann, T., et al. (2004) Versatile gene-specific sequence tags for Arabidopsis functional genomics: transcript profiling and reverse genetics applications. *Genome Res* **14**, 2176–2189.
25. Park, J. and Labaer, J. (2006) Recombinational cloning. *Curr Protoc Mol Biol* Chapter 3:Unit 3.20.
26. Finer. J., Vain, P., Jones, M., and McMullen, M. (1992) Development of the particle inflow gun for DNA delivery to plant cells. *Plant Cell Rep* **11**, 323–328.
27. Rojas, M. R., Hagen, C., Lucas, W. J., and Gilbertson, R. L. (2005) Exploiting chinks in the plant's armor: evolution and emergence of geminiviruses. *Annu Rev Phytopathol* **43**, 361–394.
28. Livak, K. J. and Schmittgen, T. D. (2001) Analysis of relative gene expression data using real-time quantitative PCR and the 2(-Delta Delta C(T)) Method. *Methods* **25**, 402–408.
29. Gutierrez, L., Mauriat, M., Pelloux, J., Bellini, C., and Van Wuytswinkel, O. (2008) Towards a Systematic Validation of References in Real-Time RT-PCR. *Plant Cell* **20**, 1734–1735.
30. Udvardi, M. K., Czechowski, T., and Scheible, W. (2008) Eleven Golden Rules of Quantitative RT-PCR. *Plant Cell* **20**, 1736–1737.
31. Xin, Z., Velten, J. P., Oliver, M. J., and Burke, J. J. (2003) High-throughput DNA extraction method suitable for PCR. *BioTechniques* **34**, 820–826.

Chapter 4

The CRE/*lox* System as a Tool for Developmental Studies at the Cell and Tissue Level

Guy Wachsman and Renze Heidstra

Abstract

Targeted gene manipulation has been used in the last few decades for better understanding of gene function. Most often mutant or overexpression genotypes are analyzed, but in many cases these are not sufficient to obtain a detailed picture on the mode of action of the corresponding protein. For example, many mutations result in pleiotropic or early phenotypic effects thereby affecting the whole organism. Conditional complementation or deletion of the gene under study in a specific cell or tissue can elucidate its exact role in a specific region within a certain time frame. Implementation of several site-specific recombination systems such as CRE/*lox* has created powerful tools to study the role of many genes at the cellular level. In this chapter, we describe in detail protocols for the application of a two-vector based CRE/*lox* system, enabling controlled timing and position of gain or loss of function clonal analyses.

Key words: Clones, recombination, CRE/*lox*, green fluorescence protein (GFP).

1. Introduction

Transgenic technology has provided a means to alter a genome and transcriptional output in a fundamental manner and is widely used for studies on developmental processes in plants and animals. However, understanding the role of many genes requires the ability to generate and visualize a knock-out event at the tissue or cell level and not only at the whole organism level. If a mutant causes alterations early in development, it is likely that secondary changes accumulate as development proceeds. These secondary effects frequently mask the primary defects making the interpretation of phenotypes not straight forward. By generating a phenotypically

L. Hennig, C. Köhler (eds.), *Plant Developmental Biology*, Methods in Molecular Biology 655,
DOI 10.1007/978-1-60761-765-5_4, © Springer Science+Business Media, LLC 2010

wild-type organism with an inducible specific loss or gain of function cell or tissue, or a mutant with gain of function wild-type regions, it is possible to pinpoint the exact role of the gene-of-interest in these specific locations. Such directed approaches utilizing temporal and/or tissue-specific over-expressing or deletion of a gene-of-interest are frequently based on the CRE/*lox* or an equivalent site-specific recombination system (1, 2).

Another application of site-specific recombination systems is to test cell autonomous vs. non-cell autonomous function of a protein, i.e., does a protein act solely in the cells where it is expressed or does it have an effect on neighboring cells, directly by movement or indirectly via signal transduction (3, 4). Furthermore, site-specific recombination systems can be exploited in gene therapy (5), cell lineage studies (6, 7), and creation of marker-free transgenic organisms (8).

There are two separately evolved families of DNA recombinases named after the amino acid residue that covalently binds the DNA (9). The Serine recombinase family is mainly present in prokaryotes (e.g., Hin in *Salmonella*) and bacteriophages while the Tyrosine family can be found in eukaryotes such as yeast and fungi as well (e.g., λ integrase, yeast Flipase (FLP), and P1 CRE recombinase). The most prominent difference between the two families is the simultaneous, i.e., double-strand break and ligation mechanism vs. sequential strands cleavage and reunion, respectively (9). The λ integrase executes the integration and excision of the phage genome to and from the *Escherichia coli* host chromosome (10) while FLP has a role in the amplification of the yeast 2 μ plasmid (11). The CRE recombinase has at least two known functions in the P1 phage life cycle (12). Initially it catalyzes the cyclization of the linear phage genome after viral infection; later, in the lysogenic phase during cell division, it enables the physical separation of P1 plasmids keeping a high frequency of infected bacteria daughter cells. The three enzymes mentioned above in addition to many other site-specific recombination systems, such as the *Ac/Ds* transposon from maize (13), are commonly used as tools for DNA manipulation.

The CRE/*lox* system is made up of two main elements, both from the bacteriophage P1: the CRE recombinase that carries out a 'cut and paste' site-specific recombination reaction of a DNA sequence placed between its DNA recognition sequences, and the *lox* sites (14). The wild-type *lox* site, *loxP* (15), consists of two inverted repeats of 13 base pairs that are separated by an eight-base pairs spacer (16 and **Fig. 4.1**). Each CRE molecule binds one repeat (17); hence, a full recombination event, which involves the exchange of two double-strand DNA sequences, is mediated by four CRE units. In case of a linear DNA substrate, the orientation of these two 34-nucleotide *lox* sequences in respect to each other determines the DNA product(s) structure. A deletion

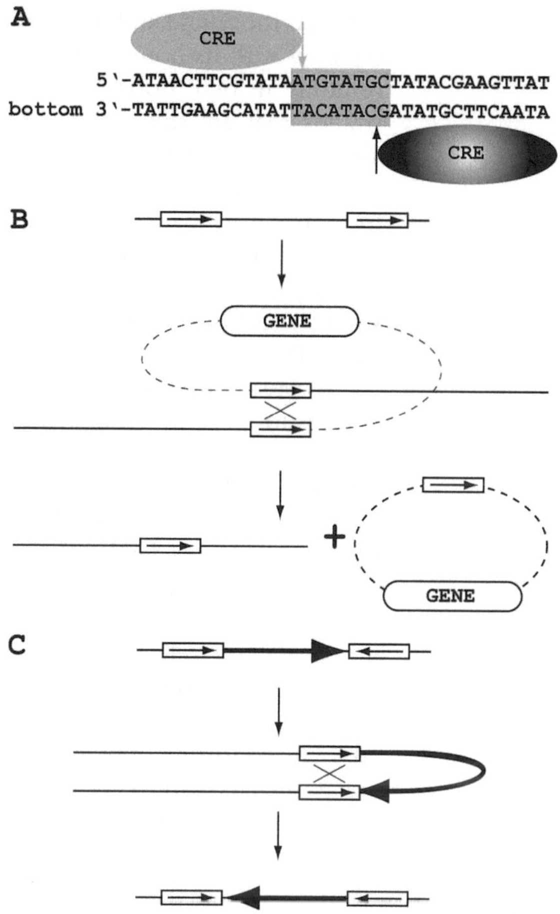

Fig. 4.1. **The CRE/*lox* recombination system.** (**A**) CRE/*lox* interaction prior to recombination. The depicted *loxP* site consists of a two 13 base-pairs repeat separated by an 8 base-pair spacer (*gray box*). The active CRE subunit (*gradient oval*) binds the "bottom" strand, subsequently cleaving it at the GC phosphodiester bond (*black arrow*) located in the *lox* spacer region. Postcleavage, ligation of "bottom" strands and isomerization of CRE units, the same process is executed on the upper strand (*see* text for details). (**B** and **C**) CRE-mediated recombination of a DNA sequence flanked by tandem *lox* sites results in deletion and cyclization (**B**, *dashed line*) or reversion of the sequence in case of inverted *lox* sites (**C**, *thick arrow*). Boxed horizontal arrow indicates *lox* site, X marks recombination.

and cyclization of the sequence flanked by the two *lox* sites occurs when they are laid as two tandem repeats and an inversion when they are positioned in a head to head orientation (*see* **Fig. 4.1**).

The sequential CRE/*lox* recombination starts with the binding of four CRE monomers to the two *loxP* sites (18). At this stage, only two of the CRE monomers are in their active conformation while the other two remain inactive. The active units first cleave the 'bottom' strand in the spacer region 3′ to the Guanine, generating a free 5′-OH end, and a covalent phosphotyrosyl bond between the 3′ end and the active CRE subunit. Each end is then rejoined with its homologue end on the parallel DNA sequence resulting in two 'bottom' recombined strands and two 'upper' strands that are still intact (18). This intermediate structure is also known as Holliday junction (19). Exchange of the 'upper' strands initiates with the isomerization of the CRE molecules from active to inactive form and vice versa. This conformation change resolves the Holliday junction by similar cleave (of the 5′ Adenosine Thymidine phosphodiester bond on the 'upper' strands) and join steps to yield a fully recombined DNA product (18).

A classical example that shows the power of applying mosaic analysis for gene function demonstrates the requirement for *Egfr* in cell proliferation in flies by way of induced deletion clones using the FLP/FRT recombination system (20). Null mutations in this gene cause embryonic lethality hampering the analysis of its function during development. Examining somatic *Egfr*$^-$ sectors in imaginal discs revealed that these contained 10-fold fewer cells compared to their wild-type twin-spot sister clones. In addition, it indicated the cell autonomous role of *Egfr* in cell proliferation.

Here we describe in detail the application, advantage, and limitations of the CRE/*lox*-based clonal system, developed in our lab (21), consisting of two vectors in which recombination events are positively marked by endoplasmic reticulum (ER)-localized green fluorescent protein (GFP_{ER}) expression.

2. Materials

2.1. Vectors

1. The *pCB1* binary vector (*see* **Fig. 4.2**) contains the recombination cassette consisting of the *CRT1* gene flanked by two direct *loxP* repeats. The *CRT1* stuffer gene prevents the induction of *GAL4VP16* by the *35S* promoter and the transactivation of GFP_{ER} while at the same time, by encoding for a nonplant phytoene desaturase, it confers resistance to bleaching herbicides such as Norflurazon (22). In between the *CRT1* gene and the second *pCB1 loxP* site, there are unique *Xba*I and *Not*I restriction sites for cloning purposes (21).
2. The *pG7CRE* construct contains a multicloning cassette in front of the CRE recombinase for cloning a promoter of choice to drive CRE expression and the option to clone a cassette with a gene-of-interest under the *UAS* promoter in the *Stu*I site next to the right border in the T-DNA region. The *pG7HSCRE* construct (*see* **Fig. 4.2**) already

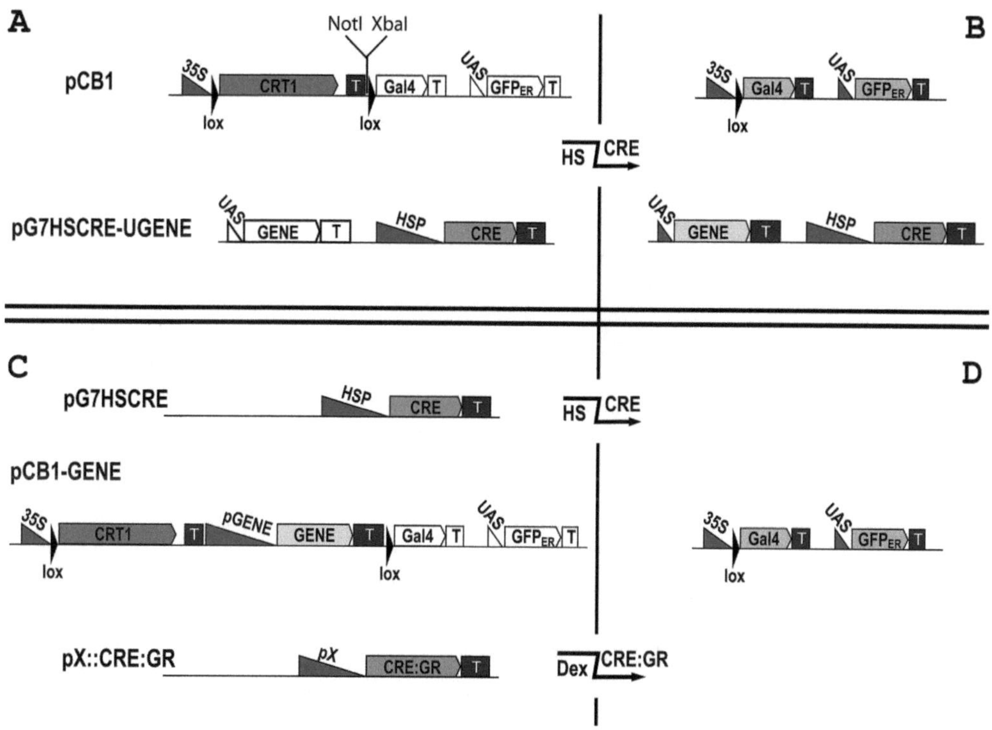

Fig. 4.2. **Constructs used for gain (A and B) and loss (C and D) of function clonal analysis prior to (A and C) and post (B and D) induction of CRE-mediated recombination.** Gray-scaled objects indicate active promoters, transcribed genes, and terminators (*see* **Section 3** for details).

contains the *Arabidopsis HSP18.2* heat-shock inducible promoter driving CRE recombinase expression while maintaining the option to clone the *UGENE* cassette in the multicloning site.

3. The *pX::CRE:GR* constructs (*see* **Fig. 4.2**) are generated by a multisite Gateway® (Invitrogen) reaction incorporating three entry clones carrying a tissue-specific promoter *X*, *CRE:GR* (23), and *NOS* terminator, respectively, in a binary destination vector. This construct can also incorporate a *UGENE* cassette next to the right border as described above.
4. The *pBnUASPTn* is used for a subcloning step to place the gene-of-interest under control of the *UAS* promoter. This vector is derived from *pB2n* (based on pCR-Script, Stratagene) and contains two *Not*I sites flanking a cassette consisting of a *6xUAS* GAL4 binding repeat fused to the -46-bp minimal *35S* promoter followed by a multicloning site and the *35S* terminator.

The binary vectors described are based on the *pGREEN* vectors (24, www.pgreen.ac.uk/) carrying a bacterial kanamycin selection marker. *pCB1* contains a plant *NOS-Basta* resistance cassette and the *pG7* vectors carry a *NOS-Hygromycin* resistance cassette. *pX::CRE:GR:T* constructs have a choice of plant resistance cassettes depending on the binary destination vector.

2.2. Microscopy

1. Forceps for seedlings handling (GGI0079, Outdoor Education, www.oe-initiatieven.nl).
2. 24×50 mm and 18×18 mm #1 (0.13–0.16 mm) cover slips (Menzel Gläser).
3. Propidium iodide (Sigma) stock solution: 10 mg/mL in water. Use 1000× dilution for confocal microscopy (*see* **Note 1**).
4. Fluorescence stereomicroscope (e.g., Leica) equipped with digital camera.
5. Confocal imaging and analysis of clones were performed using a Leica SP2 inverted microscope, and the accompanying software (version 2.61).

2.3. Media and Reagents

1. Forceps for crossing: Watchmaker forceps #5.
2. 2-(*N*-Morpholino) ethanesulfonic acid (MES) buffer (50 g/L = 100×) MES, pH 5.8, adjust with 1 M KOH. Autoclave.
3. ½ GM growth medium: Add 1.1 g of Murashinge & Skoog medium (MS + vitamins, Duchefa Biochemie), 4 g of plant agar (Duchefa Biochemie), 5 g of sucrose and 5 mL of 100× MES buffer and water to 500 ml. Autoclave. Post autoclaving 50 mg/L ampicillin may be added to inhibit microbial growth.
4. Dexametasone (Sigma) (20 mM) in DMSO.
5. Agarose (Sphaero) (0.1%) in water. Autoclave.
6. Bleach: NaClO (Sodium hypochlorite acid) <5%.
7. HCl (36–38%).
8. Tools and materials for genomic Southern blot (25).
9. Tools and materials for plant transformation (26).
10. Square petri dishes, 120×120×15 mm.
11. Urgo Pore Tape (Laboratoires Urgo, France).
12. Phosphate buffer (100 mM, pH 7.0): Prepare from 1 M stocks of Na_2HPO_4 and NaH_2PO_4.
13. Selective agent stock solutions: Ampicillin (Duchefa Biochemie), 50 mg/mL in water. Hygromycin (Duchefa Biochemie), 25 mg/mL in water. Basta (DL-Phosphinothricin, Duchefa Biochemie), 25 mg/mL in water. Norflurazon (Novartis), 10 mM in 96% EtOH. Kanamycin (Duchefa Biochemie), 50 mg/mL in water.

3. Methods

The CRE/*lox* system can be used for two main purposes, inducing cell- or tissue-specific gain or loss of function. These two objectives are achieved by using one of the two features of the system: inverting (*see* **Note 2**) or deleting a sequence flanked by two *lox* sites (*see* **Fig. 4.1**). For that purpose, we designed a set of two vectors that can be used to generate both activation and deletion clones (*see* **Fig. 4.2**). Combining the CRE/*lox* recombination system with the GAL4/*UAS* transactivation system allows positively marking the cells in which recombination has taken place with GFP_{ER}(*see* **Note 3**). The CRE is driven by either heat-shock or by a tissue-specific promoter. In the latter case, inducibility is achieved by fusion of CRE to the human glucocorticoid receptor (CRE:GR). Thereby, CRE can be targeted to the nucleus (where it executes the recombination reaction) upon Dexametasone (Dex) application. Using a heat-shock promoter has the advantage of producing random clones and therefore it can be used to analyze deletion effects in any tissue and cell at a certain moment. In addition, exposing the plants to short and longer heat-shock periods will generate a range of clone sizes from single cell up to a complete null that should acquire the mutant phenotype. Furthermore, the heat-shock promoter provides strong expression relative to often weak tissue-specific promoters. On the other hand, the tissue-specific promoter combined with *CRE:GR* allows timely gene deletion generating desirable mutant cell or tissues preferably within the wild-type expression environment of the gene-of-interest.

Below we describe protocols and examples of using the CRE/*lox* system for clonal deletion and activation analyses in *Arabidopsis thaliana* by deleting a sequence placed in between two *loxP* sites.

3.1. Loss of Function Clones

The application of the clonal-deletion system for loss of function mutations is based on the following reasoning: First, as for all CRE/*lox* mosaic systems, there are two main components, an inducible CRE and a second construct harboring the *lox* sites (*see* **Note 4**). The *CRE* is induced by either a heat shock or by Dex application in case of a tissue-specific promoter driving *CRE:GR*. Second, a knock-out mutant is complemented by a *CB1-GENE* T-DNA integration containing the corresponding wild-type allele (driven by its own promoter), which is flanked by two *loxP* sites (*see* **Fig. 4.2**). Induction of CRE and recombination results in the excision and degradation (*see* **Note 5**) of the wild-type copy, since it is a nonchromosomal unstable circular DNA molecule (*see* **Fig. 4.1**). Recombination results in a relocation of the *35S* promoter in a manner that induces the expression of GFP_{ER}, marking the loss of function cell with a green fluorescence. Third, the *CB1-GENE* T-DNA holding the wild-type allele in it, must exist as a hemizygous single copy, as such the GFP_{ER} expression is correlated to a unique credible recombination and excision event marking a null cell.

3.1.1. Cloning, Transformation, and Selection

1. Clone the wild-type complementing allele in one of the available *Not*I or *Xba*I sites of the *pCB1* plasmid, either directly or via an intermediate cloning step, e.g., by using *pB2n*. The construct must harbor the promoter region in a manner that fully complements the mutation. Optionally, fuse the sequence of the wild-type allele to a fluorescence protein (FP), but make sure to maintain complementation. This allows the visualization of reduction in protein level over time (*see* **Note 6**).
2. Transform mutant plants with *pCB1* and *pG7* constructs by floral dip and select resistant plants according to standard procedures (*see* **Note 7**).
3. Select plants with a single copy insertion of the *CB1-GENE* T-DNA by Southern blotting. The size of each hybridizing fragment on a blot is determined by the position of each T-DNA and therefore correlates directly to the number of T-DNA insertions. Use four-cutter restriction enzyme for better separation of similar sized fragments. Hybridize with a probe complementary to the sequence located between the restriction site closest to the right border and the right border itself (*see* **Notes 8** and **9**). It is not necessary to select for single copy *G7HSCRE* plants since the CRE is only used as a means to recombine DNA (*see* **Note 10**). In case of *CRE:GR*, multiple insertion lines may even be beneficiary contributing to higher efficiency of recombination upon Dex application (*see* **Note 11**).

3.1.2. Crossing

As mentioned above, the plants used to assay clonal-deletion effects should be hemizygous containing a single *CB1-GENE* T-DNA. Keeping heterozygosis is easily achieved by analyzing F1 plants from the following cross: $gene^{-/-}$; $pCB1\text{-}GENE^{+/+}$ × $gene^{-/-}$; *pG7HSCRE* where the heat-shock promoter can be replaced by any tissue-specific promoter in combination with *CRE:GR* to maintain inducibility. The dominant allele should always be used as pollen donor to ensure a successful cross. In the case of the clonal-deletion system, it means that plants bearing the *CB1-GENE* T-DNA harboring the wild-type complementing allele are used as pollen donors, while the mutant plants, carrying the *CRE* receive the pollen. By carrying out this type of cross, F1 offspring should have a wild-type phenotype indicating a successful cross and can be used directly for experimentation. This is by far the fastest and easiest option when working with viable transformable mutants. Induction and analysis of clones in embryonic or gametophytic lethal mutants are possible albeit complicated (*see* **Note 12**) and preferably requires an improved system (*see* **Section 3.6**).

It is also possible to perform a sequential transformation and generate plants that harbor both the *pCB1-GENE* and the *CRE* insertions and back cross them to the mutant plant thereby reducing the *pCB1-GENE* copy number to one. In order to increase the amount of the CRE:GR protein, one can carry out the following cross: $gene^{-/-}$; $pCB1\text{-}GENE^{+/+}$; $pX::CRE:GR^{+/+}$ × $gene^{-/-}$; *pX::CRE:GR* thereby enhancing recombination efficiency when using weak tissue-specific promoters.

3.2. Gain of Function Clones

The gain of function system also consists of two components (*see* **Fig. 4.2**), i.e., the *pCB1* and *pG7HSCRE-UGENE* constructs. Both corresponding T-DNAs should be present in the mutant background in order to analyze local gain of function or in wild-type when the goal is to test clonal ectopic expression. Induction of CRE by the heat-shock promoter leads to recombination and deletion of the stuffer fragment between the *loxP* sites and relocation of the *35S* promoter in front of *GAL4VP16* driving its transcription. Subsequently, GAL4VP16 will bind and transactivate transcription from the *UAS* promoters of both the gene-of-interest from the *G7HSCRE-UGENE* and GFP_{ER} from the *CB1* T-DNAs, marking the clone with green fluorescence. Optionally, a tissue-specific gene induction can be achieved by combined use of the *pCB1* and *pG7XCRE-UGENE* or *pX::CRE:GR-UGENE* where X is a tissue-specific promoter driving *CRE* or Dex inducible *CRE:GR*.

3.2.1. Cloning, Transformation, Selection and Crossing

1. Clone the gene-of-interest into *pBnUASPTn* between the *UAS* promoter and *35S* terminator.
2. Ligate this *UGENE* cassette into the *Not*I restriction site of *pG7HSCRE* generating *pG7HSCRE-UGENE* (*see* **Note 13**).
3. Transform constructs by floral dip and select resistant plants according to standard procedures. There is no strict need to select single-copy T-DNA insertion plants for clonal activation, but keep in mind that different and multiple insertion lines can affect expression levels.
4. Combine the *CB1* and *CRE* expressing T-DNAs by genetic crosses. In case of ectopic gain of function, there is no preferential direction of the cross. Sequential transformation is an alternative means to combine T-DNA insertions.

3.3. Induction of Root Clones

Reproducible induction of clones is highly dependent on having identical conditions throughout all experiments. Therefore, it is required to replicate the working conditions as much as possible to facilitate the prediction of clone size.

1. Plate sterile and stratified seeds (*see* **Note 14**) on $^{1}/_{2}$GM square plates, seal the edges with parafilm (*see* **Note 15**), and germinate at 20°C (*see* **Note 16**) in a 16 h light/8 h dark cycle with the plates in near vertical position.
2. Heat-shock induction of CRE is carried out at 37°C (*see* **Notes 17, 18** and **19**), 2–5 d postgermination (*see* **Notes 20** and **21**). Short induction periods will generate fewer clones, randomly appearing in the tissues treated. Using the conditions described here, a heat shock of 18–20 min will generate small clones as shown in **Fig. 4.3**. Larger clones (i.e., a group of adjacent cells showing GFP expression) can be induced by increasing heat-shock time up to 2 h (*see* **Note 22**). Outer tissue clones in root epidermis and columella are usually induced faster than inner ones. Generally, we have observed the first clones, based on GFP_{ER} expression, within 8 h after heat shock (HAHS), and up to 16 HAHS newly induced clones appeared. Roots displaying clones can readily be identified using a fluorescence stereomicroscope.
3. Activation of *CRE:GR* is carried out by flooding the roots with 20 μM Dex solution. Specifically, one row of seeds is plated and germinated 1–3 cm from the bottom of a $^{1}/_{2}$GM plate. At 2–5 d postgermination, the plate is vertically dipped

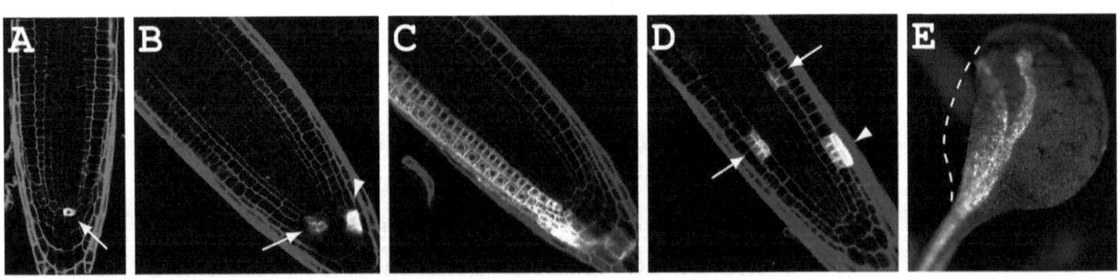

Fig. 4.3. **Examples of root and leaf clones.** (**A**) Single quiescence center (QC) cell-deletion clone induced by Dex-activated *CRE:GR* driven from the QC-specific *WOX5* promoter. (**B**) Clones induced by root cap promoter driven *CRE:GR* at 54 h after Dex application. (**C**) Same root as in (**B**) 8 days after clone induction. The clone marked by the *arrow* in (**B**) now proximally extends throughout the lateral root cap, epidermis, and cortex due to division of cells in which recombination has occurred. Note the loss of the collumela clone, marked by the *arrow head* in (**B**), as a result of cell displacement following division of nonrecombined columella stem cells and their daughter cells. (**D**) Heat shock-induced clonal *SCARECROW* activation in a *scr* mutant background leads to periclinal cell division only in the mutant ground tissue layer (*arrows*) in a cell autonomous manner, i.e., cells neighboring the ground tissue activation clone were not induced to divide. No effects of *SCARECROW* activation were detected in other tissues (*arrow head, see* also (21)). (**E**) Large $hobbit^{2311}$ deletion clone induced in leaf primordial. In the mediolateral axis, $hobbit^{2311}$ clones have strongly reduced cell numbers and appear to lack growth compensation leading to incomplete leaf blade expansion (*dashed line, see* also (27)). GFP_{ER}-marked clones in *white*.

in a sterile tray containing liquid $^1/_2$GM supplemented with Dex, thereby completely submerging the roots for 8–16 h. Following liquid removal, the first clones marked with GFP_{ER} will generally be visible within 2 h under a fluorescence stereomicroscope depending on the promoter strength and Dex incubation time. These roots should be placed on a $^1/_2$GM plate devoid of Dex to avoid formation of new clones.

3.4. Imaging, Analysis, and Confocal Settings of Root Clones

1. Approximately 16 h after induction of CRE-mediated recombination by heat shock or Dex, preselect plants with GFP_{ER} expressing clones of the desired size using a fluorescence stereomicroscope for subsequent analysis. This way the number of plants for labor-intensive confocal microscopy is considerably reduced.
2. For confocal microscopy of seedling roots, pipet ~30 μL of propidium iodide solution on a 24×50 #1 coverslip and mount the seedling in a way that only the root touches the liquid. Cover the root with a 15×15 #1 coverslip and keep the shoot and hypocotyl parts wet (*see* **Note 23**).
3. GFP and propidium iodide are visualized by excitation at 488 nm and collection at 498–523 nm and around 600 nm, respectively (*see* **Fig. 4.3**). In case the complementing protein is fused with another FP, it is recommended to adjust the settings to collect both fluorescence emissions either simultaneously or sequentially so that the two images perfectly overlap (*see* **Note 24**).
4. Return the seedling to the $^1/_2$GM plate, in case of a time course experiment (*see* **Note 22**), by dripping approximately 100 μL of water between the two coverslips. This makes the 15×15 #1 coverslip slide down easily without damaging the root. Carefully place the seedling back on a $^1/_2$GM plate in the exact same position it had on the coverslip during imaging (*see* **Notes 25** and **26**).

   Dependent on parameters such as protein and mRNA half-life times, altered phenotypes should be apparent 3 h–3 d after clone formation using a time course confocal analysis procedure.

3.5. Induction, Imaging, and Analysis of Leaf Clones

Loss and gain of function analyses in *Arabidopsis* leaves using *pCB1* and heat shock-driven CRE are performed essentially as described above with a few modifications (27). In general, leaf clone size can be also manipulated depending on the time of heat shock, but in addition depends on the developmental stage of the emerging leaf and the number of divisions that remain for the cell(s) in which recombination has taken place, until full maturation of the leaf.

1. Plate sterile and stratified seeds on $^{1}/_{2}$GM plates at given density of around 75 seeds per plate and seal around with Urgo Pore tape. Germinate and grow at 20°C in a horizontal position for 9–12 d.
2. Induce CRE-mediated recombination with a 20–60 min heat shock at 37°C. Return to the growth chamber for an additional 3–10 d before imaging and analysis (*see* **Note 27**).
3. Select the leaf with the desired clone under the fluorescence stereomicroscope based on GFP expression (*see* **Note 28**).
4. Cut the leaf or a section of it and maintain in phosphate buffer (pH 7.0) until the time of assay, e.g., in a 24-well microtiter plate. Mount as described above between two coverslips and a drop of water or phosphate buffer for confocal imaging. Propidium iodide is not required as it will interfere with chlorophyll fluorescence.
5. Clones are easily visualized in the adaxial epidermis, which does not contain functional chloroplasts, i.e., excited epidermal cells in clones show only GFP_{ER} emission in green without interference of chlorophyll autofluorescence in red. Abaxial epidermis is normally wrinkled and collapses in between the coverslips (*see* **Note 29**). Upon identification of an epidermal clone, image down into the palisade mesophyll to determine whether the clone expands into this L2 layer or only affects L1 (*see* **Note 30**).

3.6. Conclusions and Future Improvements

We have described in detail the conditions to perform and analyze clonal activation and deletion of a gene-of-interest. The protocols are based on our experience with the CRE/*lox*-based site-specific recombination system developed in our lab consisting of two vectors, in which recombination events are positively marked with GFP_{ER}(*see* **Note 31**). Much of the methods will also apply to different recombination systems, constructs, and analysis of other cells and tissues than mentioned here. Nevertheless a number of improvements can be implemented. Most obvious of all is combining the CRE recombinase, the *lox* sites, and the gene to activate or delete, all in a single vector (28, 29). However, the clonal-deletion system described is particularly suited for viable, easily maintained, transformable mutants. Combining both vectors by crossing allows analysis of F1 seedlings with the insurance of working with hemizygotes.

Equally important are improvements to allow the induction and analysis of deletion clones in an embryonic and, in particular, gametophytic lethal background for which the current system is inapplicable. The simple reason being that the mutant individuals are unviable and thus cannot be used for transformation with the *CRE* construct and later crosses. One way to circumvent these problems is to design a clonal-deletion system historically used in animal mosaic studies that relies on deletion of a marker together with the gene-of-interest. Upon recombination, a homozygous deletion clone will no longer be marked. A complication may be the identification of these clones in a background of positively marked cells. Another possibility is to utilize two incompatible *lox* pair variants (e.g., 30). Starting with a single-insertion homozygote plant, each recombination event would excise the complementing wild-type allele by one of two *lox* pairs translocating the *35S* promoter in front of either a CFP or RFP coding sequence. Consequently, a double CFP and RFP fluorescence in a given cell indicates a complete deletion of both wild-type copies. A clonal-deletion system based on these principles is currently under construction in our lab and should extend the application of mosaic studies to analysis of lethal genes during and post-embryogenesis.

4. Notes

1. The membrane impermeable fluorescent dye propidium iodide is used to visualize the outlines of cells in roots.
2. CRE recombination results in an inversion of a sequence when it is flanked by two oppositely directed *lox* sites (*see* **Fig. 4.1**). This type of recombination creates two identical new *lox* sites that can recombine again in the presence of residual CRE protein and reinvert the defined sequence into its original orientation. It is not recommended to use this feature of the CRE/*lox* system to generate clones without having an accurate indication (e.g., an FP) for the correct orientation.
3. The presence of many cells in living tissues would hamper clone identification due to fluorescence penetration from surrounding cells. Therefore, we chose a system based on an FP-marked recombination event as opposed to a whole organism expressing an FP with nonfluorescent clones. In addition, the GFP is anchored to the ER preventing movement to wild-type cells.
4. There are a few reports in which the *CRE* and the *lox* sites-carrying plasmids can be combined to a single vector (28). Such a setup may be particularly useful for gain of function studies. In these cases, there is no requirement for an out cross such as the one needed in the clonal-deletion system described here. However, upon merging *CRE* recombinase and *lox* sites in one vector, we and others have observed CRE activity and subsequent recombination in *E. coli*, which warrants additional measures to

prevent *CRE* transcription and translation, e.g., to include an intron within the *CRE*-coding sequence (29).

5. Although the gene-of-interest is deleted, the protein and mRNA may remain for some time in the cell. Functional protein in a null clone can delay any changes in phenotypes for several days.
6. In that case, one should choose a fluorescence protein that does not have an emission overlap with the GFP_{ER} that marks the deletion, e.g., RFP. Naturally, it is possible to use another FP for marking the deletion but ultimately, and although GFP overlaps with a wide range of FPs, it has a significant advantage over other FPs since its brightness permits the detection of a single GFP_{ER} marked cell by a simple fluorescence stereomicroscope. This enables screening for the right individual prior to confocal microscope analysis.
7. *pCB1* carries both Basta (select on 20 μg per 1 L) and Norflurazon (select on 200 nM – 1 μM) resistance under *NOS* and *35S* promoter control, respectively. *pG7*-based vectors carry a hygromycin resistance under the control of the *NOS* promoter (select on 15 μg per 1 L). Positive transformants can also be selected on the basis of phenotype and/or the expression of an FP in case it is fused to the analyzed gene.
8. In case of *CB1-GENE* T-DNA insertions, *Hpa*II makes an excellent four-cutter restriction enzyme for plant DNA digestion. Then, a GFP probe can be used for identification of transformants carrying a single *CB1-GENE* copy. One should take into account the high homology between GFP and a few other FPs, which might yield an additional invariable hybridizing fragment, e.g., a translational fusion of *YFP* to the gene-of-interest results in a strong hybridization signal even under stringent hybridization conditions.
9. The left border of the T-DNA integrates with less accuracy than the right border (31). Therefore, performing the Southern blot analysis using a right border adjacent probe is preferred since it allows a precise estimation with respect to the minimal size of the hybridizing fragment.
10. Different *HS::CRE* lines have a varied induction response regarding the number of clones generated per given heatshock. Working with a single *HS::CRE* line in all experiments is recommended in order to be able to reasonably predict clone size and number.
11. We have found that Dex activation of *CRE:GR* is not very efficient when using relative weak promoters. Therefore,

CRE::GR should be driven by a strong promoter and one needs to select for the best clone inducing line, i.e., the one that shows clones formation upon shortest and lowest Dex-induction conditions. By using these strong CRE drivers, induction period in those nonoptimal conditions is shortened. Alternatively, elevating the CRE:GR protein levels in the tissue-of-interest by introducing multiple copies of the corresponding T-DNAs may help to sensitize these cells for recombination.

12. An alternative for working with an F1 offspring is to generate clones in a segregating population (i.e., from a heterozygous plant) and genotype them later.
13. Cloning of the *UGENE* cassette into *pG7XCRE* is similar to cloning in *pG7HSCRE*, but cloning into *pX::CRE:GR* requires blunt-end ligation into the *Stu*I restriction site adjacent to the right border.
14. Seeds in open tubes are sterilized using chlorine gas in a dry-sealed jug containing a small beaker with 100 mL of bleach and 3 mL of 37% HCl for 5–12 h. After release of gas for at least 1 h, 200 μL of 0.1% agarose is added to each tube and stratification is carried out in the dark, at 4°C for 2–5 d.
15. To ensure reproducibility of heat-shock induction of clones in relation to induction time, exactly 50 mL of medium is used per plate and plates are dried for equal time periods. The location of seedlings on the plate and of the plate in the 37°C chamber affects clones formation as well, i.e., seedlings placed towards the edges of the plate are induced faster than the center ones.
16. Precise temperature conditions are critical to avoid recombination and clone formation by suboptimal induction conditions. Therefore, plates should always be placed on lower shelves far from any heat source such as lamps.
17. Optionally, cool the plates for about 5 min on ice before retuning them to the growth chamber.
18. Heat induction of CRE-mediated recombination should preferably take place in a large 37°C room to minimize the temperature drop while opening the door.
19. Heat shock can also be applied by directing the microscope laser towards specific cells (32, 33). The system described here seems unsuitable for laser-induced clones as we have not been able to reproducibly obtain GFP_{ER}-marked clones by this technique beyond columella and lateral root cap tissues. However, it cannot be excluded that different lasers or confocal microscopes can give better results.

20. It is recommended to induce root clones as early as possible keeping in mind that one needs to follow them, in many cases for days, by transferring the seedlings back and forth the coverslips and $^{1}/_{2}$GM plates.
21. For both gain and loss of function analysis, it is required to perform the two following controls. First, the seedlings must be analyzed prior to *CRE* induction to ensure no clones are spontaneously induced and second, no clones are present in the absence of a *CRE*-expressing T-DNA. These controls confirm that recombination and GFP_{ER}-expressing clones occur specifically by *CRE* induction.
22. We have not seen any effect on plant development when implementing these long heat shocks.
23. Shoot parts need to be kept wet to prevent them from drying out. This is particularly important when following clones in time and seedlings need to be maintained viable and transferred back to the $^{1}/_{2}$GM plate.
24. For example, the Leica SP2 confocal microscope has the option to implement a sequential scanning in order to collect fluorescence emission from more than one FP as well as the propidium iodide emission. For sequential scanning, adjust the desired setting, e.g., one for GFP and propidium iodide visualization and the other for YFP and load them in the open window after pressing "seq". On the right panel of the same window choose "between frames" and scan by pressing "series".
25. It is tricky to keep the plates and seedlings sterile but it is possible, up to a period of 3 weeks. One needs to work as clean as possible and be sure to use sterile water and propidium iodide, both supplemented with 50 μg/mL ampicillin.
26. It must be taken into account that the position and shape of the clone are continuously changing as a result of cell division, growth, and differentiation. Clones can be reidentified by first, placing the seedling at the exact orientation to the one it had on the coverslip and second, by comparing their shape, size, and relative position with other clones in previous images.
27. Typically, a short heat shock of ~20 min will result in small clones in already-emerged leaves restricted to a number of cell diameters and affecting only one cell layer of the leaf, as judged by GFP_{ER} expression. Longer heat shock results in sectors in leaves still to emerge indicating these have been induced in the emerging leaf primordia or the shoot apical meristem (SAM) and also that these tissues are less accessible for heat-shock induction. The extension of the clones in these subsequent fully developed leaves reflects the layers in

the SAM in which recombination took place ranging from clones restricted to only one cell layer to clones spanning all cell layers within the leaf. Notably, these GFP_{ER}-marked clones extend along the entire proximo-distal axis and often include the midvein.

28. By using a regular GFP filter, leaf clones are shown in green, on a red background formed by chlorophyll emission. A custom GFP filter allowing passage of wavelength with a bandwidth between 500 and 530 nm blocks this red auto-fluorescence.
29. This is not a problem when analyzing early stages of leaf development. If one is interested in analysis of later stages of leaf development, the leaf can be fixed before analysis.
30. To visualize epidermal morphologies, it is possible to make a leaf cast using Eukitt (Sigma-Aldrich). Leaf casting is not straightforward since the Eukitt needs to be almost solidified, but not too much, before making the cast. Subsequently, the leaf is removed carefully to leave behind an impression of the leaf surface. The leaf casts are visualized with Nomarsky light microscopy.
31. Vector sequences are available upon request.

Acknowledgments

We are indebted to René Benjamins for critical reading, Jose Manuel Perez-Perez and Olivier Serralbo for valuable discussions, critical reading, and sharing data.

References

1. Kolb, A. F. (2002) Genome engineering using site-specific recombinases. *Cloning Stem Cells* **4**, 65–80.
2. Wirthl, D., Gama-Norton, L., Riemer, P., Sandhul, U., Schucht, R., and Hauser, H. (2007) Road to precision: recombinase-based targeting technologies for genome engineering. *Curr Opin Biotechnol* **18**, 411–419.
3. Sieburth, L. E., Drews, G. N., and Meyerowitz, E. M. (1998) Non-autonomy of AGAMOUS function in flower development: use of a Cre/loxP method for mosaic analysis in Arabidopsis. *Development* **125**, 4303–4312.
4. Becraft, P. W., Kang, S., and Suh, S. (2001) The Maize CRINKLY4 receptor kinase controls a cell-autonomous differentiation response. *Plant Physiol* **127**, 486–496.
5. Maedaa, M., Namikawaa, K., Kobayashia, I., Ohbaa, N., Takaharaa, Y., Kadonoa, C., Tanakab, A., and Kiyamaa, H. (2006) Targeted gene therapy toward astrocytoma using a Cre/*loxP*-based adenovirus system. *Brain Res* **1081**, 34–43.
6. Zinyk, D. L., Mercer, E. H., Harris, E., Anderson, D. J., and Joyner, A. L. (1998) Fate mapping of the mouse midbrain–hindbrain constriction using a site-specific recombination system. *Curr Biol* **8**, 665–668.
7. Saulsberry, A., Martin P. R., O'Brien, T., Sieburth, L. E., and Pickett, F. Bryan. (2002) The induced sector *Arabidopsis* apical

embryonic fate map. *Development* **129**, 3403–3410.
8. Srivastava, V. and Ow, D. W. (2004) Marker-free site-specific gene integration in plants. *Trends Biotechnol* **22**, 627–629.
9. Craig, N. L. (1988) The mechanism of conservative site-specific recombination. *Annu Rev Genet* **22**, 77–105.
10. Nash, H. A. (1981). Integration and excision of bacteriophage λ: The mechanism of conservative site specific recombination. *Annu Rev Genet* **15**, 143–167.
11. Volkert, F. C. and Broach, J. R. (1980) Site-Specific Recombination promotes plasmid amplification in yeast. *Cell* **46**, 541–550.
12. Sternberg, N. and Hoess, R. (1983) The molecular genetics of bacteriophage P1. *Annu Rev Genet* **17**, 123–154.
13. Osborne, B. I. and Baker, B. (1995) Movers and shakers: maize transposons as tools for analyzing other plant genomes. *Curr Opin Cell Biol* **7**, 406–413.
14. Sternberg, K. and Hamiltons, D. (1981) Bacteriophage Pl site-specific recombination. *J Mol Boil* **150**, 467–486.
15. Hoess, R. H., Ziese, M., and Sternberg, N. (1982) P1 site-specific recombination: Nucleotide sequence of the recombining sites. *Proc Natl Acad Sci USA* **79**, 3398–3402.
16. Hoess, R. H. and Abremski, K. (1985) Mechanism of strand cleavage and exchange in the Cre-lox site-specific recombination system. *J Mol Biol* **181**, 351–362.
17. Mack, A., Sauer, B., Abremski, K., and Hoess, R. (1992) Stoichiometry of the Cre recombinase bound to the lox recombining site. *Nucleic Acids Res* **20**, 4451–4455.
18. Lee, L. and Sadowski, P. D. (2003) Sequence of the loxP Site determines the order of strand exchange by the Cre recombinase. *J Mol Biol* **326**, 397–412.
19. Holliday, R. A. (1964) Mechanism of gene conversion in fungi. *Genet Res* **5**, 282–304.
20. Xu, T. and Rubin, G. M. (1993) Analysis of genetic mosaics in developing and adult *Drosophila* tissues. *Development* **117**, 1223–1237.
21. Heidstra, R., Welch, D., and Scheres, B. (2004) Mosaic analyses using marked activation and deletion clones dissect *Arabidopsis* SCARECROW action in asymmetric cell division. *Genes Dev* **18**, 1964–1969.
22. Misawa, N., Yamano, S., Linden, H., de Felipe, M. R., Lucas, M., Ikenaga, H., and Sandmann, G. (1993) Functional expression of the Erwinia uredovora carotenoid biosynthesis gene crtl in transgenic plants showing an increase of beta-carotene biosynthesis activity and resistance to the bleaching herbicide Norflurazon. *Plant J* **4**, 833–840.
23. Brocard, J., Feil, R., Chambon, P., and Metzger, D. (1998) A chimeric Cre recombinase inducible by synthetic, but not by natural ligands of the glucocorticoid receptor. *Nucleic Acids Res* **26**, 4086–4090.
24. Hellens, R. P., Edwards, E. A., Leyland N. R., Bean, S., and Mullineaux, P. M. (2000) pGreen: a versatile and flexible binary Ti vector for *Agrobacterium*-mediated plant transformation. *Plant Mol Biol* **42**, 819–832.
25. Sambrook, J., Fritsch, E. F., and Maniatis, T. (1989) Molecular Cloning: A Laboratory Manual, 2nd edition, Volume 1, 2, 3. Cold Spring Harbor Laboratories Press, New York, NY.
26. Clough, S. J. and Bent, A. F. (1998) Floral dip: A simplified method for *Agrobacterium*-mediated transformation of *Arabidopsis thaliana*. *Plant J* **16**, 735–743.
27. Serralbo, O., Pérez-Pérez, J. M., Heidstra, R., and Scheres, B. (2006) Non-cell-autonomous rescue of anaphase-promoting complex function revealed by mosaic analysis of *HOBBIT*, an *Arabidopsis CDC27* homolog. *Proc Natl Acad Sci USA* **103**, 13250–13255.
28. Hoff, T., Schnorrl, K. M., and Mundy, J. (2001) A recombinase-mediated transcriptional induction system in transgenic plants. *Plant Mol Biol* **45**, 41–49.
29. Mlynarova, L. and Nap, J. (2003) A self-excising Cre recombinase allows efficient recombination of multiple ectopic heterospecific *lox* sites in transgenic tobacco. *Transgenic Res* **12**, 45–57.
30. Livet, J., Weissman, T. A., Kang, H., Draft, R. W., Lu, J., Bennis, R. A., Sanes, J. R., and Lichtman, J. W. (2007) Transgenic strategies for combinatorial expression of fluorescent proteins in the nervous system. *Nature* **450**, 56–62.
31. Gheysen, G., Van Montagu, M., and Zambryski, P. (1987) Integration of Agrobacterium tumefaciens Transfer DNA (T-DNA) Involves Rearrangements of Target Plant DNA Sequences. *Proc Natl Acad Sci USA* **84**, 6169–6173.
32. Harris, J., Honigberg, L., Robinson, N., and Kenyon, C. (1996) Neuronal cell migration in C. elegans: regulation of Hox gene expression and cell position. *Development* **12**, 3117–3131.
33. Kurup, S., Runions, J., Köhler, U., Laplaze, L., Hodge, S., and Haseloff, J. (2005) Marking cell lineages in living tissues. *Plant J* **42**, 444–453.

Chapter 5

Inducible Gene Expression Systems for Plants

Lorenzo Borghi

Abstract

Several systems for induction of transgene expression in plants have been described recently. Inducible systems were used mainly in tobacco, rice, Arabidopsis, tomato, and maize. Inducible systems offer researchers the possibility to deregulate gene expression levels at particular stages of plant development and in particular tissues of interest. The more precise temporal and spatial control, obtained by providing the transgenic plant with the appropriate chemical compound or treatment, permits to analyze also the function of those genes required for plant viability. In addition, inducible systems allow promoting local changes in gene expression levels without causing gross alterations to the whole plant development. Here, protocols will be presented to work with five different inducible systems: AlcR/AlcA (ethanol inducible); GR fusions, GVG, and pOp/LhGR (dexamethasone inducible); XVE/OlexA (β-estradiol inducible); and heat shock induction.

Key words: Dexamethasone, β-estradiol, ethanol, heat shock, inducible system, GVG, XVE, OlexA, pOp, LhGR, AlcA, AlcR, GR fusion.

1. Introduction

Inducible systems are usually based on two components. The first component is typically a chimeric transcription factor that can specifically bind to tightly controlled promoters. The chimeric transcription factor can activate the promoter only after induction, and it is commonly called *driver* or *activator*. The second component contains the binding sites for the driver to control the expression of the gene of interest (the *target*) and is commonly referred to as *reporter* or *effector*.

The driver can be cloned under control of the *CaMV35S* promoter to obtain strong and ubiquitous expression. Alterna-

L. Hennig, C. Köhler (eds.), *Plant Developmental Biology*, Methods in Molecular Biology 655,
DOI 10.1007/978-1-60761-765-5_5, © Springer Science+Business Media, LLC 2010

tively, tissue-specific promoters can be used for local induction experiments. The reporter is usually flanked by a minimal 35S promoter, therefore allowing the binding of the basal transcription machinery but minimizing its activation by endogenous transcriptional activators. Driver and reporter can be part of the same cassette or split onto two different vectors for plant transformation (*see* **Fig. 5.1**).

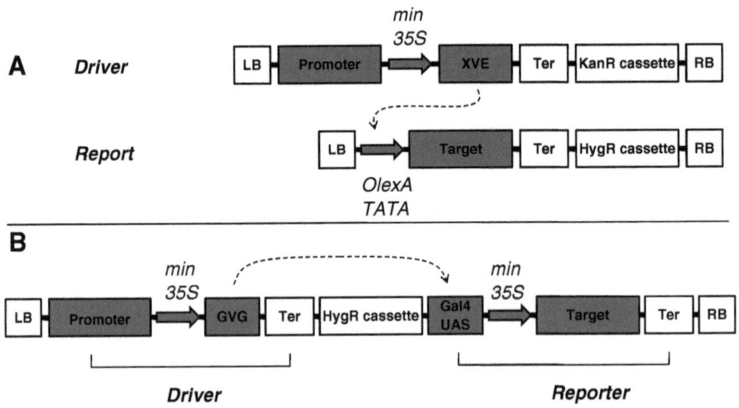

Fig. 5.1. **Inducible transgene expression.** (**A**) Two components system for β-estradiol-inducible expression. The chimeric transcription factor XVE cloned on the *driver* vector binds, after β-estradiol treatment, to the *OlexA* promoter placed on the reporter vector, thus activating the transcription of the gene of interest (33). (**B**) Single T-DNA vector for dexamethasone-inducible expression. The chimeric transcription factor GVG binds, after dexamethasone treatment, to six copies of the *Gal4 UAS*, thus activating the transcription of the gene of interest (20).

Single cassette systems speed up transformation procedures. However, binary systems give the possibility to better check for system functionality after integration in the plant genome. In fact, constitutive and ubiquitous expression of the driver can occasionally cause the so-called *squelching* (i.e., sequestration of general transcriptional factors) (1). Position effects can alter the activity of the driver as well the reporter (2). High transcriptional levels of the driver can lead to posttranscriptional gene silencing (PTGS) (3) or nonspecific binding, thus promoting the inactivation of the system or causing unpredictable side effects on plant development, as shown in (4–7). The choice of a binary system permits to select for functional driver and reporter lines without undesired side effects. Subsequently, these lines can be crossed to obtain the complete inducible system. In addition, binary systems assure great versatility by combining different driver and reporter lines, thus making it easy to generate lines with inducible expression confined to various specific tissues or bearing multiple reporters. It is always advisable to carry out experiments with several (at least two) driver and target lines, thus to isolate and discard those lines showing any side effects as previously mentioned.

Activation of inducible systems can be carried out by pouring, spraying, or adding the activator compound respectively to soil, plant surfaces, or plant growth media. Ways of actions, preparations, and treatments, being widely various depending on each case, will be described for each of the explained method in the following sections.

1.1. Ethanol-Inducible Expression with the AlcR/AlcA System

The AlcR transcription factor (the activator) and the *AlcA* promoter (driving the target) were isolated from *Aspergillum nidulans* (8–10). AlcR is activated by acetaldehyde, which is produced in plants by ethanol metabolism. The *AlcA* promoter is not activated by Arabidopsis endogenous transcription factors (9) even under anoxic conditions (i.e., excessive watering, growth on agar plates), which are known to promote ethanol production. Driver and promoter are split in two different plant transformation vectors. This system is very suitable for pulse and local inductions (11), as AlcR expression can be restricted to precise patterns using specific promoters and ethanol quickly evaporates (12, 13). However, due to ethanol toxicity, pulse inductions have to be carefully calibrated to assure the maximum effect in the minimal time.

1.2. Dexamethasone-Inducible Expression with – GR Fusions, GVG/UAS or pOp/LhGR Systems

All these three systems are based on the interaction properties of the ligand-binding domain of the glucocorticoid receptor (GR) from rat (14). This domain has a size of 93 amino acids. In the absence of steroids, GR interacts with cytosolic complexes containing heat shock proteins 90 (HSP90). Transcription factors eventually fused to GR (the drivers) are inactive, because they bind to HSP90 anchored in the cytosol and thus cannot enter the cell nucleus (15). After treatment with the synthetic steroid hormone dexamethasone, the GR–HSP90 interaction is disrupted and the GR transcription factor fusion-protein is free to enter the cell nucleus where it can bind and activate its target(s). The GR domain can be directly combined with the transcription factor of interest (16, 17). If the gene under analysis is not a transcription factor, chimeric transcription factors containing the GR domain are available, as described below.

The GVG driver consists of a yeast Gal4 DNA-binding domain fused to the strong transcription activator VP16 (from the *Herpes virus*) and to the GR receptor (18). The target is cloned under control of an Upstream Activation Sequence (UAS). This system has been successfully used (18–20), but results need to be interpreted with great care because high GVG expression levels were reported to cause strong developmental defects in Arabidopsis and several other plants (4–7). The choice of the binary GVG version (instead of the single cassette containing both driver and target) is highly recommended, as strict controls must be conducted on the driver line: high GVG expression levels can even lead to plant death, as experienced in our group.

The driver LhGR consists of a high-affinity mutant of the *Escherichia coli* lac repressor fused to the yeast Gal4 transactivation domain and GR (21). The target can be cloned downstream of six lac operator copies. The system, both available as single and binary cassette, is strongly and quickly inducible and no side effects have been reported to date (22–25). Versions of this system optimized for tobacco are available (25). Several constitutive tissue-specific driver lines, which lack the GR domain, are also available (2, 21, 26, 27).

1.3. β-estradiol-Inducible Expression with the XVE/OlexA System

The XVE driver consists of a *lexA* repressor domain fused to the VP16 transcription activation domain and the human estrogen receptor ER (28). In the presence of estrogens like 17-β-estradiol, XVE binds to the eight copies of the *lexA* domain, thus activating the transcription of the downstream target. Phytoestrogens are unable to activate the system in Arabidopsis, while in soy bean their high concentrations are reported to cause unspecific activation (29). High levels of exogenous β-estradiol up to 25 μM do affect neither Arabidopsis development nor endogenous gene expression (28, 30). The laboratory of Dr. Mark Curtis generated a GATEWAY®-compatible β-estradiol two-components system for both drivers and reporters (31).

1.4. Heat Shock-Inducible System

The promoter of the gene for heat shock protein 18.2 (*HSP18.2*) from Arabidopsis (32) has been successfully used in several plants to induce gene expression after heat shock (33–35). The cis-activation is carried out by cloning the gene of interest downstream of the *HSP18.2* promoter and by inducing the transgenic plant with heat shocks at 37°C. In absence of heat stress, the *HSP18.2* promoter is known to be repressed (36).

2. Materials

2.1. Ethanol-Inducible Expression with the AlcR/AlcA System

1. Microcentrifuge tubes.
2. Closed containers for plant growth. Very suitable are trays of pots where the trays can be well closed with transparent hoods.
3. Ethanol. Pure ethanol (99.8%) is used at different dilutions, depending on the strength of activation required (*see* **Note 1**). Ethanol is diluted in water when poured on soils or vaporized. Most published studies use ethanol concentration between 0.5 and 1% (*see* **Note 2**), both for soil drenches and agar plates (9). 70–100% ethanol concentrations can be used to induce the system by ethanol vapors.

2.2. Dexamethasone-Inducible Expression with – GR Fusions, GVG or pOp/LhGR Systems

1. Dexamethasone (Sigma) stock: 20 mM dexamethasone in DMSO or ethanol (*see* **Note 3**). Stocks can be stored at 4°C or at –20°C for 2 weeks. Avoid repeated freeze/thaw.
2. Silwet L-77 (GE Silicones).
3. Dexamethasone-containing lanolin paste. Place approximately 1 mL of lanolin into a 2-mL microcentrifuge tube and heat it to 60°C (e.g., in a water bath). Add dexamethasone to obtain a concentration of 1–2 μM, vortex the tube and return to 60°C.
4. Dexamethasone-containing MS plates: 1×MS salts (Sigma,) 1 μM dexamethasone, and 1% plant agar (Duchefa). Add dexamethasone to medium when temperature is around 40°C (i.e., when you can touch the bottle after autoclaving).
5. Spray nozzle.

2.3. β-estradiol-Inducible Expression with the XVE/OlexA System

1. 17-β-estradiol (Sigma) stock: 20–40 mM β-estradiol in DMSO or 70% ethanol (*see* **Note 3**). Stocks can be stored in the dark at 4°C or at –20°C for 1 week. Avoid repeated freeze/thaw. β-estradiol is light- and temperature-sensitive: stocks should be wrapped with aluminum foils.
2. Silwet L-77 (GE Silicones).
3. β-estradiol-containing lanolin paste. Place approximately 1 mL of lanolin into a tube and warm it at 60°C. Add the proper amount of β-estradiol to obtain a concentration of 10–25 μM, vortex the tube and put it back at 60°C.
4. β-estradiol-containing MS plates: 1×MS salts (Sigma,) 5 μM estradiol, and 1% plant agar (Duchefa). When preparing plates containing estradiol (*see* **Note 4**), add β-estradiol to 1% MS agar medium when temperature is around 40°C (i.e., when you can touch the bottle after autoclaving). Plates containing β-estradiol should be not older than 1 week.
5. Spray nozzle.

2.4. Heat Shock-Inducible System

Preinduction and induction can be carried out in incubators tightly set, respectively, at 38 and 37°C (*see* **Note 5**).

3. Methods

3.1. Ethanol-Inducible Expression with the AlcR/AlcA System

1. Grow the plants to be induced in trays that can be closed with transparent hoods (*see* **Note 6**).
2. For ethanol vapor inductions, fill 1.5- or 2-mL microcentrifuge tubes with 0.5–1 mL of 70–100% ethanol and force

them into the soil (8–16 tubes for 27×27-cm trays, 16–28 tubes for 30×50-cm trays). Distribute the open tubes in a way to allow ethanol vapors to evenly saturate the tray (*see* **Note 7**).

3. Keep the tray closed overnight or for shorter times (*see* **Note 8**), depending on the strength of induction needed.
4. Remove hoods and tubes. Keep the trays open for at least half a day and then repeat the procedure for further inductions.

Additionally or alternatively, the soil can be soaked with 0.5–1% ethanol. Also in this case, alternate inductions times with aeration times, the latter being not shorter than 12 h, thus avoiding growth of fungi and moulds. Repeat the procedure for prolonged inductions.

Ethanol induction on plates can be tricky. Even 0.1% ethanol may inhibit growth of Arabidopsis seedlings (22, 25). Anoxia (*see* **Note 9**) is also frequent in roots of seedlings grown on agar plates and cell culture (37).

3.2. Dexamethasone-Inducible Expression with – GR Fusions, GVG or pOp/LhGR Systems

GR fusions and GVG systems are fully induced with 1 μ*M* dexamethasone (18). pOp/LhGR systems are maximally induced with 2 μ*M* and are 50% induced with 0.2 μ*M* dexamethasone dilutions (38). Treatments should always be carried out under fume hoods or equivalent delimited zones.

1. Grow the plants to be induced on soil.
2. Spray plants with dexamethasone solution (*see* **Note 10**). Adding 0.01% Silwet L-77 supports dexamethasone penetration into plant tissues by lowering water surface tension. Avoid inducing neighboring plants: let sprayed droplets dry before returning plants into their growth chamber. Although single treatments are known to be sufficient for induction, repeated sprayings every 24 h increase expression levels of the target of interest (39).

Beads or lanolin soaked with dexamethasone can be used for local inductions (*see* **Note 11**). Prewarmed pipette tips can be used to transfer lanoline–dexamethasone drops from the tube to the plant to be induced.

For induction on plates, seeds can be either germinated directly on dexamethasone-containing medium, or established seedlings can be transferred from noninducing to inducing plates at an age of 3–10 days.

3.3. β-estradiol-Inducible Expression with the XVE/OlexA System

β-estradiol-inducible systems are fully activated with 5 μ*M* β-estradiol (31), even if concentrations up to 25 μ*M* have been used without causing side effects to plant development (30). Exogenous estrogens are toxic for human health. Although β-estradiol does not evaporate (30), wear protective gloves (*see* **Note 12**) whenever working with it.

1. Grow the plants to be induced on soil.
2. Spray plants with β-estradiol solution (*see* **Note 10**) under a fume hood. Adding 0.01% Silwet L-77 supports β-estradiol penetration into plant tissues by lowering water surface tension (31). Avoid inducing neighboring plants: let sprayed droplets dry before returning plants into their growth chamber. Repeat sprayings daily to increase expression levels of the target of interest.

Local inductions (*see* **Note 13**) can be carried out by applying lanolin paste containing β-estradiol on the tissue of interest. Prewarmed pipette tips can be used to transfer drops of lanoline containing β-estradiol from the tube to the plant (*see* **Note 14**).

For induction on plates, seeds can be either germinated directly on β-estradiol-containing medium, or established seedlings can be transferred from noninducing to inducing plates at an age of 3–5 days. The physical contact between β-estradiol MS medium and the tissue of interest greatly enhances the system induction strength.

3.4. Heat Shock-Inducible System

1. If plants are grown on soil, water them properly before starting the induction procedure (*see* **Note 15**).
2. Preincubate at 38°C for 15 min to activate Arabidopsis thermotolerance (40) and return plants to 22°C for 1 h.
3. Induce the plants at 37°C every other hour for 3–5 times (*see* **Note 16**): during no-induction times return plants to 22°C (*see* **Note 17**).
   Local inductions can be obtained by incubating part of the plants in 37°C water for 5–15 min.

4. Notes

1. Ethanol dilutions can be freshly prepared before starting induction.
2. 2% ethanol v/v soil is the optimal concentration to obtain full system induction (9). However, as 5% ethanol v/v soil is toxic for plant development, lower ethanol concentrations are used for system induction.
3. DMSO is preferred to ethanol for dilutions, as 0.1% v/v ethanol is sufficient to alter root development in seedlings grown on Petri dishes (22). However, as DMSO tends to accumulate into soil, dilutions in ethanol are preferred to DMSO for soil drenches.
4. Submerging 3- to 5-day old seedlings on β-estradiol-containing plates with 2–5 mL of 5 μ*M* estradiol solution

for short times (3–6 h) permits to obtain high induction levels similar to spraying (unpublished observation).

5. The incubators have to be kept closed during the experiment, as temperature must be constant to obtain maximum induction.
6. Ethanol is toxic to plants: proper system calibrations are required to obtain the maximum induction at the lowest toxicity. Toxicity changes along with plant developmental stages: young seedlings are far more sensitive to ethanol than adult plants.
7. Ethanol vaporizes and penetrates also internal cells (9, 12). Strong precautions have to be kept to avoid inductions of neighboring plants (10, 41). Growth chambers/greenhouses where the experiment takes place must be 100% ethanol-free (cleaning products containing ethanol can be sufficient to activate the system).
8. A few hours are sufficient to induce the system, as reported in (9, 11, 12) and (42). Prolonged inductions can promote moulds and fungal growth in pots and can be detrimental to plant fertility.
9. Avoid anoxic conditions on soil as well on plates. Do not water in excess and use high concentrations of agar (around 1.5%) to prevent roots growing into media.
10. Use spray bottles with adjustable spraying nozzles to minimize droplet size, thus obtaining a more even distribution of the inductor.
11. Because dexamethasone is systematically transported throughout the plant (22), systems induced by β-estradiol are preferable for local induction experiments.
12. Properly clean the area where induction occurred with ethanol. Avoid breathing vapors from agar containing β-estradiol.
13. β-estradiol is not easily transported in adult plants (30): applications on leaves do not affect newly produced organs. Brushing plant surfaces with 5–10 μ*M* estradiol solutions can be used to promote local inductions (31).
14. Prepare new lanolin β-estradiol mixes every time you need them.
15. Both seedlings in sealed petri dishes and plants on soil can be induced.
16. Three to five cycles of inductions of 1-h each promote stronger system activity in comparison to single inductions, without altering plant development (Y. Laizet, internal communication). The HSP promoter stays active for approximately 24 h after induction (36).

17. High temperatures are detrimental to pollen development: carry on parallel experiments with wild-type plants to recognize and discard false positive effects.

Acknowledgment

I would like to thank Dr. Yec'han Laizet for improving the activation of the heat shock system, Dr. Mark Curtis for the support provided with the β-estradiol-inducible system, and Prof. Dr. Rüdiger Simon for the advice in calibrating the ethanol-inducible system.

References

1. Rutherford, S., Brandizzi, F., Townley, H., Craft, J., Wang, Y., Jepson, I., Martinez, A., and Moore, I. (2005) Improved transcriptional activators and their use in mis-expression traps in Arabidopsis. *Plant J* **43**, 769–788.
2. Baroux, C., Blanvillain, R., Betts, H., Batoko, H., Craft, J., Martinez, A., Gallois, P., and Moore, I. (2005) Predictable activation of tissue-specific expression from a single gene locus using the pOp/LhG4 transactivation system in Arabidopsis. *Plant Biotechnol J.* **3** 91–101.
3. Schubert, D., Lechtenberg, B., Forsbach, A., Gils, M., Bahadur, S., and Schmidt, R. (2004) Silencing in Arabidopsis T-DNA transformants: the predominant role of a gene-specific RNA sensing mechanism versus position effects. *Plant Cell* **16**, 2561–2572.
4. Kang, H. G., Fang, Y., and Singh, K. B. (1999) A glucocorticoid-inducible transcription system causes severe growth defects in Arabidopsis and induces defense-related genes. *Plant J* **20**, 127–133.
5. Andersen, S. U., Cvitanich, C., Hougaard, B. K., Roussis, A., Gronlund, M., Jensen, D. B., Frokjaer, L. A., and Jensen, E. O. (2003) The glucocorticoid-inducible GVG system causes severe growth defects in both root and shoot of the model legume *Lotus japonicus*. *Mol Plant Microbe Interact* **16**, 1069–1076.
6. Ouwerkerk, P. B., de Kam, R. J., Hoge, J. H., and Meijer, A. H. (2001) Glucocorticoid-inducible gene expression in rice. *Planta* **213**, 370–378.
7. Amirsadeghi, S., McDonald, A. E., and Vanlerberghe, G. C. (2007) A glucocorticoid-inducible gene expression system can cause growth defects in tobacco. *Planta* **226**, 453–463.
8. Caddick, M. X., Greenland, A. J., Jepson, I., Krause, K. P., Qu, N., Riddell, K. V., Salter, M. G., Schuch, W., Sonnewald, U., Tomsett, A. B. et al. (1998) An ethanol inducible gene switch for plants used to manipulate carbon metabolism. *Nat Biotechnol* **16**, 177–180.
9. Roslan, H. A., Salter, M. G., Wood, C. D., White, M. R., Croft, K. P., Robson, F., Coupland, G., Doonan, J., Laufs, P., Tomsett, A. B., and Caddick, M. X. (2001) Characterization of the ethanol-inducible alc gene-expression system in *Arabidopsis thaliana*. *Plant J* **28**, 225–235.
10. Salter, M. G., Paine, J. A., Riddell, K. V., Jepson, I., Greenland, A. J., Caddick, M. X., and Tomsett, A. B. (1998) Characterisation of the ethanol-inducible *alc* gene expression system for transgenic plants. *Plant J* **16**, 127–132.
11. Muller, R., Borghi, L., Kwiatkowska, D., Laufs, P., and Simon, R. (2006) Dynamic and compensatory responses of Arabidopsis shoot and floral meristems to CLV3 signaling. *Plant Cell* **18**, 1188–1198.
12. Deveaux, Y., Peaucelle, A., Roberts, G. R., Coen, E., Simon, R., Mizukami, Y., Traas, J., Murray, J. A., Doonan, J. H., and Laufs, P. (2003) The ethanol switch: a tool for tissue-specific gene induction during plant development. *Plant J.* **36**, 918–930.
13. Maizel, A. and Weigel, D. (2004) Temporally and spatially controlled induction of gene expression in *Arabidopsis thaliana*. *Plant J* **38**, 164–171.

14. Schena, M., Picard, D., and Yamamoto, K. R. (1991) Vectors for constitutive and inducible gene expression in yeast. *Meth Enzymol* **194**, 389–398.
15. Picard, D. (1993) Steroid-binding domains for regulating the functions of heterologous proteins in cis. *Trends Cell Biol* **3**, 278–280.
16. Kirch, T., Simon, R., Grunewald, M., and Werr, W. (2003) The *DORNROSCHEN/ENHANCER OF SHOOT REGENERATION1* gene of Arabidopsis acts in the control of meristem cell fate and lateral organ development. *Plant Cell* **15**, 694–705.
17. Jasinski, S., Piazza, P., Craft, J., Hay, A., Woolley, L., Rieu, I., Phillips, A., Hedden, P., and Tsiantis, M. (2005) KNOX action in Arabidopsis is mediated by coordinate regulation of cytokinin and gibberellin activities. *Curr Biol* **15**, 1560–1565.
18. Aoyama, T. and Chua, N. H. (1997) A glucocorticoid-mediated transcriptional induction system in transgenic plants. *Plant J* **11**, 605–612.
19. Aoyama, T. (1998) Glucocorticoid-inducible gene expression in plants. In: Inducible Gene Expression in Plants, Reynolds, P., ed. CAB International, Wallingford, CT, pp. 44–59.
20. van Dijken, A. J., Schluepmann, H., and Smeekens, S. C. (2004) Arabidopsis trehalose-6-phosphate synthase 1 is essential for normal vegetative growth and transition to flowering. *Plant Physiol* **135**, 969–977.
21. Moore, I., Galweiler, L., Grosskopf, D., Schell, J., and Palme, K. (1998) A transcription activation system for regulated gene expression in transgenic plants. *Proc Natl Acad Sci USA* **95**, 376–381.
22. Craft, J., Samalova, M., Baroux, C., Townley, H., Martinez, A., Jepson, I., Tsiantis, M., and Moore, I. (2005) New pOp/LhG4 vectors for stringent glucocorticoid-dependent transgene expression in Arabidopsis. *Plant J* **41**, 899–918.
23. Reddy, G. V. and Meyerowitz, E. M. (2005) Stem-cell homeostasis and growth dynamics can be uncoupled in the Arabidopsis shoot apex. *Science* **310**, 663–667.
24. Wielopolska, A., Townley, H., Moore, I., Waterhouse, P., and Helliwell, C. (2005) A high-throughput inducible RNAi vector for plants. *Plant Biotechnol J* **3**, 583–590.
25. Samalova, M., Brzobohaty, B., and Moore, I. (2005) pOp6/LhGR: a stringently regulated and highly responsive dexamethasone-inducible gene expression system for tobacco. *Plant J* **41**, 919–935.
26. Schoof, H., Lenhard, M., Haecker, A., Mayer, K. F., Jurgens, G., and Laux, T. (2000) The stem cell population of Arabidopsis shoot meristems in maintained by a regulatory loop between the CLAVATA and WUSCHEL genes. *Cell* **100**, 635–644.
27. Segal, G., Song, R., and Messing, J. (2003) A new opaque variant of maize by a single dominant RNA-interference-inducing transgene. *Genetics* **165**, 387–397.
28. Zuo, J., Niu, Q. W., and Chua, N. H. (2000) Technical advance: An estrogen receptor-based transactivator XVE mediates highly inducible gene expression in transgenic plants. *Plant J* **24**, 265–273.
29. Zuo, J. and Chua, N. H. (2000) Chemical-inducible systems for regulated expression of plant genes. *Curr Opin Biotechnol* **11**, 146–151.
30. Tornero, P., Chao, R. A., Luthin, W. N., Goff, S. A., and Dangl, J. L. (2002) Large-scale structure-function analysis of the Arabidopsis RPM1 disease resistance protein. *Plant Cell* **14**, 435–450.
31. Brand, L., Horler, M., Nuesch, E., Vassalli, S., Barrell, P., Yang, W., Jefferson, R. A., Grossniklaus, U., and Curtis, M. D. (2006) A versatile and reliable two-component system for tissue-specific gene induction in Arabidopsis. *Plant Physiol* **141**, 1194–1204.
32. Takahashi, T. and Komeda, Y. (1989) Characterization of two genes encoding small heat-shock proteins in *Arabidopsis thaliana*. *Mol Gen Genet* **219**, 365–372.
33. Shinmyo, A., Shoji, T., Bando, E., Nagaya, S., Nakai, Y., Kato, K., Sekine, M., and Yoshida, K. (1998) Metabolic engineering of cultured tobacco cells. *Biotechnol Bioeng* **58**, 329–332.
34. Matsuhara, S., Jingu, F., Takahashi, T., and Komeda, Y. (2000) Heat-shock tagging: A simple method for expression and isolation of plant genome DNA flanked by T-DNA insertions. *Plant J* **22**, 79–86.
35. Masclaux, F., Charpenteau, M., Takahashi, T., Pont-Lezica, R., and Galaud, J. P. (2004) Gene silencing using a heat-inducible RNAi system in Arabidopsis. *Biochem Biophys Res Commun* **321**, 364–369.
36. Yoshida, K., Kasai, T.Garcia, M. R., Sawada, S., Shoji, T., Shimizu, S., Yamazaki, K., Komeda, Y., and Shinmyo, A. (1995) Heat-inducible expression system for a foreign gene in cultured tobacco cells using the *HSP18.2* promoter of *Arabidopsis thaliana*. *Appl Microbiol Biotechnol* **44**, 466–472.
37. Roberts, G. R., Roberts, G. R., Garoosi, G. A., Koroleva, O., Ito, M., Laufs, P., Leader, J., Caddick, M. X., Doonan, J. H., and Tomsett, A. B. (2005) The alc-GR system: a modified alc gene switch designed for use

in plant tissue culture. *Plant Physiol* **138**, 1259–1267.
38. Moore, I., Samalova, M., and Kurup, S. (2006) Transactivated and chemically inducible gene expression in plants. *Plant J* **45**, 651–683.
39. Borghi, L., Bureau, M., and Simon, R. (2007) Arabidopsis JAGGED LATERAL ORGANS is expressed in boundaries and coordinates KNOX and PIN activity. *Plant Cell* **19**, 1795–1808.
40. Hong, S. W. and Vierling, E. (2000) Mutants of *Arabidopsis thaliana* defective in the acquisition of tolerance to high temperature stress. *Proc Natl Acad Sci USA* **97**, 4392–4397.
41. Sweetman, J. P., Chu, C., Qu, N., Greenland, A. J., Sonnewald, U., and Jepson, I. (2002) Ethanol vapor is an efficient inducer of the alc gene expression system in model and crop plant species. *Plant Physiol* **129**, 943–948.
42. Laufs, P., Coen, E., Kronenberger, J., Traas, J., and Doonan, J. (2003) Separable roles of UFO during floral development revealed by conditional restoration of gene function. *Development* **130**, 785–796.

Chapter 6

Trichome Development in *Arabidopsis*

Joachim F. Uhrig and Martin Hülskamp

Abstract

Arabidopsis trichomes are giant single epidermal cells that are easily accessible for genetic, genomic and cell-biological analysis. They have therefore become a convenient model system to study developmental and physiological processes. Trichome studies are greatly facilitated by methods specifically applicable for this particular cell type. In addition, it is very important to use conventions and definitions that have been developed to make studies comparable and capture the relevant aspects. This chapter will highlight these two aspects of trichome analysis.

Key words: Pattern formation, cell cycle, endoreduplication, expression profiling.

1. Introduction

Arabidopsis trichomes are single epidermal cells that are found on various organs including rosette and cauline leaves and the stem. Trichome development is mostly studied on rosette leaves. Here, incipient trichomes are found in a regular spacing pattern at the base of young developing leaves (1). Due to cell divisions of the intervening epidermal pavement cells, trichomes become further separated. The differentiating trichomes undergo typically four endoreduplication cycles and form three or four branches (1). The systematic search for trichome mutants has revealed a large number of mutants interfering with various developmental processes (2–5). Based on these mutants, trichomes are used to study pattern formation, cell-cycle regulation and cell morphogenesis. In addition, trichomes have been established as a model system for single cell type-specific biochemical and genomic approaches (6–8). In the following, we will describe selected methods that are central for the analysis of the above-mentioned processes.

L. Hennig, C. Köhler (eds.), *Plant Developmental Biology*, Methods in Molecular Biology 655,
DOI 10.1007/978-1-60761-765-5_6, © Springer Science+Business Media, LLC 2010

2. Materials

2.1. Plant Material and Growth Conditions

1. Plant material: Commonly used ecotypes for trichome analyses are *Arabidopsis thaliana* var. Columbia (Col-0) and *A. thaliana* var. Landsberg *erecta* (L*er*) (*see* **Note 1**), and any controlled growth conditions may be suitable for trichome analyses (*see* **Note 2**).
2. Murashige Skoog (MS) medium (9): 0.5×–1× MS salts (*Duchefa*, Haarlem, Niederlande), supplemented optionally with 10–30 g/L sucrose and/or with 0.5× vitamins–glycine mix. Adjust pH to 5.8 with KOH, optionally add 8 g/L agar and autoclave.
3. Vitamins–glycine mix (100×): 100 mg/L nicotinic acid, 1 g/L thiamine-HCl, 100 mg/L, pyridoxine-HCl, and 400 mg/L glycine. Filter sterilize and store at −20°C.

2.2. Analysis of the Cell Cycle

1. 4′,6-Diamino-2-phenylindole (DAPI).
2. PBS buffer: 8 g NaCl, 0.2 g KCl, 1.44 g Na_2HPO_4, 0.24 g KH_2PO_4 in 800 mL of distilled H_2O. Adjust the pH to 7.4 with HCl. Add water to 1L. Sterilize by autoclaving.
3. PBT buffer: PBS buffer + 0.1% Triton X100.
4. Formaldehyde.
5. Ethanol (70%).
6. Glycerol.
7. Fluorescence microscope (e.g. DMR Fluorescence Microscope, *Leica* Camera AG, Solms, Germany).
8. CCD camera with analysis software (e.g. DISKUS, Carl H. Hilgers-Technisches Büro, Königswinter, Germany).

2.3. Analysis of Cell Morphogenesis

1. Particle gun, e.g. PDS-1000 He Biolistic Particle Delivery System (Biorad).
2. Gold particles.

2.4. Trichomes as a Single Cell System for Biochemical and Genomics Analyses

1. Borosilicate glass capillaries (WPI, Berlin, Germany).
2. List pipette puller (Darmstadt, Germany) with a tip aperture of 1–10 μm.
3. Stereo zoom microscope (e.g. Optiphot 2 microscope, Nikon, Duesseldorf, Germany) with micromanipulator.
4. Liquid nitrogen.
5. Extra-fine forceps.
6. Trypsin buffer: 5% acetonitrile, 100 mM NH_4HCO_2, 5 mM $CaCl_2$.

7. Guanidinium isothiocyanate buffer: 4 M guanidinium isothiocyanate, 25 mM sodium citrate, 0.5% sarkosyl, 0.1% beta-mercaptoethanol; or commercially available buffers such as RNeasy^{TM} RLT buffer (Qiagen, Hilden, Germany).
8. 60/80 μm glass beads.
9. Screen door mesh (Home Depot).
10. 100-μm cell strainer (Falcon-Becton Dickinson).
11. Diethyl pyrocarbonate (DEPC).
12. RNase inhibitor (e.g. RNAsin, Promega, Heidelberg, Germany).

3. Methods

3.1. Analysis of Pattern Formation

Trichomes on rosette leaves develop at the base of developing leaves such that trichomes at the tip of a leaf are older than more basal ones (1). During leaf stages when new trichomes are still formed, a region called patterning zone can be defined that is characterized by the presence of only unbranched trichomes (*see* **Fig. 6.1**, region A). In this region, new trichomes are formed basal to the already existing ones. In region B, trichomes become separated by cell divisions of the epidermal cells.

In principle, three aspects of pattern formation can be analysed: first, the establishment of the initial pattern from equivalent epidermal cells in region A, second the placement of trichomes between already existing mature trichomes in region B, third, the separation of trichomes by epidermal cell divisions in region B. The analysis of patterning by genetic or pharmacological manipulations relies on a meaningful quantification of mature trichomes on rosette leaves and the ability to trace the development on growing leaf primordia.

3.1.1. Quantification of Trichomes on Rosette Leaves

It is imperative to use plants of similar developmental stage and the same leaf number throughout a given study as trichome number can vary considerably. In particular, the first two leaves have significantly less trichomes than leaves initiated later. A meaningful analysis should distinguish two major aspects of trichome patterning: the overall density and cluster formation, which does not occur in wild-type *Arabidopsis*. Genetic evidence indicates that these are separate regulatory events (3, 10).

3.1.1.1. Trichome Density

For the determination of trichome density, it is in most cases sufficient to simply compare the number of trichomes on mature rosette leaves provided that the leaf size is not altered. Due to considerable variations in trichome numbers, it is necessary to

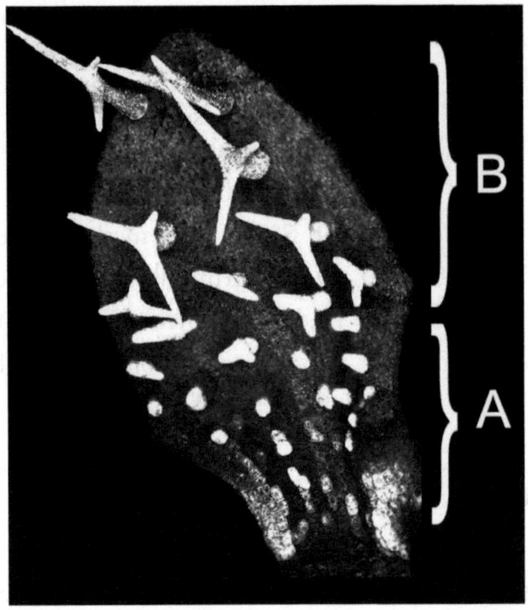

Fig. 6.1. **Overview of a young Arabidopsis leaf depicting two trichome patterning regions.** Leaf of a transgenic plant expressing GFP under the trichome-specific *GL2* promoter reconstructed from Confocal Laser Scanning Microscope sections. Region A harbours young trichomes with no more than two branches and new trichomes emerge in areas basal to the already existing trichomes. Region B has three- and more branched trichomes that become separated by epidermal cell divisions. In some situations, new trichomes can emerge between already existing older trichomes.

analyze a sufficient number of leaves (at least ten) from different individuals to permit statistically sound conclusions.

3.1.1.2. Trichome Cluster

During the analysis of mutants with trichome clusters, the determination of trichome number may be misleading as density and local patterning effects around a trichome are mixed. In such cases, determine three values:

First, the number of trichome initiation sites (TIS) (11). One trichome cluster with two or more trichomes is considered one TIS and therefore equivalent to a single trichome. This value can be directly compared to the trichome density of leaves showing no cluster formation.

Second, the frequency of clusters among the trichome initiation sites.

Third, the average number of trichomes in trichome clusters.

3.1.2. Tracing Trichome Development on Growing Leaves

The quantification of trichome density and clustering does not allow judging, when and how the pattern is established. For example, a higher density can be caused by a higher initiation density in region A, by more intercalary trichome initiation in region B, or by differences in leaf growth. To resolve this, it is possible to monitor the development of a single trichome from the earliest stages on.

1. Dissect soil-grown young plants with no more than four rosette leaves using a hypodermic needle as a micro knife under a dissecting microscope. Select a young leaf (about 2 mm long) and remove all other leaves. Make sure not to hurt stem and hypocotyl.
2. Cut off the hypocotyl to produce a 5-mm section.
3. Place the hypocotyl end in 1% Murashige Skoog (MS) solid medium in a Petri dish. Take care to prevent the leaf from touching the surface of the medium.
4. Keep the Petri dish closed except during microscopic investigation (*see* **Note 3**). With this procedure it is easily possible to monitor the development of trichomes for several days. The use of appropriate fluorescent markers greatly facilitates the analysis as it enables the use of a confocal laser-scanning microscope and thereby a cellular and even sub-cellular resolution (*see* **Fig. 6.2**).

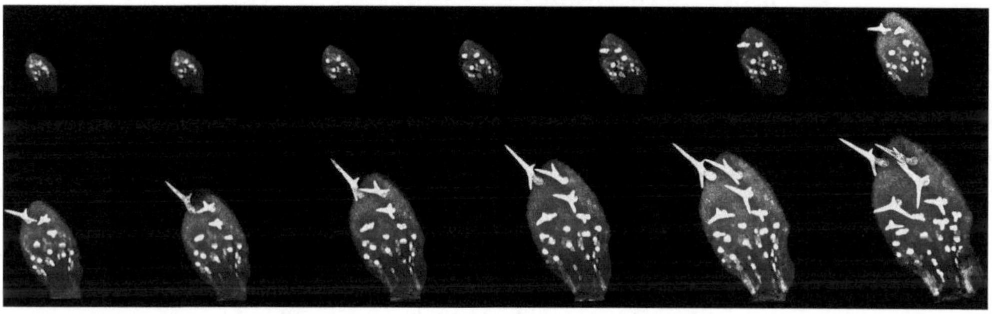

Fig. 6.2. **Time series of developing trichomes on a developing leaf.** Pictures of one young leaf were taken in intervals of about 6 h. Individual pictures were reconstructed from Confocal Laser Scanning Microscope sections of a plant expressing GFP under the trichome-specific *GL2* promoter.

3.2. Analysis of the Cell Cycle

Wild-type leaf trichomes undergo on average four endoreduplication cycles, that is DNA replication cycles in the absence of nuclear and cellular divisions (1). Various mutants have been identified that change the trichome cell cycle including mutants resulting in more or fewer endoreduplication cycles (12, 13), mutants that undergo nuclear but not cellular divisions (14) and mutants with complete cell divisions (15). In addition, overexpression studies have been an extremely powerful tool to study the role of specific cell-cycle genes (16–18). Below, two methods are described that are most relevant for the analysis of the cell cycle in *Arabidopsis* trichomes.

3.2.1. Measurement of the DNA Content

The DNA content of trichomes is typically measured using DAPI to label the DNA.

1. Collect rosette leaves at the desired stage and fix for 4 h in 3.7% formaldehyde in PBT (PBS + 0.1% Triton X-100) buffer.
2. Wash leaves three times for 10 min in PBT.
3. Stain leaves for 15 min with DAPI (5 mg/mL in PBT).
4. Wash leaves three times for 10 min in PBT.
5. Wash leaves two times for 15 min in 70% ethanol.
6. Mount leaves on slides in glycerol.

Measure fluorescence with a fluorescent microscope equipped with a CCD camera and a software enabling the quantitative analysis of pixel intensities in selected areas.

1. Select a single trichome and mark the area of the nucleus to determine the total fluorescence intensity (*see* **Note 4**).
2. Measure background fluorescence in an area of equal size outside the nucleus. Subtract background intensity from the nuclear intensity.
3. Measure nuclei of stomata cells on the same leaf in parallel. Stomata have a DNA content of 2*C* and serve as a reference to calculate the *C*-value of trichomes.

3.2.2. Manipulating the Cell Cycle in Trichomes

Plants devoid of trichomes are perfectly viable under laboratory conditions. Therefore, trichome-specific manipulations of gene activities are possible even if lethal for the trichome cells. To do this, genes of interest are expressed under the control of a trichome-specific promoter. Typically, the GLABRA2 (*GL2)* promoter is used that mediates gene expression in selected cell types in roots, hypocotyl and leaves including the trichomes (16–19). Although not perfectly trichome specific, this promoter was successfully used to express a number of cell-cycle genes without causing detrimental effects to the whole plant. In case a more specific trichome promoter is desired, several trichome specific promoters are now available including the *CAPRICE* and *TRIPTYCHON* promoters (11).

3.3. Analysis of Cell Morphogenesis

Trichome morphogenesis in *Arabidopsis* involves a sequence of defined stages that are commonly used as a reference (19). During the first stage, the trichome initial undergoes the first endoreduplication round and increases in size by radial expansion. Stage 2 is characterized by growth perpendicular to the leaf surface. Subsequently, two or more branching events occur (stage 3). After branch initiation, cell elongation occurs with the tips still blunted (stage 4). Cell growth continues and eventually the tip becomes pointed (stage 5). The mature cell develops papillae on the surface (stage 6). The mature trichome has a remarkably regular architecture with characteristic branch positions that are aligned with respect to the apical basal leaf axis. In principle, two aspects of cell morphogenesis can be studied using trichomes as a model system: the establishment of polarity with several axes and the expansion growth of the stem and the branches. Depending on the specific question, careful quantification of the trichome architecture is required. Below, we describe the relevant aspects for the quantitative analysis of trichome morphogenesis and a method enabling the single trichome transformation with genes of interest. Numerous techniques to visualize intracellular components with antibodies or GFP variants are available but are not considered here as trichome-specific techniques.

3.3.1. Quantitative Analysis of Trichome Morphogenesis

A large number of trichome morphogenesis mutants have revealed that most parameters of trichome form are genetically controlled and are therefore potentially relevant for a quantitative analysis (12, 20). **Figure 6.3** shows a typical trichome in which all parameters are highlighted that have been shown to

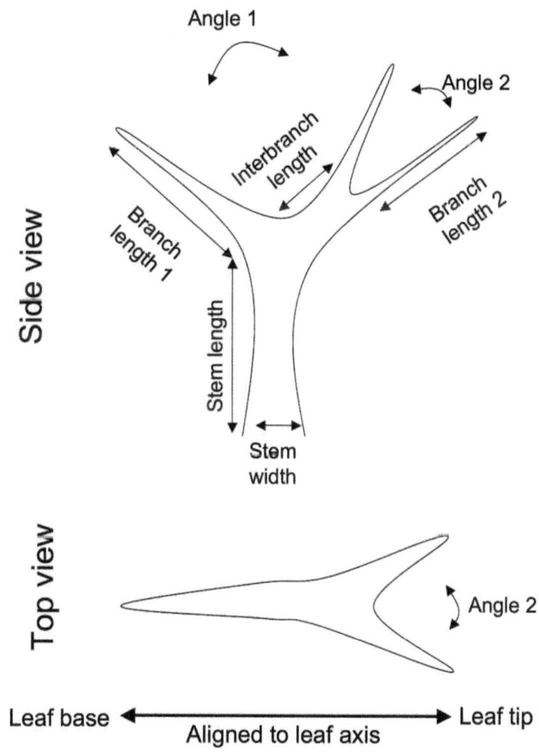

Fig. 6.3. **Possible quantification of relevant parameters of Arabidopsis trichome architecture.** A typical three-branched Arabidopsis trichome is sketched from the side view (*top*) and a top view (*bottom*). All parameters depicted here have been shown to be altered in mutants and are therefore genetically controlled. The first branching event gives rise to two branches that point to the basal part and the tip of the leaf. The second branching event occurs on the tip pointing branch and the branching plane is perpendicular to the first one. Note that the angles between the lower and the upper branches are different.

be under genetic control. Depending on the particular question, these parameters can be used for a quantitative analysis. Conventional light microscopy or Scanning Electron Microscopy can be used for measurements (*see* **Note 5**).

3.3.2. Transient Expression by Particle Bombardment

Frequently it is desired to analyze gene function in developing trichomes in a short time frame. The localization of the GFP-tagged protein, for example, shall be analyzed, the ability of a candidate gene to rescue a mutant phenotype or the effect of overexpression shall be tested. One elegant way to greatly accelerate this type of analysis is transient transformation of young trichomes by particle bombardment.

1. Prepare young leaves as described in **Section 3.1.2**.
2. Arrange about 15–20 leaves on solid 1% MS medium in a Petri dish.
3. Optimize DNA concentrations, distances and rupture disks to determine the region in which the gold particles are delivered without destroying the tissue. Using a PDS-1000 He Biolistic Particle Delivery System, the following parameters work best in our hands: gold particle size: 1 μm; amount of DNA: 300 ng; distance of plant material from rupture disk: 8–10 cm; rupture disk specification: 900 psi. One example showing a single GFP-labelled trichome is shown in **Fig. 6.4**.

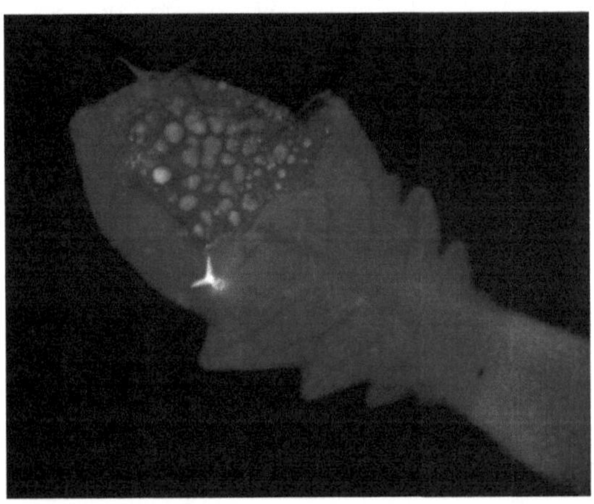

Fig. 6.4. **Single trichome transformation.** Using the particle-bombardment method, a single trichome cell was transformed early in development. After 12 h, this trichome is recognized as a GFP-labelled cell on an otherwise unlabelled leaf.

3.4. Trichomes as a Single Cell System for Biochemical and Genomics Analyses

Size and characteristic shape of trichomes allow direct macroscopic inspection and measurements, and there are a number of quite simple procedures to isolate and purify the cellular content of trichomes and even intact trichome cells to homogeneity. This gives rise to the exceptional possibility to perform single cell type-specific biochemical analyses, transcription profiling and proteomics approaches to investigate trichome metabolism and cell morphogenesis.

Specific cell sampling and the application of nano LC/MS/MS shotgun peptide sequencing allow cell type-specific protein profiling. Using the cellular content of 1000–2000 individually sampled trichome cells, more than 60 unique proteins have been identified, including several proteins involved in sulphur metabolism and a glutathione S-conjugate translocator (21). This finding is in agreement with trichome-specific gene expression data and supports the role of trichomes as a sink during detoxification processes (22).

Manual collection, microcapillary sampling and isolation of intact trichome cells have been successfully applied for expression profiling (6–8). Mature trichome cells contain approximately 0.1 ng of total RNA (6). Therefore, a few hundred cells yield sufficient material for microarray or filter array hybridization experiments. These expression profiles confirmed the trichome-specific expression of transcription factors such as GL2, ETC1, TRY, CPC and TTG2 (8), and generated information about trichome differentiation and endocycling (7). Furthermore, trichome-specific expression profiling has been instrumental to reveal the identity of the *NOECK* gene, underlying the over-branched phenotype of the *noeck* mutants (6, 12).

Additionally, isolated intact trichome cells have been used for the analysis of DNA content, for immunolocalization studies and to probe the biochemical composition of trichome cell walls (8, 23).

3.4.1. Microcapillary Sampling of Trichome Content

Trichomes have been used as model cells to investigate developmental processes with the ultimate spatial resolution of single cells. Kryvych et al. (7) used the individually sampled cellular content of as few as ten trichome cells for the generation of probes for gene expression profiling. The extraordinary size and easy accessibility of trichome cells on the leaf epidermis allow the direct sampling of cellular contents using microcapillaries. Borosilicate glass capillaries with a tip aperture of 1–10 μm have been found suitable for trichomes (7, 24). Young rosette leaves are fixed under a microscope equipped with a remote-controlled micromanipulator. The capillary is inserted into a single trichome cell, and immediately after withdrawal, the content of the capillary was released into a sterile, 0.5-mL reaction vial containing 1 mL of DEPC-treated water with 5 U of RNase inhibitor. Each capillary was used for only one sampling process. Pooling the samples of ten individually probed trichome cells provides sufficient material for subsequent cDNA synthesis, quantitative PCR analyses and probe generation for transcription profiling approaches by array hybridization.

3.4.2. Manual Collection of Trichomes

The exposed position and the large size of trichome cells allow rather simple experimental approaches to collect sufficient quantities of homogeneous cell-specific material to perform metabolite, protein and transcript analyses. Several groups have used clipping off cells from the epidermis of frozen leaves using extra-fine forceps with immediate transfer to buffers suitable for either protein or transcript analysis. Each wild-type trichome cell contains on average of 0.1 ng of total RNA and approximately 1 ng of total protein. Therefore, a few hundred cells provide sufficient material for Affymetrix microarray hybridizations (~100 ng starting material) or for protein profiling using, for example, a combination of HPLC separation with subsequent tandem mass spectrometry (LC/MS/MS).

3.4.3. Collection of Biochemically Intact Trichome Cells

Trichome cells are only rather loosely attached to the leaf surface mainly by the cross-linked pectins in the middle lamellae. These cross-linkings can be destabilized by chemicals removing Ca^{2+} ions. Such treatment leads to the detachment of biochemically intact trichome cells in large quantities suitable for a range of different downstream analyses.

A rather simple method to isolate intact trichome cells has been described by Zhang and Oppenheimer (23). Leaves are treated with EGTA in the presence of 0.05% Triton X-100 and trichomes are detached from the leaf surface by gentle rubbing using a small paint brush. The cells are then collected by low-speed centrifugation and washed several times. These isolated trichomes remained intact and suitable for immunolocalization studies and DNA content analyses. The method has recently been improved and further developed for isolating trichome cells on a larger scale to facilitate biochemical analyses and proteomic and transcriptomic approaches (8).

1. Add approximately 1.5 g of 4 week-old seedlings to a 50-mL test tube containing 15 mL of modified PBS buffer with 50 mM EGTA, pH 7.5, and 50 mg of 60/80-μm glass beads.
2. Mix at maximum speed on a vortexer with four cycles of 30 s on and 30 s rest on ice.
3. Wash detached trichomes from the plant material by filtering through a coarse-meshed sieve.
4. Rinse the plant material several times with modified PBS buffer without EGTA.
5. Collect trichomes from this filtrate by sieving through a 100-μm cell strainer.
6. Wash with modified PBS.
7. Wash trichomes out of the inverted cell strainer using approximately 10 mL PBS solution.

8. Transfer the trichome suspension to a 15-mL centrifuge tube and centrifuge for 1.5 min at 150×*g*.
9. Remove supernatant. Trichomes are now ready for downstream applications. In this way, up to 3000 trichomes per gram plant material can be isolated.

4. Notes

1. Due to the variability in trichome density and morphogenesis between different *Arabidopsis thaliana* accessions, comparative experiments should be carried out always within the same ecotype.
2. *A. thaliana* plants can be grown in a greenhouse or in growth chambers, on soil or under aseptic conditions on solid Murashige Skoog (MS) medium (9) and with continuous illumination or in different photoperiods. For the profiling experiments described here, plants were usually grown for 2–4 weeks.
3. If during microscopy the leaf gets wet, the water needs to be removed before further culturing because insufficient aeration greatly reduces the fitness of the detached leaves.
4. It is important to avoid saturation of pixel intensities.
5. It is extremely important to view the trichome from the correct angle to obtain the right values.

References

1. Hülskamp, M., Misra, S., and Jürgens, G. (1994) Genetic dissection of trichome cell development in Arabidopsis. *Cell* **76**, 555–566.
2. Szymanski, D. B., Lloyd, A. M., and Marks, M. D. (2000) Progress in the molecular genetic analysis of trichome initiation and morphogenesis in Arabidopsis. *Trends Plant Sci* **5**, 214–219.
3. Larkin, J. C., Brown, M. L., and Schiefelbein, J. (2003) How do cells know what they want to be when they grow up? Lessons from epidermal patterning in Arabidopsis. *Annu Rev Plant Biol* **54**, 403–430.
4. Hülskamp, M. (2004) Plant trichomes: a model for cell differentiation. *Nat Rev Mol Cell Biol* **5**, 471–480.
5. Smith, L. G. and Oppenheimer, D. G. (2005) Spatial control of cell expansion by the plant cytoskeleton. *Annu Rev Cell Dev Biol* **21**, 271–295.
6. Jakoby, M. J., Falkenhan, D., Mader, M. T., Brininstool, G., Wischnitzki, E., Platz, N., Hudson, A., Hülskamp, M., Larkin, J., and Schnittger, A. (2008) Transcriptional profiling of mature Arabidopsis trichomes reveals that *NOECK* encodes the MIXTA-like transcriptional regulator MYB106. *Plant Physiol* **148**, 1583–1602.
7. Kryvych, S., Nikiforova, V., Herzog, M., Perazza, D., and Fisahn, J. (2008) Gene expression profiling of the different stages of *Arabidopsis thaliana* trichome development on the single cell level. *Plant Physiol Biochem* **46**, 160–173.
8. Marks, M. D., Betancur, L., Gilding, E., Chen, F., Bauer, S., Wenger, J. P., Dixon, R. A., and Haigler, C. H. (2008) A new method for isolating large quantities of Arabidopsis trichomes for transcriptome, cell wall and other types of analyses. *Plant J* **56**, 483–492.

9. Murashige, T. and Skoog, F. (1962) A revised medium for rapid growth and bio assays with tobacco tissue cultures. *Physiologia Plantarum* **15**, 473–497.
10. Pesch, M. and Hülskamp, M. (2004) Creating a two-dimensional pattern de novo during Arabidopsis trichome and root hair initiation. *Curr Opin Genet Dev* **14**, 422–427.
11. Schellmann, S., Schnittger, A., Kirik, V., Wada, T., Okada, K., Beermann, A., Thumfahrt, J., Jürgens, G., and Hülskamp, M. (2002) TRIPTYCHON and CAPRICE mediate lateral inhibition during trichome and root hair patterning in Arabidopsis. *Embo J* **21**, 5036–5046.
12. Folkers, U., Berger, J., and Hülskamp, M. (1997) Cell morphogenesis of trichomes in Arabidopsis: Differential control of primary and secondary branching by branch initiation regulators and cell growth. *Development* **124**, 3779–3786.
13. Perazza, D., Herzog, M., H¨ülskamp, M., Brown, S., Dorne, A. M., and Bonneville, J. M. (1999) Trichome cell growth in *Arabidopsis thaliana* can be derepressed by mutations in at least five genes. *Genetics* **152**, 461–476.
14. Spitzer, C., Schellmann, S., Sabovljevic, A., Shahriari, M., Keshavaiah, C., Bechtold, N., Herzog, M., Müller, S., Hanisch, F. G., and Hülskamp, M. (2006) The Arabidopsis *elch* mutant reveals functions of an ESCRT component in cytokinesis. *Development* **133**, 4679–4689.
15. Walker, J. D., Oppenheimer, D. G., Concienne, J., and Larkin, J. C. (2000) *SIAMESE*, a gene controlling the endoreduplication cell cycle in *Arabidopsis thaliana* trichomes. *Development* **127**, 3931–3940.
16. Schnittger, A., Schobinger, U., Bouyer, D., Weinl, C., Stierhof, Y. D., and Hülskamp, M. (2002) Ectopic D-type cyclin expression induces not only DNA replication but also cell division in Arabidopsis trichomes. *Proc Natl Acad Sci USA* **99**, 6410–6415.
17. Schnittger, A., Schobinger, U., Stierhof, Y. D., and Hülskamp, M. (2002) Ectopic B-type cyclin expression induces mitotic cycles in endoreduplicating Arabidopsis trichomes. *Curr Biol* **12**, 415–420.
18. Schnittger, A., Weinl, C., Bouyer, D., Schobinger, U., and Hülskamp, M. (2003) Misexpression of the cyclin-dependent kinase inhibitor ICK1/KRP1 in single-celled Arabidopsis trichomes reduces endoreduplication and cell size and induces cell death. *Plant Cell* **15**, 303–315.
19. Szymanski, D. B., Jilk, R. A., Pollock, S. M., and Marks, M. D. (1998) Control of GL2 expression in Arabidopsis leaves and trichomes. *Development* **125**, 1161–1171.
20. Zhang, X., Grey, P. H., Krishnakumar, S., and Oppenheimer, D. G. (2005) The *IRREGULAR TRICHOME BRANCH* loci regulate trichome elongation in Arabidopsis. *Plant Cell Physiol* **46**, 1549–1560.
21. Wienkoop, S., Zoeller, D., Ebert, B., Simon-Rosin, U., Fisahn, J., Glinski, M., and Weckwerth, W. (2004) Cell-specific protein profiling in *Arabidopsis thaliana* trichomes: identification of trichome-located proteins involved in sulfur metabolism and detoxification. *Phytochemistry* **65**, 1641–1649.
22. Gutierrez-Alcala, G., Gotor, C., Meyer, A. J., Fricker, M., Vega, J. M., and Romero, L. C. (2000) Glutathione biosynthesis in Arabidopsis trichome cells. *Proc Natl Acad Sci USA* **97**, 11108–11113.
23. Zhang, X. and Oppenheimer, D. G. (2004) A simple and efficient method for isolating trichomes for downstream analyses. *Plant Cell Physiol* **45**, 221–224.
24. Brandt, S., Kehr, J., Walz, C., Imlau, A., Willmitzer, L., and Fisahn, J. (1999) Technical Advance: A rapid method for detection of plant gene transcripts from single epidermal, mesophyll and companion cells of intact leaves. *Plant J* **20**, 245–250.

Chapter 7

Phenotyping the Development of Leaf Area in *Arabidopsis thaliana*

Sarah J. Cookson, Olivier Turc, Catherine Massonnet, and Christine Granier

Abstract

The study of leaf expansion began decades ago and has covered the comparison of a wide range of species, genotypes of a same species and environmental conditions or treatments. This has given rise to a large number of potential protocols for today's leaf development biologists. The final size of the leaf surface of a plant results from the integration of many different processes (which may be quantified by various developmental variables) at different organizational levels, such as, the duration and the rate of leaf production by the plant, the duration and the rate of individual leaf expansion, and also cell production and expansion in the leaf. There is much evidence to suggest that the magnitude of a variable at one organizational scale cannot be inferred to another scale because of different feedbacks from one scale to another. This chapter offers a series of protocols, which are the most commonly used in plant developmental biology, to assess quantitatively leaf expansion both at the scale of the shoot and the individual leaf. The protocols described here are for the comparison of *Arabidopsis thaliana* genotypes, but can be easily adapted to compare leaf expansion under different environmental conditions and in other dicotyledonous plants.

Key words: leaf expansion, leaf production, rate, duration, *Arabidopsis thaliana*.

1. Introduction

Leaf growth and development are affected by hormone, nutritional and environmental cues; combined with this regulation are the internal genetic differences in leaf size and form between species, subspecies and ecotypes of a species. These factors give rise to the extraordinary diversity in leaf shape and size found in nature. Identifying the genetic control of leaf size and

L. Hennig, C. Köhler (eds.), *Plant Developmental Biology*, Methods in Molecular Biology 655,
DOI 10.1007/978-1-60761-765-5_7, © Springer Science+Business Media, LLC 2010

form and the developmental regulators involved, requires a precise characterisation of leaf growth phenotypes with quantitative variables.

In general, the initial investigation of leaf development is made by comparing the number and sizes of leaves of different plants at a given time or developmental stage. Nonetheless, it is clear from growth studies (where the component of time is added to the analysis of size and/or form) that identical leaf size or number can be reached via different patterns of growth (3, 1). Consequently, the next level of leaf development investigation is the characterisation of the patterns of growth, which is done by quantifying the dynamics of expansion of individual leaves of a plant. Here we will describe how the dynamics of individual leaf development can be broken down into quantifiable variables. However, an important consideration is the type of leaf being studied (e.g. juvenile or adult) and its position, as well as the interactions between individual leaf development and the development of the shoot as a whole (2, 3, 4). Therefore, we also describe methods to compare the timing and rate of individual phases of leaf development as part of the whole shoot system.

2. Materials

2.1. Plant Culture

1. The amount of seeds necessary for each genotype (= 120 minimum per genotype).
2. A micro-centrifuge tube per genotype and a pipette.
3. Twenty appropriately labelled cylindrical pots per genotype (approximately 9 cm height and 4.5 cm diameter) with holes or slits at the bottom to prevent water logging.
4. Horticultural soil mixture rich in compost (*see* **Note 1**).
5. Balance and nutrient solution to maintain soil water content of individual pots at identical values.
6. Trays and aluminium foil.
7. Water spray device.
8. Drawing pins of different colours and scaling labels of known length (*see* **Fig. 7.1A**).

2.2. Control of Micro-meteorological Conditions

1. Sensors to measure air temperature and air humidity (e.g. HMP35A Vaisala, Oy, Helsinki, Finland).
2. Light sensor over the waveband of 400–700 nm to measure incident light (e.g. LI-190SB; LICOR, Lincoln, NE, USA).

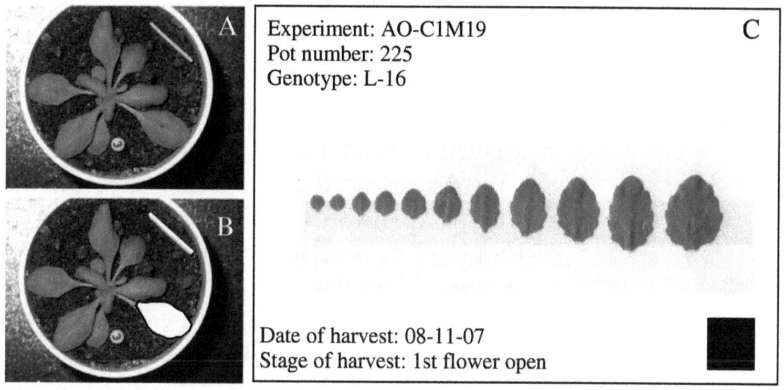

Fig. 7.1. Example of a picture of an *Arabidopsis thaliana* plant marked with a drawing pin and scaled with a label of known length (**A**), a picture of the same plant analysed with an image-analysis software (**B**). Experimental data sheet used for the analysis of the distribution of final leaf area according to the nodal position (**C**). The black square is a scaling mark.

3. Data-logger to record environmental conditions during the experiment (e.g. Campbell Scientific, LTD-CR10 Wiring panel, Sepshed, Leicestershire, UK).

2.3. Leaf Initiation and Expansion of Hidden Leaves

1. Petri dishes and a fridge or cold room at 2–4°C
2. Two scalpels, fine forceps, microscope slides and water.
3. Stereomicroscope with a magnification of at least × 160 (e.g. Leica stereomicroscope, Wild MZ8, Wetzlar, Germany) attached to a CCD camera, which is connected to a computer.
4. Image-analysis software, e.g. Optimas version 5.2 (Bioscan-Optimas V 4.10, Edmonds, WA, USA) or Image J (5).

2.4. Leaf Emergence and Expansion of Emerged Leaves

1. Digital camera connected to a tripod with a height of approximately 50 cm and a spirit-level to ensure that the digital camera is perfectly horizontal.
2. Image-analysis software.
3. Curve-fitting software, e.g. Sigma Plot (Systat Software Inc. San Jose, CA, USA) or R (6).

2.5. Distribution of Final Leaf Area According to Nodal Position

1. A pair of fine scissors, 1 scalpel, paintbrush, double-sided adhesive and A4 sheet of paper or experimental data sheet (*see* **Fig. 7.1C**).
2. Scanner with a minimum resolution of 150 dpi, black and white.

2.6. Statistical Analyses of Data

1. Data-analysis software, e.g. Microsoft Excel.
2. Computer package SPSS 11.0 for Windows (SPSS Inc., Chicago, USA) or R.

3. Methods

During shoot development, leaves are initiated at regular intervals as primordia at specific sites on the shoot apical meristem (7). Expansion of each leaf starts as soon as the leaf is initiated, and as a consequence, leaf expansion from initiation to emergence is hidden in the apical bud. Many studies have ignored this hidden phase of leaf expansion because leaf expansion during this phase is technically difficult to study. The exponential growth of this period of expansion, in dicotyledonous leaves, implies that changes during this time have a strong impact on the growth and development of the leaf as a whole (8, 9). It is therefore necessary to gain insights into this early phase of leaf expansion with destructive measurements, which can be followed by non-destructive measurements after the leaf has emerged.

In this chapter, we will describe how to measure leaf area during the whole development of a given leaf at a given position on the plant, with destructive measurements from leaf initiation to leaf emergence, and with non-destructive measurements from leaf emergence to the completion of leaf expansion. From the resultant leaf growth curves, we will describe how to calculate the dynamic variables required to compare leaf expansion between different genotypes. Because of the interaction between individual leaf and whole shoot development, the explanation of dynamic variables at the scale of the individual leaf will be complemented by the description of a protocol to assess whole shoot development.

3.1. Plant Culture

1. 24 h before sowing, put the necessary quantity of seeds (= 120 minimum per genotype) in a micro-centrifuge tube and cover them with water. Keep the seeds covered by water in the tubes at 4°C for 24 h.
2. Fill each pot with a known and equal amount of soil by weighing each pot on the balance and adjusting its weight with soil when necessary (*see* **Note 2**).
3. Aspirate seeds and water with a suitable pipette and sow the seeds on the soil: one or two seeds in the centre of each pot and three or four additional seeds towards the edges of the pot. The total amount of seeds depends on the germination rate of the seeds and their homogeneity of development.
4. After sowing, arrange the pots randomly in trays and spray some water on the soil surface. Then cover the pots in the

trays with aluminium foil. Leave them covered for 48 h at 21°C.

5. After 48 h, remove the aluminium foil and move the trays into the growth chamber. From this moment (removal of the aluminium foil), do not hesitate to spray water on the pot surface once or twice a day depending on the air circulation in the growth chamber (*see* **Note 3**). Until the end of the experiment, weigh and water your plants daily with nutrient solution to ensure that all plants are grown at comparable soil water content.
6. When more than two thirds of the plants in each pot have reached the stage 1.02, two cotyledons and two visible leaves (10), discard the plants that are not of the correct developmental stage. Leave just two or three healthy plants in each pot with sufficient space between them to ensure that the plants are not in contact with each other.
7. At this stage, 40 plants per genotype are required for the rest of the protocol (*see* **Note 4**).
8. Add markers to the pots, e.g. by using drawing pins of different colours to identify the genotype on photographs. The leaf studied can also be identified by positioning the drawing pin next to the leaf of interest or by a mark of India ink on it (called leaf Z hereafter).
9. Add a scaling label of known length to each pot, at the level of the rosette, to calibrate photographs of the plants (*see* **Fig. 7.1A**, *see* **Note 5**).

3.2. Imposition of Homogeneous Micro-meteorological Conditions (see Note 6)

1. Measure light intensity at the level of the plants in different parts of the growth chamber by displacing the light sensor from one position to another. If necessary, light intensity can be homogenised by changing lamps, reducing or increasing plant distance from the light source, adding neutral filters or increasing the number of reflective surfaces.
2. Measure air humidity and air temperature at the level of the plants in different parts of the growth chamber by displacing the humidity and temperature sensors from one position to another. Both variables can be homogenised by fine-tuning of the air circulation in the chamber.
3. If possible, use a data logger to record air temperature, air humidity and incident light for the full duration of the experiment or at least for selected days (i.e. around harvest time) to check that there is no deviation during the experiment.

3.3. Leaf Initiation and Expansion of Hidden Leaves

3.3.1. Leaf Initiation Rate, the Plastochron and the Date of initiation of a given leaf (leaf Z)

1. Harvest five rosettes per genotype at intervals of 2–3 d (*see* **Note 7**) from stage 1.02, two cotyledons and two visible leaves, until stage 5.01, bolting (10). Cut the rosettes from the pot surface with a pair of fine scissors and put them in a labelled Petri dish with some water. Store at 2–4°C until dissection (maximum 2–3 d later). Dissect the harvested rosettes in a drop of water using the microscope at magnification ×160. Remove the two cotyledons and subsequent leaves one by one with the two scalpels until all initiated leaves are visible on the apex.
2. Count the number of removed leaves and leaf primordia visible on the apex (the smallest leaf is visible when its area is approximately 0.001 mm^2 at this magnification).
3. Plot the total number of counted leaves against time (*see* **Fig. 7.2A**).
4. Estimate leaf initiation rate by the slope of the relationship between the number of initiated leaves against time (*see* **Fig. 7.2A**).
5. Calculate plastochron, the time necessary to initiate a new leaf, by the reciprocal of leaf initiation rate (*see* **Fig. 7.2A**).
6. The date of leaf initiation can be precisely calculated for each leaf Z by reversing the above relationship between the number of leaves and time (*see* **Fig. 7.2A**).

3.3.2. Area of a Given Leaf (Leaf Z) from Initiation to Emergence

1. At each sampling date, excise leaf Z and flatten it by placing it under a cover slip on a microscope slide in a drop of water. If the leaf primordium is too small to be excised without breaking, manipulate the apex so that leaf Z is horizontal and place the apex under a coverslip on a microscope slide in a drop of water.
2. Place the slide under the microscope at the appropriate magnification and measure the area *A* of the leaf Z with image-analysis software, after calibration with a micrometer slide.

3.4. Leaf Emergence and Expansion of Emerged Leaves

3.4.1. Leaf Emergence Rate, Phyllochron and Date of emergence of a given leaf (leaf Z)

1. Count the number of visible leaves on 10 plants per genotype (the smallest leaf is visible when the area is approximately 0.5 mm^2) at intervals of 2–3 days (*see* **Note 7**) from

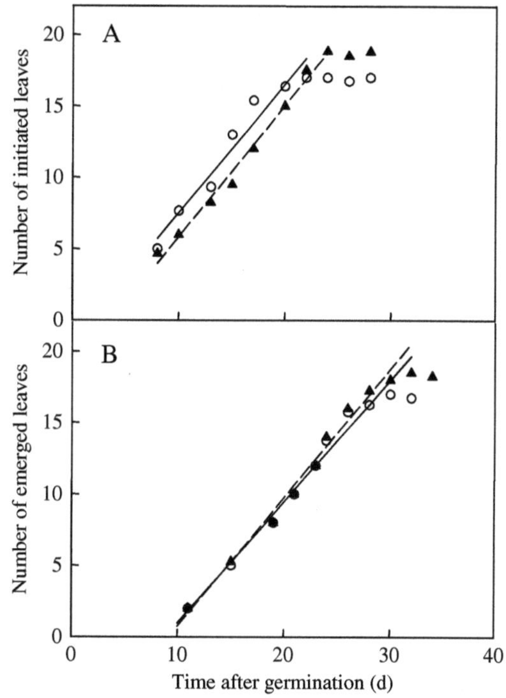

Fig. 7.2. Number of initiated (**A**) and emerged (**B**) leaves as function of time for two *Arabidopsis thaliana* genotypes, namely L*er* (*circles, straight line*) and *ron2-1* (*triangles, dashed line*) grown in long days. L*er* is the genetic background of the *ron2-1* EMS-induced leaf mutant described in (11). *ron2-1* was classified in the rotunda class, with the representative trait of broad and rounded lamina. The date of leaf initiation can be calculated for each leaf by the linear regression of the relationship between the number of initiated leaves and time. In the example shown, $y = 0.90\ d^{-1} \times t - 1.47$ for L*er* and $y = 0.92\ d^{-1} \times t - 3.36$ for *ron2-1*. Leaf 8 ($y = 8$) therefore is initiated at $t = 10.5$ d and at $t = 12.3$ d for L*er* and *ron2-1*, respectively. The leaf initiation rate (i.e. the slope of these relationships) is 0.9 leaves d^{-1} for L*er* and 0.92 leaves d^{-1} for *ron2-1*. The plastochron (calculated as the reciprocal of the leaf initiation rate) is 1.11 d for L*er* and 1.09 d for *ron2-1*. The date of leaf emergence can be calculated for each leaf by the linear regression of the relationship between the number of emerged leaves and time. In the example shown, $y = 0.85\ d^{-1} \times t - 7.55$ for L*er* and $y = 0.90\ d^{-1} \times t - 8.26$ for *ron2-1*. Leaf 8 ($y = 8$) therefore emerges at $t = 18.3$ d and at $t = 18.1$ d for L*er* and *ron2-1*, respectively. The leaf emergence rate (i.e. the slope of these relationships) is 0.85 leaves d^{-1} for L*er* and 0.90 leaves d^{-1} for *ron2-1*. The phyllochron (calculated as the reciprocal of the leaf emergence rate) is 1.18 d for L*er* and 1.11 d for *ron2-1*.

stage 1.02, two cotyledons and two visible leaves until stage 5.01, bolting (10).

2. Plot the total number of visible leaves against time (*see* **Fig. 7.2B**).
3. Calculate leaf emergence rate by the slope of the relationship between the number of leaves and time (*see* **Fig. 7.2B**).
4. Calculate phyllochron, the time necessary for a new leaf to appear/emerge, by the reciprocal of leaf emergence rate (*see* **Fig. 7.2B**).

5. Calculate the date of leaf Z emergence by the linear regression of the above relationship between the number of leaves and time (*see* **Fig. 7.2B**).

3.4.2. Area of a Given Leaf (Leaf Z) from Emergence to End of Expansion

1. At a time interval of 2–3 d (*see* **Note 7**), from leaf emergence to end of leaf expansion (*see* **Note 8**), take a photograph of 10 individual plants per genotype (*see* **Note 9**). Genotypes and/or leaves can be identified in the photographs by the markers added to the pots (*see* **Fig. 7.1A**).
2. Measure the area of the leaf Z with image-analysis software after calibration with the scaling label of known dimensions (*see* **Fig. 7.1B**).

3.4.3. Absolute Leaf Expansion Rate, Relative Leaf Expansion Rate and Duration of Leaf Expansion from Growth Curve of a Given Leaf (Leaf Z)

1. Calculate mean values of A ($n = 5$) for leaf Z at each sampling date j from leaf initiation to leaf emergence (*see* **Note 10**).
2. Calculate mean values of A for leaf Z on each picture from leaf emergence to end of expansion (n=10).
3. Plot mean leaf Z area against time during the whole growing period from leaf initiation (calculated before but *see* also **Note 11**) to end of expansion (*see* **Fig. 7.3A**).
4. Calculate leaf absolute expansion rate, LER, at any given time from leaf initiation to completion of leaf expansion as the local slope of the relationships between the leaf area A and time t: $\mathrm{LER}_j = [\mathrm{d}A/\mathrm{d}t\,]_j$
   Calculate it by linear regression on the three values of A and t corresponding to sampling dates j–1, j and j+1.
5. Calculate leaf relative expansion rate, RER, at any given time from leaf initiation to completion of leaf expansion as the local slope of the relationships between the logarithm of leaf area ln (A) and time t (*see* **Fig. 7.3B**). $\mathrm{RER_j} = [\mathrm{d}(\ln A)/\mathrm{d}t]_j$
   Calculate it by linear regression on the three values of ln(A) and t corresponding to sampling dates j–1, j and j+1.
6. Plot absolute leaf expansion rate and relative leaf expansion rate against time for each genotype to compare the dynamics of their leaf Z expansion (*see* **Fig. 7.3C**, D).

3.4.4. Final Leaf Area, Maximal Absolute Leaf Expansion Rate and Duration of Leaf Expansion from Parameters of a Fitted Curve for a Given Leaf (Leaf Z) (see ***Note 12****)*

1. Plot leaf area (A) expansion of leaf Z against time (t) from leaf initiation to the end of leaf expansion. This should produce a sigmoid curve. Then use a curve-fitting software (e.g. Sigma Plot or R) to fit the equation: $A = A_{\mathrm{end}}/[1 + \exp(-(t-t_0)/B)]$ where B represents the steepness or curvature of the sigmoid curve, t_0 represents the inflection point, i.e. the time when leaf area reaches the half of its final value and A_{end} corresponds to the final leaf area (*see* **Fig. 7.3A**).
2. Estimate the final leaf area of leaf Z with the value of A_{end}.

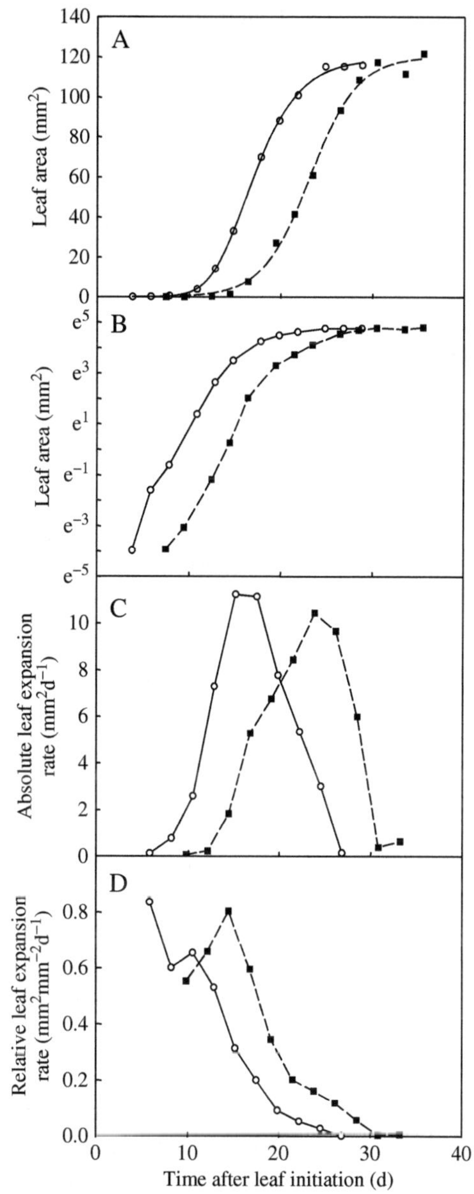

Fig. 7.3. Area of leaf 6 as function of time on either a linear (**A**) or a logarithmic (**B**) scale and corresponding changes over time in absolute (**C**) and relative (**D**) leaf expansion rate for two *Arabidopsis thaliana* genotypes, L*er* (*circles*) and *elo-1* (*squares*) ($n = 10$). L*er* is the genetic background of the *elo-1* EMS-induced leaf mutant described in (11). *elo-1* was classified in the elongata class, with the representative trait of narrow, elongated lamina and long petioles. The 3 parameter sigmoid curves (*see* **Section 3.4.4**) fitted in panel A are $A = 116.56\ \text{mm}^2/(1 + \exp-((t- 17.11)/2.26\ \text{d}))$ for L*er* and $A = 120.23\ \text{mm}^2/(1 + \exp-((t- 23.18)/2.58\ \text{d}))$ for *elo-1*. Note that the two genotypes have achieved the same final leaf 6 area (116.5 mm^2 for L*er* and 120.2 mm^2 for *elo*-1) but with different dynamics of leaf expansion (maximal absolute leaf expansion rate of 12.9 $\text{mm}^2\ \text{d}^{-1}$ for L*er* and 11.6 $\text{mm}^2\ \text{d}^{-1}$ for *elo-1*, and a duration of expansion of 23.8 d for L*er* and 30.8 d for *elo-1*).

3. Calculate the maximum absolute leaf expansion rate (LER_{max}) at the point of inflection of the fitted sigmoid curve by the equation (12): $LER_{max} = A_{end}/(4B)$
4. Calculate the duration of leaf expansion (t_{95}) as the time interval between the date of leaf initiation and the date when leaf area reaches 95% of its final size (A_{end}), from the sigmoid curve : $A_{95} = 0.95 \times A_{end} = A_{end}/[1 + \exp(-(t_{95}-t_0)/B)]$ Which gives, $t_{95} = t_0 - B \times \ln((1/0.95)-1)$.

3.5. Distribution of Leaf Expansion Variables and Developmental Stages According to Nodal Position

3.5.1. Distribution of Final Leaf Area According to Nodal Position

1. Appropriately label an A4 sheet of paper with, for example, the name of the experiment, name of the genotype, the pot number and other necessary information (*see* **Fig. 7.1C**).
2. Stick on to it a given length of double-sided adhesive tape (the length depends on the number and size of leaves on the rosette).
3. Cut the whole rosette from the pot surface with a pair of scissors when rosette area has stopped expansion (*see* **Note 8**). Clean it with a paintbrush rapidly to remove all soil.
3. Detach all leaves (only lamina, without the petiole) in their order of emergence and stick the leaves on the sheet prepared with double-sided adhesive (*see* **Fig. 7.1C**).
4. Scan the sheet of paper with the leaves stuck on it. A black and white scan with a 150 dpi resolution is sufficient.
5. Measure the area of each individual leaf with image-analysis software after calibration with the scaling mark of known dimensions (black square in the right-hand bottom corner on the example in **Fig. 7.1C**).
6. Plot individual leaf area against its nodal position on the plant (*see* **Fig. 7.4A**).

3.5.2. Distribution of Leaf Expansion Variables of All the Leaves of the Shoot

1. The methods described above under **Section 3.4** can be expanded to quantify leaf area development of all the leaves of the plant; instead of measuring just leaf *Z*, measure all leaves of the plant.
2. For each leaf, plot mean leaf area against time during the whole growing period from leaf initiation to end of expansion and calculate the variables as described under **Section 3.4**.

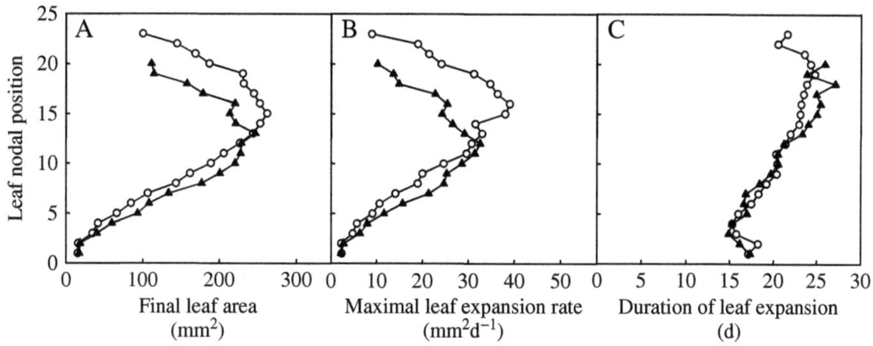

Fig. 7.4. Final leaf area (**A**), maximal leaf expansion rate (**B**) and duration of leaf expansion (**C**) as function of leaf nodal position for two *Arabidopsis thaliana* genotypes, L*er* (*circles*) and *ron2-1* (*triangles*) grown in short days.

3. Plot both maximal absolute leaf expansion rate (*see* **Fig. 7.4B**) and the duration of expansion (*see* **Fig. 7.4C**) as a function of the nodal position of the leaf in the plant.

3.5.3. Distribution of Leaf Developmental Stages According to Nodal Position

1. Plot the number of initiated primordia, the number of emerged leaves and the number of fully expanded leaves as calculated before against time on the same plot.
2. For each leaf, this plot indicates the duration of the early exponential phase, when leaf is hidden in the apical bud, and the duration of the growth phase after leaf emergence (*see* **Fig. 7.5**).
3. For each date, this plot indicates the number of leaves in the early exponential phase of their development, the number of leaves in the late phase of their development or the number of leaves that have stopped growing.

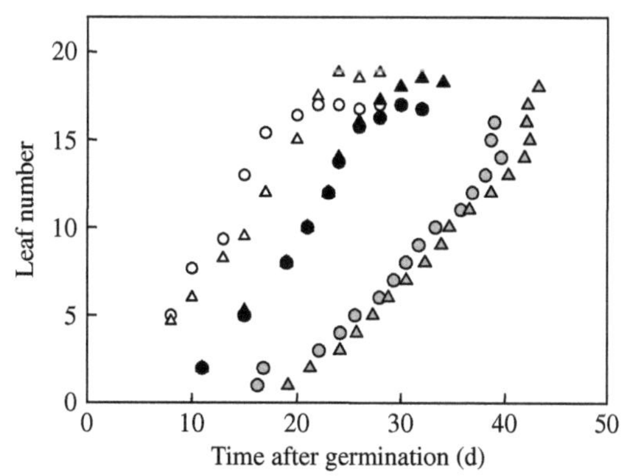

Fig. 7.5. Number of initiated leaves (*white*), emerged leaves (*black*) and fully expanded leaves (*grey*) as function of time for the *Arabidopsis thaliana* genotypes L*er* (*circles*) and *ron2-1* (*triangles*) grown in long days.

3.6. Statistical Analyses

1. Compare means of leaf growth variables using a two-way ANOVA where the assumptions of the ANOVA are met (General Linear Model with genotype and treatment as factors (also termed fixed effects)).
2. For comparison of a large number of genotypes, a Tukey Post Hoc test can be done on the genotype factor to identify sub-groups of genotypes.
3. Compare slopes of growth curves by covariance analyses.
4. Procedures included in the curve-fit software (calculation of standard errors and confidence intervals) can be used to compare parameters generated by curve fitting.

4. Notes

1. Plants can be grown in soil, in hydroponic culture or on vertical agar plates, but here the protocol for plants grown in soil will be outlined.
2. The volume and weight of soil necessary to fill in each pot depend both on soil composition and soil water content. Generally, the optimal soil water content is around 70% of the soil water content at retention capacity. The soil water content at retention capacity (SWC_R) can be deduced from a preliminary experiment in which pots are filled with soil, fully wetted and allowed to drain freely. It is then calculated by weighing the soil before (W_R) and after drying 4 d at 120°C (W_D) from the equation : $SWC_R = (W_R - W_D)/W_D$ (soil water content is expressed in g of water per g dry soil). Soil water content can then be adjusted at a given value identical for all genotypes, during the whole experiment (as described in (13)) if the weight of empty pots is known, if soil aliquots are dried to estimate the amount of dry soil and water in each pot at the time of filling, if the weight of soil in each pot at filling is known, and by calculating changes in soil water status in each pot from subsequent changes in pot weight.
3. In *A. thaliana*, the week following sowing is crucial: the germination stage is very sensitive to drought and, as the root is very short, do not hesitate to spray water on the pot surface once or twice a day depending on the air circulation in the growth chamber. Do not allow the soil surface to dry out.
4. The number of plants necessary for the whole assay depends on environmental conditions that will affect the duration of the experiment and the number of leaves

produced by the rosette. In *A. thaliana*, the number of leaves and their duration of expansion are increased by short day length. The number of 40 plants suggested here is given for Col-0 plants grown under a 12 h day length.

5. Depending on the digital camera used for the analysis; it could be useful to calibrate both in length and width (*X* and *Y*) with a square scale. Check that there is no deformation of the pixel shape by your camera.
6. Leaf expansion is very sensitive to small changes in the environmental conditions surrounding the plants during their culture. To compare genotypes on a rigorously identical basis, the growth conditions have to be homogeneous throughout the plant culture area. In addition, the different genotypes studied should be randomised and mixed together, and if the conditions surrounding the plants are not homogeneous, their position in the plant culture chamber should be changed every 1–2 d.
7. The interval between the harvests or dynamic measurements depends considerably on temperature, as plants develop more quickly as the temperature increases (within certain limits), and this influences the choice of interval used to study the plants. In general, when *A. thaliana* plants are grown at 21°C, an interval of every 2 or 3 d is appropriate. However, a daily interval would be necessary for plants grown at 26°C.
8. In most *A. thaliana* genotypes, whole rosette leaf area expansion is achieved when the first flower is visible. However, for a few of them such as the L*er* accession, rosette leaf area continues to expand until the first silique is formed.
9. This part of the protocol using a digital camera is only applicable to the study of genotypes with more or less flat rosette leaves. The study of plants with curled leaves or leaves with uneven surfaces requires all measurements of leaf area to be made destructively; the leaves must be flattened between a microscope slide and coverslip or glued on a sheet of paper with double-sided adhesive tape.
10. Because leaf growth is exponential from primordium initiation to leaf emergence, it is preferable to use the geometric mean instead of arithmetic mean during early developmental stages.
11. The projected area of the youngest visible leaf primordium, with the techniques described here, measures around 0.001 mm^2. Because the growth of leaf primordia is generally exponential between initiation and emergence, the date of initiation of leaf Z can be deduced by plotting leaf Z area against time. Fit experimental data measured at early

stages ($A < 1\ mm^2$) to an exponential function of time and calculate the time at which the leaf size is 0.001 mm^2.

12. Except when plants are grown in disturbed environmental conditions (transient periods of stress), the time course of leaf area fits to a sigmoid curve. The parameters of the sigmoid can be used to deduce leaf growth variables and they differ according to the nodal position of the leaf, the genotype or environmental conditions (3, 14, 15).

Acknowledgements

The authors acknowledge the European Commission for the financial support to analyse and model the plasticity of leaf growth in *A. thaliana* first with the Community Human Potential Program via the DAGOLIGN Research Training Network (HPRN-CT-2002–00267) and then the Framework Programme 6 via the AGRON-OMICS Integrated Project (LSHG-CT-2006-037704).

References

1. Aguirrezabal, L., Bouchier-Combaud, S., Radziejwoski, A., Dauzat, M., Cookson, S. J., and Granier, C. (2006) Plasticity to soil water deficit in *Arabidopsis thaliana*: dissection of leaf development into underlying growth dynamic and cellular variables reveals invisible phenotypes. *Plant Cell Environ* **29**, 2216–2227.
2. Tisné, S., Reymond, M., Vile, D., Fabre, J., Dauzat, M., Koornneef, M., and Granier, C. (2008) Combined genetic and modeling approaches reveal that epidermal cell area and number in leaves are controlled by leaf and plant developmental processes in Arabidopsis. *Plant Physiol* **148**, 1117–1127.
3. Cookson, S., Chenu, K., and Granier, C. (2007) Day-length affects the dynamics of leaf expansion and cellular development in *Arabidopsis thaliana* partially through floral transition timing. *Annals Bot* **99**, 703–771.
4. Dosio, G. A. A., Rey, H., Lecoeur, J., Izquierdo, N. G., Aguirrezábal, L. A. N., Tardieu, F., and Turc, O. (2003) A whole-plant analysis of the dynamics of expansion of individual leaves of two sunflower hybrids. *J Exp Bot* **54**, 2541–2552.
5. Abramoff, M. D., Magelhaes, P. J., and Ram, S. J. (2004) Image Processing with ImageJ. *Biophot Internat* **11**, 36–42.
6. R Development Core Team (2007) R: A language and environment for statistical computing. R Foundation for Statistical Computing, Vienna, Austria. ISBN 3-900051-07-0, URL http://www.R-project.org.
7. Turc, O. and Lecoeur, J. (1997) Leaf primordium initiation and expanded leaf production are co-ordinated through similar response to air temperature in pea (*Pisum sativum* L.). *Annals Bot* **80**, 265–273.
8. Lecoeur, J., Wery, J., Turc, O., and Tardieu, F. (1995) Expansion of pea leaves subjected to short water deficit: cell number and cell size are sensitive to stress at different periods of leaf development. *J Exp Bot* **46**, 1093–1101.
9. Granier, C. and Tardieu, F. (1999) Water deficit and spatial pattern of leaf development. Variability in responses can be simulated using a simple model of leaf development. *Plant Physiol* **119**, 609–620.
10. Boyes, D. C., Zayed, A. M., Ascenzi, R., McCaskill, A. J., Hoffman, N. E., Davis, K. R., and Gorlach, J. (2001) Growth stage-based phenotypic analysis of Arabidopsis: A model for high throughput functional genomics in plants. *Plant Cell* **13**, 1499–1510.

11. Berná, G., Robles, P., and Micol, J. L. (1999) A mutational analysis of leaf morphogenesis in *Arabidopsis thaliana*. *Genetics* **152**, 729–742.
12. Torres, M. and Frutos, G. (1989) Analysis of germination curves of aged fennel seeds by mathematical models. *Env Exp Bot* **29**, 409–415.
13. Granier, C., Aguirrezabal, L., Chenu, K., Cookson, S. J., Dauzat, M., Hamard, P., Thioux, J. J., Rolland, G., Bouchier-Combaud, S., Lebaudy, A., Muller, B., Simonneau, T.. and Tardieu, F. (2006) PHENOPSIS, an automated platform for reproducible phenotyping of plant responses to soil water deficit in *Arabidopsis thaliana* permitted the identification of an accession with low sensitivity to soil water deficit. *New Phytol* **169**, 623–635.
14. Cookson, S. J. and Granier, C. (2006) A dynamic analyses of the shaded-induced plasticity in *Arabidopsis thaliana* rosette leaf development reveals new components of the shade-adaptative response. *Annals Bot* **97**, 443–452.
15. Cookson, S. J., Radziejwoski, A., and Granier, C. (2006) Cell and leaf size plasticity in Arabidopsis: what is the role of endoreduplication? *Plant Cell Env* **28**, 1355–1366.

Chapter 8

Analyzing Shoot Apical Meristem Development

Cristel C. Carles, Chan Man Ha, Ji Hyung Jun, Elisa Fiume, and Jennifer C. Fletcher

Abstract

The shoot apical meristem of *Arabidopsis thaliana* contains a reservoir of pluripotent stem cells that functions as a continuous source of new cells for organ formation during development. The SAM forms during embryogenesis, when it becomes stratified into specific cell layers and zones that can be delineated based on morphological and molecular criteria. The primary SAM produces all the aerial structures of the adult plant, and alterations in SAM organization or function can have profound effects on vegetative and reproductive plant morphology. Such SAM-specific defects can be identified, evaluated, and quantified using specialized microscopic and histological techniques.

Key words: Shoot apical meristem, confocal laser scanning microscopy, histology, in situ hybridization.

1. Introduction

Leaves, stems, and flowers are the products of the shoot apical meristem (SAM), which is located at the tip of the growing plant (1). Throughout development, the SAM sustains a stem cell reservoir at its apex, which is maintained by signals from the underlying niche cells, while forming new organ primordia on its flanks (2). Developmental defects in growth rate, organ initiation, organ number, and/or stem size displayed by many different Arabidopsis mutants can be traced back to perturbations in SAM structure or function (3, 4). However, the embedded location

Cristel C. Carles, Chan Man Ha, Ji Hyung Jun, Elisa Fiume, have contributed equally to this work.

L. Hennig, C. Köhler (eds.), *Plant Developmental Biology*, Methods in Molecular Biology 655,
DOI 10.1007/978-1-60761-765-5_8, © Springer Science+Business Media, LLC 2010

and small size of the Arabidopsis SAM make it difficult to access, rendering analysis challenging. Thus specific methodologies have been developed or optimized to study the morphological characteristics of embryonic and vegetative SAMs, measure their parameters, and visualize the mRNA transcript domains within them. In our chapter, we will explain methods and approaches for the analysis of SAM development: (i) Study the morphology of Arabidopsis embryonic SAMs using confocal laser scanning microscopy that permits the ready imaging of mature embryonic meristem organization, cells and cell layers from whole mount samples (5) (*see* **Section 3.1**). These images can be reconstructed three-dimensionally with computer software. (ii) Histological sectioning of resin-embedded samples allows the analysis of vegetative SAM tissues, which are inaccessible by other methods. Sections are stained to colorize the cellular components, permitting high-resolution imaging of the internal cellular morphology (*see* **Section 3.2**). (iii) Meristem size is a key indicator of stem cell activity in Arabidopsis (6). The use of publicly available image-analysis software provides a fast, convenient, and highly accurate method of measuring the height and width of vegetative meristem sections (*see* **Section 3.3**). (iv) In situ hybridization uses a labeled probe to localize a specific DNA or RNA sequence within a whole-mount or tissue section (7). First introduced in the late 1960s for RNA detection in Xenopus oocytes (8), this technique was adapted to plants 20 years later (9, 10). Arabidopsis vegetative SAMs are surrounded by leaf primordia, making it very difficult to achieve full tissue penetration and obtain a satisfactory signal to noise ratio. A modified RNA in situ hybridization protocol has been shown to be effective for detecting and localizing mRNA transcripts to specific regions within vegetative shoot apices (11–13) (*see* **Section 3.4**).

2. Materials

Arabidopsis thaliana seedlings are grown at 21°C on plates containing Murashige and Skoog (MS) medium (14) under cool-white fluorescent lights (100–140 mmol/m^2 s) for 4–7 days.

MS medium: 4.33 g MS salts, 3 g sucrose, and 900 mL distilled water. Adjust pH to 6.0 with 1 M KOH. Add 9 g bactoagar and adjust volume to 1 L with distilled water.

2.1. Confocal Laser Scanning Microscopy of the Embryonic Meristem

1. Glass scintillation vials or similar containers.
2. Pasteur pipettes or micropipettes.
3. Microscope depression slides.
4. Coverslips (24 × 60 mm).

5. Fine forceps (number 5, Ted Pella Inc.).
6. Fine needles.
7. Formaldehyde acetic acid (FAA) fixation solution: 3.7% formaldehyde, 50% ethanol, and 5% acetic acid (*see* **Note 1**). Prepare a fresh mixture of ~10 mL per vial.
8. Graded ethanol series: Prepare 15, 30, 50, 70, 85, 90, and 95% ethanol in distilled water.
9. Graded Histoclear series: Prepare 500 mL solutions of 25:75 Histoclear (National Diagnostics):Ethanol, 50:50 Histoclear:Ethanol, and 75:25 Histoclear:Ethanol.
10. Propidium iodide stock solution: Prepare 100 μg/mL propidium iodide in 0.1 M L-arginine and adjust the pH to 12.4 with 5 M NaOH (*see* **Note 2**). This solution is deep red and can be stored for months at 4°C. When it turns dark orange it cannot be used anymore and should be disposed of in the hazardous waste.
11. Prepare propidium iodide staining solution by diluting the 100 μg/mL propidium iodide stock solution to 5 μg/mL in 0.1 M L-arginine and adjust the pH to 12.4 with 5 M NaOH.
12. Rinsing solution: Prepare a solution of 0.1 M L-arginine buffer in distilled water and adjust the pH to 8 with HCl.
13. Immersion oil (Cargille Laboratories).
14. Nail polish.

2.2. Histological Sectioning of the Seedling Meristem

1. Single-edge razor blades.
2. Scintillation vials.
3. Vacuum bell.
4. Microtome.
5. 42–55°C slide-warmer.
6. Histoform S (Teflon embedding mould; Heraeus Kulzer, Wehrheim/Ts., Germany).
7. Histobloc (Heraeus Kulzer, Wehrheim/Ts., Germany).
8. Tungsten Carbide Microtome Knife.
9. FAA fixation solution: 3.7% formaldehyde, 50% ethanol, and 5% acetic acid. Prepare a fresh mixture of ~10 mL per vial.
10. Graded ethanol series: Prepare 50, 60, 70, 80, 90, and 95% ethanol in distilled water.
11. Technovit Glycol Methacrylate (GMA) Kit 7100: Each kit contains 500 mL of GMA monomer, five 1 g packs of Hardener I, and 40 mL of Hardener II (*see* **Note 3**).

12. Graded Technovit series: Prepare 50% (v/v), 70% (v/v), and 90% (v/v) Technovit 7100 resin in 100% ethanol.
13. Toluidine blue solution: Prepare 0.1% Toluidine blue O in 0.1% aqueous sodium tetraborate (*see* **Note 4**).
14. Neutral red solution: Prepare 0.01% solution by melting neutral red powder in Technovit 7100 Hardener I solution.

2.3. Meristem Size Measurement

1. Computer equipped with NCBI Image J software for image processing and analysis (*see* **Note 5**).
2. Digital images of shoot apical meristem sections (*see* **Note 6**).

2.4. Seedling RNA In Situ Hybridization

1. Single-edge razor blades.
2. Scintillation vials.
3. Vacuum bell.
4. 56°C oven for flasks.
5. Bench-top microscope slide warming table.
6. Weighing boats (8 cm × 8 cm).
7. Microtome.
8. Paraffin block holder.
9. Positively charged adhesion slides (Probe-on plus slides, FisherBrand).
10. 42°C water bath for slides.
11. Wooden applicator sticks.
12. 42–55°C bench for slides (slide-warmer).
13. Racks for slides or Copeland jars: Glassware (racks or Copeland jars) that will hold the slides must be baked and cooled very slowly or else the glass will crack (*see* **Note** 7).
14. 37–60°C incubator with a shaker for the slide racks/ Copeland jars.
15. UV cross-linker.
16. Dark boxes for slides.
17. Stereo-microscope.
18. Eosin.
19. RNA molecular weight marker.
20. T7/ T3 RNA polymerase.
21. RNAse out/ RNAsin.
22. DNaseI.
23. Boehringer blocking reagent (*see* **Note 8**): A 5 × Blocking reagent (2.5%) stock can be prepared in advance and stored at −20°C.

24. Nitroblue tetrazolium (NBT) (*see* **Note 8**).
25. 5-Bromo-4-chloro-3-indolyl phosphate (BCIP) (*see* **Note 8**).
26. Stock of 10 × phosphate buffer saline (PBS): Prepare 1 L of 10 × PBS (1.3 M NaCl, 70 mM Na_2HPO_4, 30 mM NaH_2PO_4, pH to 7 with NaOH) and autoclave it.
27. 4% Paraformaldehyde fixation solution (*see* **Note 9**): Take 100 mL of 1 × PBS and adjust pH to 11 with 1 M NaOH. Heat to 60°C, add 4 g of fresh paraformaldehyde, and stir to dissolve. Cool on ice and adjust pH to 7 with 1 M H_2SO_4. Add 1 mL of Triton X-100 and 1 mL of dimethyl sulfoxide (DMSO) (*see* **Note 10**). Make this solution fresh each time and keep on ice for fixation.
28. Prepare 30, 50, 75, and 95% ethanol in distilled water.
29. 2 × Probe hydrolysis solution: Make fresh each time by mixing 30 μL of 21 mg/mL Na_2CO_3 and 20 μL of 16.8 mg/mL $NaHCO_3$ per probe.
30. Proteinase K buffer: Prepare 50 mL of 100 mM Tris-HCl, pH 8.0, 50 mM EDTA.
31. Proteinase K stock: 5 mg/mL proteinase K in 1 mL of proteinase K buffer. The stock can be aliquoted and stored at –20°C for subsequent use.
32. 20 × Sodium salt citrate (SSC) stock solution: Mix 3 M NaCl and 0.3 M $Na_2C_6H_5O_7$ $2H_2O$. Prepare 1 L and autoclave solution.
33. Hybridization buffer: To prepare 4 mL of hybridization buffer, mix 2 mL of pure deionized formamide, 1.2 mL of 20 × SSC, 600 μL of 20% SDS, 40 μL of 10 mg/mL yeast tRNA, and 158 μL of RNAse-free Milli-Q water. Leftover buffer can be frozen at -20°C for further use (*see* **Note 11**). The hybridization buffer precipitates at room temperature. Keep it at hybridization temperature (55°C) before aliquoting it on the slides.
34. 1 × Tris buffer saline (TBS) solution: Prepare 2 L of 0.4 M NaCl, 0.1 M Tris-HCl, pH 7.5 and autoclave the solution. A 10 × TBS stock can be prepared, although the powders will not dissolve easily.
35. Detection buffer: For 250 mL, mix 25 mL of 1 M Tris-HCl, pH 9.6, 6.25 mL of 4 M NaCl, 12.5 mL of 1 M $MgCl_2$, and 207 mL of RNAse-free Milli-Q water. Prepare the detection buffer with the substrates by mixing 1.6 μL of 50 mg/mL BCIP and 2.2 μL of 100 mg/mL NBT per 1 mL of detection buffer.

3. Methods

3.1. Confocal Laser Scanning Microscopy of the Embryonic Meristem

3.1.1. Embryo Dissection

1. Place two discs of Whatman or 3MM paper in a Petri dish and let them absorb as much water as they can hold.
2. Distribute mature and dry Arabidopsis seeds on the surface of the paper, close the Petri dish, and let the seeds imbibe overnight at 4°C in the dark (*see* **Note 12**).
3. The next day place a small piece of wet Whatman or 3MM paper on a slide. Transfer a few seeds onto the slide and dissect the embryos out of the seed coat under a dissecting microscope using fine forceps and needles (*see* **Note 13**). Hold the seeds with fine forceps by the micropylar end and make an incision on the other end of the seed coat with a needle or another pair of fine forceps. Apply gentle pressure on the micropylar side of the seed using the forceps slanted to one side. The embryo should pop out of the scar in the seed coat. Add more water if the paper on the slide dries.
4. With a fine needle immediately transfer the isolated embryo into a vial containing a few mL of cold water and keep the vial on ice.

3.1.2. Tissue Staining and Rinsing

1. Replace the water with 2–3 mL of the 5 μg/mL propidium iodide staining solution (*see* **Note 14**). Stain for 6 h at room temperature. After the staining incubation, the tissue appears pale orange.
2. Replace the staining solution with 2–3 mL of the L-arginine buffer rinsing solution. Leave in L-arginine buffer at 4°C for 4 days, changing the rinsing solution once per day.

3.1.3. Tissue Clearing

1. Dehydrate the samples through an ethanol series (15, 30, 50, 70, 85, 95, and 100%). Each wash should last 30 min.
2. Wash twice more with 100% ethanol.
3. Incubate the tissues in a Histoclear series (75:25, 50:50, 25:75 Ethanol:Histoclear) to 100% Histoclear. Leave the samples in each solution for at least 2 h to completely clear the tissue.
4. Change the 100% Histoclear three times and leave the tissue overnight in the last change.

3.1.4. Mounting and Imaging

1. Pipet three or four separate drops of immersion oil on a microscope slide.
2. Pipet some embryos with a P100 micropipet (*see* **Note 15**) into the first drop of immersion oil.
3. Transfer the embryos from one oil drop to the next using fine forceps, in order to remove the Histoclear (*see* **Note 16**).
4. Place the embryos in the center of a coverslip. Each embryo should be lying on its side with the root and the two cotyledons touching the coverslip and be fully covered with immersion oil to prevent the tissue drying out.
5. Flip the coverslip over a depression slide, such that the embryo side is down and seal the four corners of the coverslip with nail polish (*see* **Note 17**).
6. Visualize each slide using a Confocal Laser Scanning Microscope (*see* **Fig. 8.1**). Propidium iodide can be excited by a 514 nm argon laser beam and emits between 580 and 610 nm.

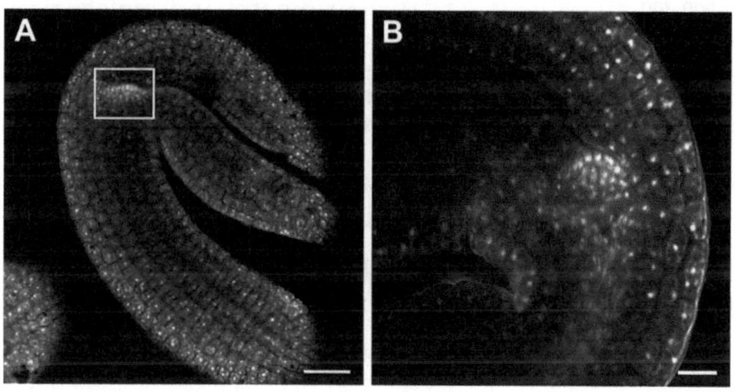

Fig. 8.1. **Confocal laser scanning micrographs of mature Arabidopsis embryo and embryonic shoot apical meristem.** (**A**) Optical longitudinal section of a wild-type Landsberg *erecta* (L*er*) mature embryo. The shoot apical meristem is boxed. (**B**) Optical longitudinal section of an embryonic shoot apical meristem from a mature wild-type L*er* embryo. The apical meristem cells stain very brightly with propidium iodide. Scale bars, 50 μm in (**A**) and 20 μm in (**B**).

3.2. Histological Sectioning of the Seedling Meristem

3.2.1. Tissue Dissection and Fixation

Plant materials for histological analysis should be grown on slanted agar plates. Depending on the growing conditions, 7- to 9-day-old seedlings are appropriate for vegetative SAM observation.

1. Remove the roots of each seedling at the base of the hypocotyl using clean forceps or small scissors.
2. Immediately place each dissected sample into a glass scintillation vial containing FAA fixation solution in a fume hood (*see* **Note 18**).
3. Loosen the caps of the scintillation vials and place the vials in a vacuum chamber.
4. Pull the vacuum slowly to 25 psi for 20–30 min. This step removes air bubbles from the samples to allow the penetration of the fixative into the tissue. The samples will begin to sink. Slowly release the vacuum to return the samples to air.
5. Repeat the above step once to permit all the samples to sink to the bottom of the vials (*see* **Note 19**).
6. Slowly release the vacuum and remove the vials from chamber. Keep the sample vials in the fume hood overnight to complete the fixation process.

3.2.2. Tissue Dehydration and Infiltration

1. The next day, remove the FAA fixation solution using a Pasteur pipette and replace with 50% ethanol. Incubate for 30 min.
2. Dehydrate the samples through an ethanol series (60, 70, 80, 90, and 95%). Each wash should last for 30 min.
3. Wash twice with 100% (v/v) ethanol for 1 h each.
4. Transfer the dehydrated samples into 50% (v/v) Technovit 7100 resin. Keep the samples at room temperature for 30 min.
5. Repeat the above step for 70% (v/v) and 90% (v/v) resin.
6. Replace with 100% resin (v/v) twice for 1 h each. Store the samples at least overnight at 4°C (*see* **Note 20**).

3.2.3. Tissue Staining and Embedding

1. Place a drop of neutral red on a slide and place the sample in it for 3–4 s (*see* **Note 21**).
2. Add together Technovit 7100 Hardener II to Hardener I at a ratio of 1–15, mix the solution and pour it into the embedding mold. Place the neutral red stained samples into the mold (*see* **Note 22**).
3. Allow polymerization for 1 h.
4. Place Histobloc on the top of each mold in an upside-down position.
5. Store the samples at least overnight at room temperature to allow polymerization.

3.2.4. Tissue Sectioning

1. After removing the sample from the mold, place it on a rotary microtome at an 8° knife inclination angle. Use a very sharp microtome knife.
2. Section the tissue at 1–4 μm thickness. The sections will be released by the microtome one by one, unattached to each other.
3. Check continuously for sections that contain the SAM tissue (*see* **Note 23**).
4. Remove each SAM tissue section from the microtome with tweezers and float it in a Petri dish filled with distilled water (*see* **Note 24**).
5. Put one section on a coverslip and dry it on a slide warmer at 42°C for 10–15 min (*see* **Note 25**).

3.2.5. Toluidine Blue Staining

1. Soak the coverslip with ribbon in Toluidine blue solution for 1–2 min.
2. Soak the coverslip with ribbon in distilled water for 3–4 min to remove the excess stain. Check the sections under the microscope. Repeat step 1 if the sections are too lightly stained, and repeat step 2 if they are too strongly stained (*see* **Note 26**).
3. Place the coverslip with the ribbon on the slide warmer at 42°C for 10–15 min.

3.2.6. Mounting and Visualization

1. Place a drop of Permount onto a glass slide (*see* **Note 27**).
2. Place the coverslip with the completely dried ribbon upside-down on top of the Permount, ribbon side down (*see* **Note 28**).
3. Place the glass slide on the slide warmer overnight to allow complete fixation of the coverslip to the glass slide.
4. Visualize the stained sections using a stereo-microscope (*see* **Fig. 8.2**).

3.3. Meristem Size Measurement

1. Launch the Image J software and open a meristem image file.
2. Set the scale by choosing the 'straight line selections' tool from the tool bar below the main menu and drawing a straight line along the scale bar in the raw image (*see* **Note 29**). Go to the Analyze scroll down menu and select 'set scale'. In the window that appears, fill in the 'known distance' and 'unit of length' customizable space. After entering the correct values click 'ok' to close the calibration window.

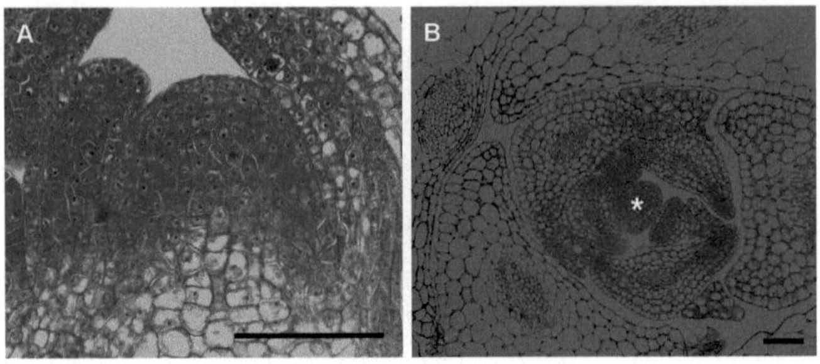

Fig. 8.2. **Histological sections of Arabidopsis vegetative shoot apical meristems.** (**A**) Longitudinal section. (**B**) Transverse section. The shoot apical meristem is marked (∗). Ten-day-old seedlings were fixed in FAA solution and infiltrated with Technovit 7100 resin before sectioning at 2-μm thickness. Scale bars, 50 μm.

3. Draw lines on the image in order to make the desired measurements. Two parameters can be measured on longitudinal shoot apical meristem images: width and height (**Fig. 8.3**). The width is measured by drawing a horizontal line between the interior boundaries of the visible ridges that will form the next primordia. The height is measured by drawing a perpendicular line from the point equidistant along the width up to the tip of the meristem.

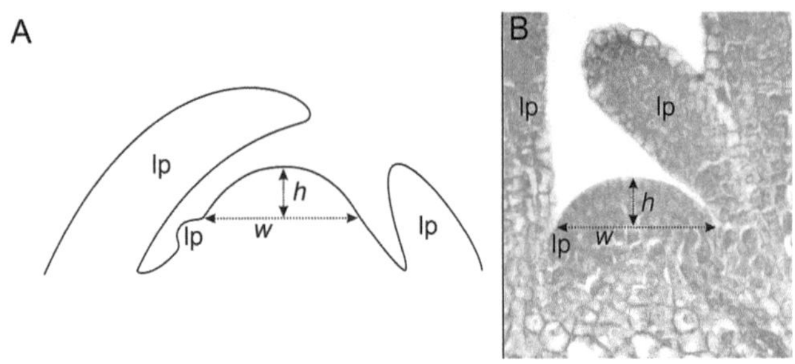

Fig. 8.3. **Measurement of the Arabidopsis vegetative shoot apical meristem.** (**A**) Schematic of the shoot apex showing the meristem as a dome of cells between the developing leaf primordia (lp). (**B**) Longitudinal section through a 7-day-old shoot apex. The meristem width (*w*) and height (*h*) measurements are made as shown.

4. Once an accurate line is drawn, use the Ctrl+M keyboard combination to open a results window where all the measurements will be saved.
5. To make a line permanent on the image (useful, for example, to measure SAM height in longitudinal sections), click the 'Edit' scroll down menu in the menu bar and select 'draw.'
6. Open the next image file and repeat steps 2, 3, and 4. The new measurement will be saved in the same results window, which can be saved later as a Microsoft Excel table (*see* **Note 30**).

7. Perform statistical analysis using a program such as Microsoft Excel or OpenOffice SpreadSheet.

3.4. Seedling RNA In Situ Hybridization

3.4.1. Fixation of Tissue Sections

For section preparation, the seedling tissues must be embedded in paraffin. To preserve morphology, the tissues should first be fixed. Next they are dehydrated and stained with a dye to facilitate sectioning. For tissue embedding, an organic solvent gradually replaces the ethanol present in the tissue. The nontoxic solvent Histoclear is now commonly used in paraffin-based embedding procedures. Finally, paraffin is slowly introduced into the solvent solution until it reaches 100% concentration. Numerous solution changes are needed in order for the paraffin to fully penetrate the tissue.

Fixation is one of the most critical steps for successful in situ hybridization, for which a compromise must be found between tissue morphology preservation and good probe penetration. A 4% paraformaldehyde fixation solution penetrates 2 mm of tissue in about 1 h at room temperature. Because of the structure formed by the cotyledons, a bubble of air is usually trapped on the top of the shoot apex, preventing full penetration of the fixative into the SAM tissue. Moreover, the stacking of leaf primordia over the SAM creates a layer of tissue for the fixative to pass through before reaching the meristematic cells. Chopping the tip of the seedling greatly enhances the SAM fixation process and allows all tissues to be cross-linked at the same pace.

1. Prepare the 4% paraformaldehyde fixation solution fresh. 100 mL of solution should be enough for the independent fixation of 5 different genotypes or batches. Aliquot ~10 mL of the 4% paraformaldehyde fixation solution into each scintillation vial kept on ice.
2. Pull a seedling from its culture plate (*see* **Note 31**). Lay the seedling sideways on a slide under a dissecting microscope. Using a sharp single-edge razor blade, trim the cotyledons and oldest leaf primordia and then cut midway along the hypocotyls to remove the roots (**Fig. 8.4**).
3. Immediately place the dissected tissue in the fixative and swirl. The dissecting process should be prompt to preserve tissue integrity. If the process is taking too long, mount seedling in few drops of fixative while trimming to avoid drying of the sample (work under a chemical hood). To achieve a good representation of the expression pattern of the gene of interest, fix at least 50 individuals per genotype or experimental batch. Up to 50 dissected seedlings can be placed in a same scintillation vial containing 10 mL of fixative.

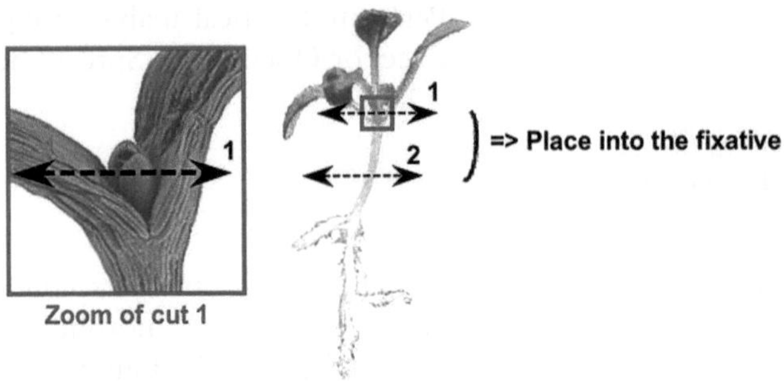

Fig. 8.4. **Dissection of the Arabidopsis vegetative shoot apex for in situ hybridization**. Lay a seedling sideways on a slide under a dissecting microscope. Holding the base of the hypocotyl firmly, make a sharp cut to remove the cotyledons and the oldest leaf primordia (*arrow 1*). Make another cut midway along the hypocotyl to remove the root system (*arrow 2*). Immediately place the dissected tissue in cold fixation solution.

4. Loosen the caps of the scintillation vials and place them on ice in the vacuum apparatus.
5. Pull the vacuum slowly to 25 psi for 10 min. This step removes air bubbles from the samples to allow the penetration of the fixative into the tissue. The samples will begin to sink. Slowly release the vacuum to return the samples to air.
6. Repeat the above step once to permit all the samples to sink to the bottom of the vials.
7. Pipet the fixation solution into a hazardous waste bottle and replace with fresh cold fixation solution.
8. Place tissue in vials at 4°C for 16 h under constant rotation.

3.4.2. Tissue Dehydration

Following fixation, the tissue is embedded in paraffin for sectioning. Tissue embedding in an organic solvent requires prior tissue dehydration. Ethanol solutions used for dehydration should be kept at 4°C.

1. Pipet the fixation solution into a hazardous waste bottle and replace it with the same volume of cold 15% ethanol. Incubate at 4°C with rotation for 30 min.
2. Dehydrate through an ethanol series (30, 50, 75, and 95%), incubating at 4°C with rotation for 30 min each.
3. Pour off the 95% ethanol solution and replace with 100% ethanol. Incubate at 4°C with rotation for 30 min.
4. Pour off the 100% ethanol solution and replace with 100% ethanol containing 0.1% Eosin to stain the tissue (*see* **Note 32**). Let the vials rotate overnight at 4°C.

3.4.3. Tissue Embedding in Paraffin

1. Remove the vials from 4°C and allow the solution to come to room temperature.

2. Remove the 100% ethanol solution, replace with room temperature 100% ethanol, and incubate for 30 min.
3. Remove 1/4 of the volume of ethanol and replace with 1/4 volume of Histoclear; incubate for 30 min (*see* **Note 33**). Repeat this step four times.
4. Pour off the remaining solution and replace with 100% Histoclear. Repeat this step once.
5. Pour off the solution and fill the vial 1/3 full with 100% Histoclear. Add an equal volume of paraplast X-tra paraffin chips. Incubate overnight at room temperature.
6. Prepare a 500-mL beaker of paraplast chips and leave it in a 56°C oven for the chips to melt overnight.
7. In the morning, place vials at 42°C to melt the paraplast chips. Keep adding paraplast chips and allow them to melt at 42°C. Repeat until the vial is full. The rest of the chips should melt within an hour or two.
8. When all chips have melted, transfer the vial to the 56°C oven.
9. Remove the Histoclear/paraffin mixture, replace with pure molten paraplast from the 500-mL beaker, and incubate at 56°C.
10. Replace the molten paraplast every 8–10 h. Make at least 6 changes of molten paraplast.
11. On a bench-top microscope slide warming table, pour the paraplast solution containing the tissues into weighing boats. Split the contents of each vial into two weighing boats. Align and orient the samples using a pipet or needle (*see* **Note 34**). The embedded tissues can be stored at 4°C for several months.

3.4.4. Tissue Sectioning and Mounting

1. Cut into the paraffin bed with a sharp single-edge razor blade in order to isolate each embedded seedling in a block, leaving at least 4 mm of paraffin on each side of the tissue.
2. Mount the block on the microtome sample holder (*see* **Note 35**).
3. Trim the block further to a rectangular or trapezoid shape, leaving about 2 mm of paraffin around the tissue. The long edges must be parallel to one another and to the hypocotyl of the sample.
4. Insert the sample holder into the microtome.
5. Align the hypocotyl of the tissue sample parallel to the knife blade, by moving the sample holder as needed (*see* **Note 36**).

6. Section the tissue at an 8° knife inclination angle for 8 μm-thick sections. Manipulate the paraffin ribbons containing the sectioned tissue with fine paintbrushes.
7. Screen the ribbons under a dissecting microscope for the serial sections containing SAM tissue (*see* **Note 23**).
8. Trim the ribbons of interest with a razor blade to remove unwanted sections.
9. Transfer the ribbons onto an adhesion slide using a fine paintbrush, shiny side down. Orient them on the slide in parallel rows. Place the ribbons such that each is a few millimeters away from the edge of the slide.
10. Add RNAse-free water beneath the ribbons and transfer the slide onto the slide warmer set at 42°C.
11. Add more water so as to cover nearly all of the slide surface and let the ribbons expand for 10 min. Each ribbon should expand approximately 50% of its initial length.
12. Once the ribbons are fully expanded, slightly incline each slide and wick away the water using a kimwipe, always avoiding touching the ribbons.
13. Leave on a slide warmer overnight at 42°C to dry. Cover with a lid to prevent dust falling on the slides. Once fully dry, sections can be stored in a clean slide box at 4°C for a couple of months.

3.4.5. Riboprobe Preparation (see Note 37)

In situ hybridization techniques were initially established using radioactively labeled probes. Improvements in nonradioactive labeling methodologies have led to equivalent sensitivity in signal detection. The most commonly used DIG indirect labeling approach uses a hapten coupled to UTP. Riboprobe is made by run-off transcription from a linearized plasmid vector template. This requires cloning of the cDNA of interest into a suitable vector (such as pBSKII) that allows linearization with restriction enzymes leaving 5′ overhangs, because 3′ overhangs or blunt ends can lead to RNA synthesis artifacts. In our hands the T7 and T3 RNA polymerases give the best synthesis results.

1. Linearize 10 μg of plasmid DNA with the appropriate enzyme in a total volume of 100 μL. Incubate the digestion reaction at 37°C for at least 2 h. Check for completeness of the digestion by running a few μL of the digestion on a 1% agarose minigel.
2. Add 100 μL of phenol:chloroform:isoamyl alcohol (25:24:1) and vortex.
3. Incubate for 5 min at room temperature and spin at 17,900×*g* for 5 min.

4. Remove the aqueous phase and add 100 μL chloroform:isoamyl alcohol (24:1).
5. Vortex, incubate for 5 min at room temperature, and spin at 17,900×*g* for 5 min.
6. Remove the aqueous phase and precipitate the linearized DNA with 1/10 volume (10 μl) of 3 M NaOAc and 3 volume (300 μL) of 95% ethanol. Incubate at −20°C for 20 min and spin at 17,900×*g* for 10 min.
7. Remove the supernatant and wash the pellet with 80% ethanol. Spin at 17,900×*g* for 5 min.
8. Remove the supernatant, air dry the pellet, and resuspend in 50 μL of RNAse-free MilliQ water.
9. Prepare the transcription mix by combining in a 1.5-mL microcentrifuge tube 10 μL linearized template (2 μg), 2 μL DIG RNA labeling mix, 4 μL 5 × transcription buffer, 1 μL RNase-out and add RNase-free Milli-Q distilled water to a total volume of 19 μL. Add 1 μL of RNA polymerase (T7 or T3). Incubate at 37°C for 1 h. The probe synthesis protocol is adapted from the Roche DIG RNA labeling mix manual. RNA is labeled to a density of 1 digoxogenin for every 20–25 nucleotides.
10. Load 1 μL of the transcription reaction on a 1% agarose minigel to check the RNA synthesis.
11. Add 1 μL DNaseI and incubate at 37°C for 15 min.
12. Precipitate the riboprobe with 1/10 volume (2 μL) of 3 M NaOAc and 3 volumes (60 μL) of 95% ethanol. Incubate at −20°C for 20 min and spin at 17,900×*g* for 10 min. Wash the pellet with 80% ethanol and air dry. Resuspend the pellet in 50 μL RNase-free Milli-Q water.
13. Run 3 μL of the riboprobe on a 1% agarose minigel to check probe synthesis and verify DNA absence. Multiple RNA bands can be observed because the riboprobe may adopt secondary structures. Save 3 μL of the probe for the gel in the next step (*see* **Note 38**).
14. If the riboprobe is longer than 1000 nucleotides, it may be beneficial to hydrolyze it for better tissue penetration (*see* **Note 39**). Long probes tend to stick randomly to the sample and give unspecific background. Add 50 μL of 2 × probe hydrolysis solution to the 50 μL of riboprobe. Incubate at 60°C for time t where $t = (L_i - L_f)/(k \times L_i \times L_f)$, where L_i is the initial length of the RNA probe in bp, L_f is the final length of the RNA probe in bp (optimal at 200–250 bp), K is the rate constant ($K = 110$ bp/min).

15. Stop the hydrolysis with 1/20 volume (5 μL) of 10% glacial acetic acid. Load 6 μL on a 1% agarose minigel, along with the 3 μL retained from the prehydrolysis riboprobe. The posthydrolysis probe should give a fuzzy band of smaller molecular weight.
16. Precipitate the riboprobe by adding 1 μL of 10 mg/mL yeast tRNA, 1/10 volume (10 μL) of 3 M NaOAc, and 3 volume (300 μL) of 95% ethanol. Incubate at –20°C for 2 h and spin at 17,900×*g* at 4°C for 30 min. Wash the pellet with 80% ethanol and air dry. Resuspend the probe in 40 μL of 50% formamide and freeze at –20°C. This should yield enough probe for 40 slides (*see* **Note 40**).

3.4.6. Slide Prehybridization and Hybridization Treatments

Slides can be placed back to back in the rack slots (make sure to pull them apart at each wash/incubation step to get rid of the prior solution trapped in between them), allowing the use of two times less of each solution. An 18-slide experiment, corresponding to one rack, will require about 6 L of RNase-free Milli-Q water.

After rehydration of the sectioned tissue samples, incubation with a protease is conducted as a pretreatment to permeabilize the tissue and thereby facilitate penetration of the probe. Proteinase K is an effective protease, the concentration of which is critical because too extensive protein degradation results in loss of tissue morphology whereas too limited protein degradation reduces probe penetration and produces a weaker signal. To enhance tissue permeabilization, a gentle acid hydrolysis step can be added prior to the proteinase K treatment, which helps break down the cell walls and partially solubilizes highly cross-linked basic nuclear proteins.

Temperature, buffer composition and pH, blocking reagents and probe concentration are interdependent parameters that affect the ability of the riboprobe and tissue RNA to form duplexes during hybridization. Because the target RNA is embedded in the tissue sections, classical hybridization kinetics and T_m (melting point temperature, the temperature at which 50% of the probe is dissociated) cannot be calculated but must be determined empirically. Optimal hybridization temperatures vary from 45 to 60°C. Formamide, which destabilizes the hydrogen bonds between probe and target sequences, is the organic solvent of choice to reduce the melting temperature of RNA–riboprobe duplexes, thereby permitting the hybridization to take place at lower temperatures. Hybridization stringency is also determined by the concentration of monovalent cations: under high Na^+ concentrations (high stringency), only sequences with high degree of homology form stable duplexes. Yeast tRNA is used as a blocking reagent to reduce nonspecific binding of the riboprobe. Increased probe concentration leads to increased signal until a saturation point is reached. Further concentration increases result in nonspecific binding that masks the true RNA localization.

1. Dewax the tissue sections in 100% Histoclear twice for 10 min each with gentle shaking. Wash twice with 100% ethanol for 2 min each.
2. Rehydrate by dipping the slides sequentially into a freshly made graded series of ethanol dilutions (95, 90, 80, 70, 50, and 30%) with gentle shaking, for 2 min each step. Wash twice in RNase-free Milli-Q water for 2 min each.
3. Incubate the slides in 0.2 M HCl (860 μL of concentrated HCl to 50 mL RNase-free Milli-Q water) for 20 min.
4. Prepare 50 mL of proteinase K buffer and preheat to 37°C in the oven.
5. Wash the slides in RNase-free Milli-Q water for 5 min.
6. Wash the slides in 1 × PBS for 5 min to neutralize remaining acid.
7. Wash the slides in RNase-free Milli-Q water for 5 min.
8. Add 10 μL of proteinase K stock to 50 mL of proteinase K buffer, mix, and incubate the slides at 37°C for 30 min.
9. Wash the slides in 1 × PBS for 5 min.
10. Stop the digestion by washing the slides in 2 mg/mL glycine in 1 × PBS for 2 min.
11. Wash the slides in 1 × PBS for 30 s.
12. Incubate the slides in 1 × PBS containing 3.7% formaldehyde (mix 5 mL of 37% formaldehyde with 45 mL of 1 × PBS) for 20 min (*see* **Note 41**).
13. Equilibrate the tissues in 1 × PBS for 5 min.
14. Dehydrate the tissues in graded ethanol series in the reverse of step two (from RNase-free Milli-Q water to 100% ethanol).
15. Pour off all ethanol and air dry the slides. Under humid conditions, a vacuum treatment may be needed to eliminate any remaining ethanol.
16. Prehybridize the tissues by placing the slides on a 55°C slide warmer while adding 200 μL of prewarmed hybridization solution per slide. Cover with a coverslip and incubate at 55°C for at least 1 h in a box humidified with paper towels soaked in water.
17. The amount of riboprobe to use per slide ranges from 10 to 50 ng per kb. For riboprobe hybridization to an RNA target with an expected identical sequence, it is recommended

to start the hybridization with a 50% formamide solution containing 1 M Na^+ and 40 ng of probe (which should correspond to 0.5–5 μL of the 40 μL probe), at 55°C. Use a total volume of 100 μL hybridization buffer per slide.

18. Per slide (multiply by the number of slides to be hybridized with the same riboprobe): Take 1 μL of DIG-labeled riboprobe, add 5 μL of 50% formamide, and heat to 75°C for 2 min to disrupt the secondary structure.
19. Add 100 μL of warm hybridization solution per slide (multiply by the number of slides) and pipet onto the slide from one edge. Overlay tissue sections by gently applying a coverslip from one edge of the slide to the other, avoiding air bubbles. Incubate slides overnight at 55°C in a box humidified with paper towels soaked in water.
20. Prepare the washing solutions for the next day. Prewarm 2 L of 0.2 × SSC and 0.1% SDS washing solution at 55°C overnight.

3.4.7. Posthybridization Washes

Posthybridization washes with a high-stringency buffer remove unbound riboprobe and separate mismatched duplexes. An RNAse treatment between the washes degrades the nonhybridized or washed ssRNA probes. Riboprobe degradation products will then be eliminated with subsequent washes. For efficient washes, it is crucial to lay the slides flat in a glass dish and immerse them in 200 mL of washing solution under gentle agitation.

1. Dip the slides in prewarmed (55°C) 0.2 × SSC and 0.1% SDS. The slide/coverslip duplexes will separate. Remove the coverslips.
2. Lay the slides flat at the bottom of a glass dish and wash with prewarmed (55°C) 0.2 × SSC and 0.1% SDS on a shaker, for 10 min at 55°C. Repeat this step once.
3. During the washes, prepare 400 mL of 2 × SSC and warm it at 37°C for the RNAse treatment in step 5.
4. Incubate the slides in 2 × SSC for 2 min at room temperature.
5. Prepare 10 μg/mL of RNAse solution in prewarmed (37°C) 2 × SSC and incubate the slides for 30 min at 37°C, under gentle agitation (*see* **Note 42**).
6. Wash the slides in 2 × SSC for 2 min at room temperature.
7. Wash with prewarmed (55°C) 0.2 × SSC and 0.1% SDS (high-stringency wash) on a shaker for 10 min at 55°C. Repeat this step once.

3.4.8. Detection

The detection process starts with immunological detection of the DIG-labeled riboprobes. Best results are obtained using an anti-DIG antibody coupled to alkaline phosphatase (AP). The AP reaction with NBT and BCIP produces formazan, a stable blue/purple dye with bright reflective properties. Formazan does not precipitate, therefore allowing long incubation times (>24 h). Because formazan is soluble in organic solvents, it is crucial to embed the slides in aqueous mounting medium after detection. Mount the slides in a 50% glycerol solution, which gives good resolution for microscope observation and imaging and allows unlimited storage of the stained sections. Process all washes and buffer incubations in glass dishes, with the slides laying flat at the bottom. Gentle rotation should be applied during the wash steps.

1. Rinse the slides in 1 × TBS for 5 min at room temperature.
2. Incubate the slides in 0.5% Boehringer blocking reagent in 1 × TBS for 1 h at room temperature.
3. Rinse in 1 × TBS containing 0.5% bovine serum albumin (BSA) and 0.1% Triton for 30 min at room temperature.
4. Replace the solution in step 3 with 1 × TBS containing 0.5% BSA and incubate at room temperature for 5 min.
5. Remove the slides from the glass dish and add anti-DIG AP conjugate (*see* **Note 43**) in 1 × TBS containing 0.5% BSA. Use 100 μL per slide and overlay the tissue sections by gently applying a coverslip, avoiding air bubbles.
6. Place the slides flat in a humidified box (containing paper towels saturated with water) and incubate for 2 h at room temperature.
7. Dip the slide/coverslip duplexes in 1 × TBS, 0.5% BSA, and 0.1% Triton X-100. Remove the coverslips.
8. Lay the slides down flat in the glass dish and rinse them in 1 × TBS, 0.5% BSA, and 0.1% Triton X-100 for 10 min at room temperature on a shaker. Repeat this step 3 times.
9. Rinse slides in detection buffer for 15 min at room temperature with shaking. Save a few mL of this buffer for the next step.
10. Apply 100 μL of detection buffer plus BCIP and NBT substrates per slide, overlay with coverslip, and place slides in a humidified box containing paper towels soaked in water. Incubate in the dark for 4–36 h (*see* **Note 44**).
11. When a satisfactory signal is observed, stop the reaction by dipping the slides in Milli-Q water.

12. Mount the slides with 80 μL of 50% glycerol and overlay with a coverslip, being careful that no drops of glycerol solution are coming out of the sandwich.
13. Visualize the slides using a stereo-microscope and troubleshoot as necessary (*see* **Note 45**).

4. Notes

1. Formaldehyde and formaldehyde-containing solutions are toxic and should be handled wearing gloves under a fume hood. Treat all materials (pipets, tubes, etc.) that touch the solutions as hazardous waste.
2. Propidium iodide is a nucleic acid intercalating agent and should always be handled wearing gloves and suitable protecting clothing. Treat all materials (pipets, tubes, etc.) that touch the solution as hazardous waste.
3. Technovit 7100 resin is useful for working with plastic sections at room temperature. This resin hardens easily at room temperature, so it should be stored at 4°C.
4. Toluidine blue is a general purpose stain/dye. This dye stains certain cellular components with different colors, i.e., lignin/phenol is stained green/blue-green, pectins stain pink/reddish purple, and DNA stains green/blue-purplish.
5. The free software can be downloaded at http://rsbweb.nih.gov/ij/. Linux, Windows, or Mac versions are available to meet the needs of all researchers.
6. The Image J program can open, process, and save images in any format (TIFF, JPEG, PNG, GIF, BMP, and raw data). For a complete list of supported data types, refer to the software documentation Web page.
7. Glassware (cylinders, beakers, flasks, bottles, slide racks or Copeland jars) must be baked for 4 h at 180°C to inactivate any trace of RNAse. It is not necessary to treat the water with DEPC, but simply use freshly autoclaved Milli-Q water stored in clean, baked glassware.
8. The Roche RNA DIG labeling and detection system is excellent for preparation and in situ detection of riboprobes. The system contains T7/T3 RNA polymerases, RNAse out, DNaseI, blocking reagent, NBT, and BCIP. Alternative reagents are available from other commercial sources and give good results (e.g., RNA polymerases and RNAsin from Promega). To prepare the 5 × blocking reagent solution, the 1 × TBS buffer must be warmed

before adding the blocking reagent powder. The solution will look cloudy.

9. Paraformaldehyde powder and solution is toxic and should be handled wearing gloves under a fume hood. Treat all materials (pipets, tubes, etc.) that touch the powder or solution as hazardous waste. Because paraformaldehyde vapors are also toxic, verify that the vacuum system used for tissue infiltration does not exhaust into the laboratory.
10. Plant tissues have a cuticle and thus float on the surface of the fixative, preventing proper infiltration. The combination of the Triton X-100 detergent and DMSO solvent enhances the penetration of the fixative while reducing the disrupting effect of the vacuum on the tissue morphology.
11. Formamide is highly corrosive in contact with skin and eyes, so the hybridization buffer should be handled wearing gloves under a fume hood. Treat all materials (pipets, tubes, etc.) that touch the solution as hazardous waste.
12. Imbibing seeds at cold temperature prevents embryo development during this step.
13. Isolating intact embryos is a key step for high-quality imaging and morphology studies. The technique is difficult to master at the beginning; it requires patience and practice, so include extra seeds when trying to dissect embryos for the first time because inevitably some samples will be lost. Sit comfortably with your forearms resting on the bench so as to have steady hands.
14. Fixation is not necessary to prepare embryos for confocal laser scanning microscopy. To avoid tissue loss or damage, remove solutions by pipetting off the liquid with Pasteur pipets or a micropipet.
15. Use tips that have been end cut with a razor blade. This will widen the tip end and avoid any damage to the sample during the pipetting process.
16. Do not try to grasp the embryos directly with the forceps. Almost completely close the forceps around the embryo to be transferred and lift the forceps. Some liquid will be trapped within the ends of the forceps and with it the embryo, which can now be safely placed in another drop of immersion oil.
17. Use nail polish to seal the coverslips because it can be easily removed using a Q-tip dipped in acetone to access the sample, in case the embryos drop or need to be moved for better visualization of the shoot apical meristem.

18. One scintillation vial should be used per genotype. Take care not to pack the samples tightly into the vial; there should be room between them as they float on the surface of the solution. Write the name of the sample on the outside of the glass using a marker and cover it with a piece of clear tape so it does not wash off during the ethanol steps.
19. The fixed samples should remain at the bottom of the vials once the vacuum has been released. If the samples rise to the surface of the fixation solution, reapply the vacuum for another 5–10 min.
20. Resin-infiltrated samples can be kept for at least 6 months at 4°C.
21. This step will make the sample handling easier at later steps.
22. The resin will harden in 10–15 min, so it is important to maintain the correct position of the sample continuously until the hardening process is complete.
23. The SAM should appear as a dome between the leaf primordia for a longitudinal section (**Fig. 8.2A**), or as a circle in the center of the primordia for a transverse section (**Fig. 8.2B**). After reaching the SAM tissue, section very carefully so as not to lose the expected sectioning cut.
24. Perform this step quickly so that the section does not have time to roll up.
25. Use sharp tweezers to unfold the ribbon in distilled water. After putting the ribbon on the cover glass, check under the microscope again to determine whether the ribbon is completely unfolded and make any needed corrections using sharp forceps.
26. For optimal viewing, the tissues should be stained light to medium blue. If the staining is too light the tissues will not be visible against the background, and if the staining is too dark the individual cells and layers will not be resolved. Note that the addition of Permount during the mounting step tends to lighten the tissue stain a few shades.
27. Permount is very harmful if inhaled, so work in the fume hood.
28. Before placing the cover glass carrying the ribbon on the Permount, carefully blow the dust off the ribbon so as not to permanently fix debris onto the slide.
29. Include a scale bar in each raw image to serve as a reference for the Image J measurements.
30. Measure at least 10 individual meristems to obtain statistically significant data.

31. For observation of a true vegetative shoot apex, the seedling should not be more than 7 days old if grown under constant light, to ensure that the SAM has not gone through the reproductive transition.
32. The Eosin staining step is crucial for later tissue sectioning because it helps visualize and orient the tissue.
33. These solution replacements avoid the need to prepare a series of multiple dilutions because stock solutions (ethanol and solvent) will be slowly added to the previous solution, gradually bringing the content of the mixture to 100% of ethanol or solvent.
34. The subsequent step of mounting the embedded tissue is made easier by aligning the samples as the paraffin hardens around them. Orient the inflorescence apices on their sides with the stems pointing in the same direction and align in straight rows of 12–14 inflorescences each. Leave ~1 cm of paraffin between each sample.
35. Scoop a small piece of paraffin onto the tip of a metal spatula and melt it in a flame. Transfer the melted paraffin onto the top of the sample holder and affix it to the bottom of the paraffin block. Hold the two together for ~1 min until the melted paraffin seals around the block. The paraffin block should be mounted such that the microtome knife will strike the longest side.
36. This step will ensure a clean longitudinal section through the shoot apical meristem in the majority of samples.
37. For expression analysis of a gene of unknown pattern, it is important to prepare a sense probe that will serve as a negative control for specific hybridization of the antisense probe.
38. The yield of the transcription reaction can be estimated by running an aliquot of the probe on a 1.5% mini agarose gel, next to an RNA standard of known concentration (RNA Molecular Weight Marker III, 0.3–1.5 kb, Roche). For testing the DIG labeling reaction yield, 1 μL of the probe can be deposited on a piece of filter, UV cross-linked, and incubated with 5 mL of detection buffer containing 5.5 μL of NBT and 4 μL of BCIP. A dark blue spot should become visible in the place where the probe was pipetted.
39. The necessity for riboprobe hydrolysis is controversial. Some researchers hydrolyze any riboprobe greater than 1 kb in length, whereas others find that it is not required in order to obtain a good signal. If a riboprobe greater than 1 kb in length gives a weak signal, then hydrolysis is recommended.

40. When using side-by-side sense and antisense probes, it is crucial to load them on a 1.5% agarose mini gel to compare their concentrations in order to use the same amount of riboprobe per slide.
41. This step helps to refix the tissues after the destabilizing proteinase K treatment.
42. This step can be essential to eliminate any background signals. The washing temperature can also be raised to 65°C to help reduce background.
43. Dilute 1:1000 to 1:500 for low-abundance transcripts. Alternatively, the antibody can be diluted to 1:3000 for an overnight incubation at 4°C.
44. Most probes require an overnight incubation. Signal from very rare transcripts can be better observed after 48 h after adding more detection buffer plus substrate.
45. If the signal is brown, the pH of the detection solution is probably incorrect. Make sure that the pH is 9.6 and increase the washing time in the detection solution before adding the NBT + BCIP. If the signal appears as a purple haze of ubiquitous staining, there may be unspecific hybridization or antibody recognition problems. Several modifications to the protocol can be tried, such as decreasing the amount of probe or antibody, increasing the hybridization temperature, and/or increasing the duration and temperature of the posthybridization washes. If this does not solve the problem, try a probe designed from another region of the gene of interest. On the other hand, the absence of signal can have multiple origins, such as transcript abundance below the threshold of detection, poorly labeled probe, too stringent posthybridization washes, deficient anti-AP antibody, and/or old NBT/BCIP substrates. Add positive controls to troubleshoot this situation. When using a probe for the first time, it is very informative to run the hybridization experiment side-by-side with a DIG-labeled riboprobe that is already known to work.

Acknowledgments

We thank George Chuck, Harley Smith, and Sabine Zachgo for sharing protocols and giving suggestions on the in situ hybridization technique, and Helena Pires and Jinsun Kim for helpful comments. This work is supported by USDA CRIS 5335-21000-016-00D and NSF IOS-0718843.

References

1. Steeves, T. A. and Sussex, I. M. (1989) Patterns in Plant Development. Cambridge University Press, New York, NY.
2. Tucker, M. R. and Laux, T. (2007) Connecting the paths in plant stem cell regulation. *Trends Cell Biol* **17**, 403–410.
3. Bhalla, P. and Singh, M. B. (2006) Molecular control of stem cell maintenance in shoot apical meristem. *Plant Cell Rep* **25**, 249–256.
4. Williams, L. and Fletcher, J. C. (2005) Stem cell regulation in the Arabidopsis shoot apical meristem. *Curr Opin Plant Biol* **8**, 582–586.
5. Running, M. P., Clark, S. E., and Meyerowitz, E. M. (1995) Confocal microscopy of the shoot apex. *Meth Cell Biol* **49**, 217–229.
6. Fletcher, J. C. (2001) The *ULTRAPETALA* gene controls shoot and floral meristem size in Arabidopsis. *Development* **128**, 1323–1333.
7. Jin, L. and Lloyd, R. V. (1997) In situ hybridization: methods and applications. *J Clin Lab Anal* **11**, 2–9.
8. Gall, J. G. and Pardue, M. L. (1969) Formation and detection of RNA–DNA hybrid molecules in cytological preparations. *Proc Natl Acad Sci USA* **63**, 378–383.
9. Houben, A., Orford, S. J., and Timmis J. N. (2006) In situ hybridization to plant tissues and chromosomes. *Meth Mol Biol* **326**, 203–218.
10. Jackson, D. (1992) In situ hybridization in plants. In: Molecular Plant Pathology: A Practical Approach, 163–174. Bowles, D. J., Gurr, S. J., McPherson, R., eds. Oxford University Press, Oxford.
11. Ambrose, B. A., Lerner, D. R., Ciceri, P., Padilla, C. M., Yanofsky, M. F., and Schmidt, R. J. (2000) Molecular and genetic analyses of the *Silky1* gene reveal conservation in floral organ specification between eudicots and monocots. *Mol Cell* **5**, 569–579.
12. Chuck, G., Muszynski, M., Kellogg, E. A., Hake, S., and Schmidt, R. J. (2002) The control of spikelet meristem identity by the *branched silkless1* gene in maize. *Science* **298**, 1238–1241.
13. Carles, C. C., Lertpiriyapong, K., Reville, K., and Fletcher, J. C. (2004) The *ULTRAPETALA1* gene functions early in Arabidopsis development to restrict shoot apical meristem activity, and acts through *WUSCHEL* to regulate floral meristem determinacy. *Genetics* **167**, 1893–1903.
14. Murashige, T. and Skoog, F.(1962) A revised medium for rapid growth and bioassays with tobacco tissue cultures. *Physiol Plant* **15**, 473–497.

Chapter 9

Analyzing Floral Meristem Development

Elisa Fiume, Helena R. Pires, Jin Sun Kim, and Jennifer C. Fletcher

Abstract

Flowers contain the male and female sexual organs that are critical for plant reproduction and survival. Each individual flower is produced from a floral meristem that arises on the flank of the shoot apical meristem and consists of four organ types: sepals, petals, stamens, and carpels. Because floral meristems contain a transient stem-cell pool that generates a small number of organs composed of a limited number of cell types, they are excellent model systems for studying stem-cell maintenance and termination, cell fate specification, organ morphogenesis, and pattern formation.

Key words: Floral meristem, organ number, confocal laser scanning microscopy, scanning electron microscopy.

1. Introduction

Plant reproduction is unique in that plants form their floral reproductive structures de novo following their embryonic and vegetative development. In response to both endogenous and environmental cues plants undergo the transition to flowering (1), during which the shoot apical meristem becomes an inflorescence meristem that produces a distinctive architecture of flowers. Flowers contain the male (stamen) and female (carpel) reproductive organs and are themselves derived from floral meristems. An *Arabidopsis thaliana* floral meristem contains a transient stem-cell population at its apex that generates progenitor cells for the sepal, petal, and stamen primordia along its flanks before becoming consumed in the production of the central carpel primordia (2).

Elisa Fiume, Helena R. Pires, Jin Sun Kim have contributed equally to this work.

L. Hennig, C. Köhler (eds.), *Plant Developmental Biology*, Methods in Molecular Biology 655,
DOI 10.1007/978-1-60761-765-5_9, © Springer Science+Business Media, LLC 2010

During flower development each organ type arises in a stereotypical spatio-temporal sequence (3) and consists of a small number of specific cell types that can be accurately visualized and quantified using several different techniques. In this chapter, we will explain methods and approaches for the analysis of floral meristem development: (i) Floral organ number counting is a simple method to quantify and statistically analyze the phenotypes of mutants that display altered floral organ number, and/or produce mosaic floral organs. Such defects are often associated with enlarged or reduced floral meristem size (4, 5) (*see* **Section 3.1**). (ii) Confocal laser scanning microscopy (CLSM) uses sophisticated laser imaging technology to obtain optical sections of whole mount inflorescence and floral meristem samples (6) for ready examination and measurement (*see* **Section 3.2**). CLSM is a powerful method for examining internal cell patterns without the technical challenges associated with histological sectioning, and the optical sections can be combined into a three-dimensional image of the sample. (iii) Scanning electron microscopy (SEM) allows high-resolution surface-structure imaging by measuring the angle and energies of electrons scattered by atoms on the surface of a sample. The extreme level of detail that can be acquired using SEM is applicable to *Arabidopsis* for analysis of the number and arrangement of floral meristems, the position and structure of floral organ primordia, and the cell-surface morphology of individual flower organs (7) (*see* **Section 3.3**).

2. Materials

A. thaliana plants are grown in a 1:1:1 mixture of perlite:vermiculite:topsoil under cool-white fluorescent lights (100–140 mmol/m^2 s) at 21–22°C. For statistical robustness a minimum of 10 samples from each genotype should be analyzed per experiment.

2.1. Floral Organ Counting

1. Fine forceps (number 5, Ted Pella Inc).
2. Scissors with a sharp blade.
3. Dissecting microscope with 10 × objective.
4. Statistical analysis software.

2.2. Confocal Laser Scanning Microscopy of the Inflorescence Meristem

1. Glass scintillation vials or similar containers.
2. Microscope depression slides.
3. Coverslips (24 × 60 mm).
4. Fine forceps (number 5, Ted Pella Inc.).
5. Fine needles.

6. Formaldehyde acetic acid (FAA) fixation solution: 3.7% formaldehyde, 50% ethanol, and 5% acetic acid. Prepare a fresh mixture of ~10 mL per vial.
7. Prepare 15, 30, 50, 70, 85, 90, and 95% ethanol in distilled water.
8. Graded Histoclear series: Prepare 500 mL solutions of 25:75 Histoclear (National Diagnostics):Ethanol, 50:50 Histoclear:Ethanol, and 75:25 Histoclear:Ethanol.
9. Propidium iodide stock solution: Prepare 100 μg/mL propidium iodide in 0.1 M L-arginine and adjust the pH to 12.4 with 5 M NaOH. This solution is deep red and can be stored for months at 4°C. When it turns dark orange it cannot be used anymore and should be disposed of in the hazardous waste.
10. Prepare propidium iodide staining solution by diluting the 100 μg/mL propidium iodide stock solution to 5 μg/mL in 0.1 M L-arginine and adjust the pH to 12.4 with 5 M NaOH.
11. Rinsing solution: Prepare a solution of 0.1 M L-arginine buffer in distilled water and adjust the pH to 8 with HCl.
12. Immersion oil.
13. Nail polish.

2.3. Scanning Electron Microscopy of Developing Flowers

1. Glass scintillation vials.
2. Plastic conical tubes.
3. Fine forceps (number 5, Ted Pella Inc.).
4. Dissecting microscope.
5. Cylinder mount gripper (Ted Pella, Inc.).
6. Mounting bases (*see* **Note 1**).
7. Mounting stub.
8. Conductive stickers.
9. White index cards.
10. Critical point dryer.
11. Sputter coater apparatus.
12. 0.1 M Sodium phosphate buffer (PB) buffer: Combine 200 mL of 0.1 M sodium phosphate monobasic NaH_2PO_4 (12 g/L) and 800 mL of 0.1 M sodium phosphate dibasic Na_2HPO_4 (14.2 g/L). The pH should be between 7.2 and 7.4.
13. 25 mM Sodium phosphate buffer (PB) wash solution: Dilute 100 mL of 0.1 M sodium phosphate buffer (PB) to 25 mM in distilled water.

14. Glutaraldehyde fixation solution: Freshly prepare 6 mL of 0.1 M PB, 3 mL 25% glutaraldehyde, and 16 mL distilled water in a plastic conical tube in the hood (*see* **Note 2**).
15. Graded ethanol series: Prepare 30, 50, 65, 75, 89, and 95% ethanol in distilled water.

3. Methods

3.1. Floral Organ Counting

The first 10 flowers from each of 10 plants of a particular genotype are removed and all organs are counted and recorded.

3.1.1. Flower Dissection and Counting

1. Use small scissors or fine forceps to remove the first 10 flowers on the first plant (*see* **Note 3**).
2. Grip the first flower by the pedicel with forceps and move it under a dissecting microscope.
3. With the other hand, use another set of forceps to remove each sepal sequentially by drawing the organ down and away from the pedicel. Score and record the total number of sepals (*see* **Note 4**).
4. Repeat step 3 with the petals and then the stamens.
5. For the carpels that remain, make a cross-section through the intact gynoecium using a sharp blade. Count and record the number of individual carpels revealed in the cross-section (*see* **Note 5**).
6. Repeat steps 1–5 with the first 10 flowers on the next plant.

3.1.2. Data Analysis

1. Enter the data into a statistical program (e.g., Microsoft Excel) for analysis (*see* **Table 9.1**). Arrange the data according to each genotype.
2. Calculate the mean and standard error values for each dataset. A chi-square test can be performed to determine the statistical significance of values that differ between two genotypes.

3.2. Confocal Laser Scanning Microscopy of the Inflorescence Meristem

3.2.1. Tissue Fixation

1. Prepare the fixation solution and chill it on ice. Dispense ~10 mL fixation solution into each glass scintillation vial or similar container, one vial per genotype. Keep the vials on ice (*see* **Note 6**).

Table 9.1
Sample tabulation of floral organ counting raw data

Plant #	Flower #	Sepals	Petals	Stamens	Carpels	Sum
1	1	6	6	7	2	21
	2	6	7	8	3	24
	3	6	7	7	2	22
	4	7	7	8	3	25
	5	7	7	8	4	26
	6	6	7	8	3	24
	7	7	7	6	2	22
	8	5	5	7	3	20
	9	6	7	7	3	23
	10	6	6	7	2	21
2	1	7	7	8	4	26
	2	8	8	8	3	27
	3	7	9	9	3	28
	4	6	6	7	3	22
	5	6	8	8	4	26
	6	7	7	6	3	23
	7	8	8	9	3	28
	8	6	8	8	3	25
	9	5	7	7	3	22
	10	8	8	7	4	27

2. Clip off the young inflorescences when the bolting stem is a few cm tall. Remove the older flowers and leave 5–7 visible flower buds. Retain several cm of stem because it will help with later manipulations.
3. Immediately place the tissue into the fixation solution. All tissue should be submerged in the fixation solution so it is recommended not to pack too many samples into the same vial.
4. Loosen the caps of the scintillation vials and place them on ice in a vacuum chamber.
5. Pull the vacuum slowly to 25 psi for 15 min. This step removes air bubbles from the samples to allow the penetration of the fixative into the tissue. The samples will begin to sink. Slowly release the vacuum to return the samples to air.
6. Swirl the vials to redistribute the fixation solution over the tissues.
7. Apply the vacuum for another 10 min, after which the tissues should sink.

8. Remove the fixation solution and add a fresh cold aliquot (*see* **Note 7**). Place the vials at 4°C and incubate overnight.

3.2.2. Tissue Dehydration

Ethanol solutions should be kept at 4°C.

1. Pour off fixation solution and replace it with cold 50% ethanol. Incubate for 1 h.
2. Dehydrate the tissue through an ethanol series (70, 85, 95, and 100%), leaving the tissues in each solution for 1 h. Ethanol solutions should be cold and samples should be kept at 4°C.
3. Remove the 100% ethanol and replace it with a fresh aliquot, then leave it overnight at 4°C to remove the remaining chlorophyll and complete the fixation process. The following morning the tissue should be white. If some chlorophyll remains in the tissue, continue replacing the 100% ethanol at 1 h intervals until the tissue is completely white.

3.2.3. Tissue Staining and Rinsing

1. Rehydrate the samples through a decreasing ethanol series (95, 85, 70, 50, 30, 15%, and distilled water), leaving the tissues in each solution for 1 h at room temperature.
2. Wash twice briefly with distilled water.
3. Prepare a stock solution of propidium iodide (*see* **Note 8**).
4. Add 2–3 mL of 5 μg/mL propidium iodide staining solution to each vial and incubate at room temperature for 24 h. The samples should be completely submerged in the staining solution. After the staining incubation period, the tissue appears pale orange.
5. Replace the staining solution with the L-arginine buffer rinsing solution. Leave in L-arginine buffer at 4°C for 4 days, changing the rinsing solution once every day.

3.2.4. Tissue Clearing

1. Dehydrate the sample through an ethanol series (15, 30, 50, 70, 85, 95, and 100%), leaving the tissues in each solution for 1 h at room temperature.
2. Wash twice more with 100% ethanol.
3. Incubate the tissue in a Histoclear series (75:25 Ethanol:Histoclear, 50:50, 25:75) to 100% Histoclear. Leave the samples in each solution for at least 2 h to completely clear the tissue.
4. Change the 100% Histoclear three times and leave the tissue overnight in the last change.

3.2.5. Tissue Dissection

It is necessary to remove the older, larger flower buds in order to view the floral meristems. Dissecting in Histoclear is perfectly safe as this reagent is nontoxic. Nevertheless, it is strongly odorous and breathing its vapors for a long time can be unpleasant. For this reason it is preferable to dissect in immersion oil, which is also used as the mounting medium.

1. Place four individual drops of immersion oil on a microscope slide. Take one sample by the stem with fine forceps and put it in the first oil drop. Begin removing the older flower buds under the dissecting microscope (*see* **Note 9**).
2. When the first drop of oil has become too dirty to continue dissecting, grip the sample with forceps by the stem and move it into the second drop of oil. Continue dissecting, increasing the magnification if necessary. Repeat this step as many times as needed to remove most of the visible flower buds.
3. Once the sample has been dissected, cut off as much of the stem as possible.

3.2.6. Mounting and Imaging

1. Transfer the sample into the center of a coverslip. Ensure there is enough immersion oil to keep the sample in place. The top of the inflorescence meristem should lay flat against the coverslip (*see* **Note 10**). If the samples are sufficiently small, several can be transferred to a single coverslip.
2. Flip the coverslip over atop a depression slide and seal the four corners of the slide with nail polish (*see* **Note 11**).
3. Visualize each slide using a confocal laser scanning microscope (*see* **Fig. 9.1**). Propidium iodide can be excited by a 514-nm argon laser beam and emits between 580 and 610 nm.

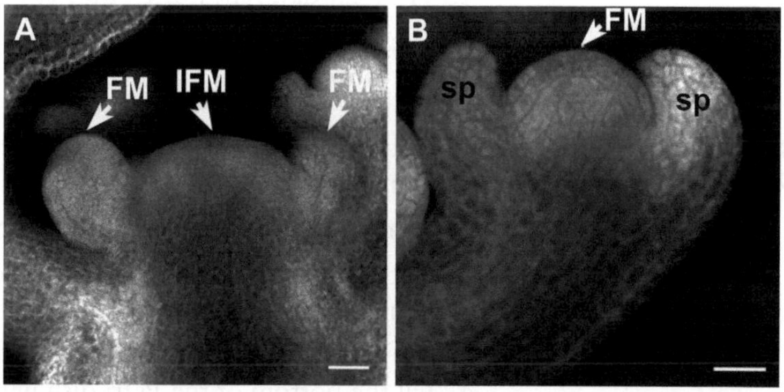

Fig. 9.1. **Confocal laser scanning micrographs of an Arabidopsis inflorescence meristem and a floral meristem.** (**A**) Optical longitudinal section of a wild-type Landsberg *erecta* (L*er*) inflorescence meristem (IFM) producing floral meristems (FM) on the flanks. (**B**) Optical longitudinal section of a wild-type L*er* stage 3 flower with sepal primordia (sp) arising from the flanks of the floral meristem (FM). Scale bars, 30 μm in (**A**) and 20 μm in (**B**).

3.3. Scanning Electron Microscopy of Developing Flowers

3.3.1. Tissue Fixation

1. Aliquot ~8 mL of glutaraldehyde fixation solution into each glass scintillation vial or similar container, one vial per genotype.
2. Using forceps or sharp scissors, gently clip off each inflorescence apex or single flower with 1 cm of the stem remaining and immediately place it into a scintillation vial. The tissue will float on the surface, so gently swirl the vial to completely cover the tissue with fixation solution.
3. Incubate overnight at room temperature under constant rotation.
4. Remove the fixation solution into a hazardous waste bottle using a Pasteur pipet.
5. Optional: Perform a secondary osmium tetroxide (OsO_4) coating step (*see* **Note 12**).

3.3.2. Tissue Rinsing and Dehydration

1. Rinse the tissues 3 × with 25 mM PB wash solution. Empty the first two washes into a hazardous waste bottle using a Pasteur pipet.
2. Dehydrate the samples through an ethanol series (30, 50, 65, 75, 89, 95, and 100%), leaving the tissues in each solution for 15–30 min.
3. Wash the tissues 3 × with 100% ethanol and leave them in the third change overnight at room temperature.
4. The next day, repeat the 100% ethanol wash twice.
5. Store the samples in 100% ethanol until ready to dry them (*see* **Note 13**).

3.3.3. Critical Point Drying

1. Choose the appropriate size specimen basket to fit the samples.
2. Remove the basket lid and place the basket in a Petri dish.
3. Cut small pieces of paper and on each write the genotype or sample name with a pencil. Using the forceps transfer each piece of paper into a separate chamber of the basket.
4. Fill the bottom of the Petri dish with 100% ethanol.
5. Quickly pour the samples from one scintillation vial into the Petri dish. Use forceps to gently transfer the samples into the respective chamber(s), minimizing their exposure to air. Close the lid over the specimen basket.
6. Dry the samples in the critical point dryer, following the manufacturer's instructions (*see* **Note 14**).

7. Use forceps to transfer the dried samples to clean, dry scintillation vials for storage.

3.3.4. Tissue Mounting

1. Place the two mounting bases on the dissecting microscope base.
2. On top of one mounting base, place a white index card that has been folded in the middle. On the other mounting base, place a mounting stub.
3. Use forceps to lift a conductive sticker by the edge and place it over the mounting stub (*see* **Note 15**). Place the two mounting bases side by side.
4. Carefully tip a specimen from the first vial onto the folded white index card.
5. View the specimen under the microscope. For an inflorescence meristem, grip it by the stem with one pair of forceps. With the other pair of forceps gently break off each older flower by pushing it very carefully down away from the stem (*see* **Note 16**), until the inflorescence apex and floral primordia are exposed. For a single flower, hold it by the stem and break off two sepals and petals from the tip downward in order to be able to see the internal floral organs.
6. Once the dissection is finished, use forceps to transfer the specimen onto the mounting stub covered by the sticker. For an inflorescence, carefully affix it by the base of the stem such that it sits perpendicular to the mounting stub. For a flower, affix the side that still maintains the sepals and petals to the mounting stub (*see* **Note 17**).
7. Coat the samples using a sputter coating apparatus (*see* **Note 18**) and visualize them with a scanning electron microscope (*see* **Fig. 9.2**), following instructions specific to the apparatus.

4. Notes

1. For inflorescences and single flowers, use a 10 × 5 mm specimen mount and corresponding mounting bases and conductive stickers.
2. Glutaraldehyde is highly toxic so it should always be handled wearing gloves under a fume hood. Treat all materials (pipets, tubes, etc.) that touch the solution as hazardous waste.
3. This procedure generally takes several days to complete. Begin when the first arising flowers on the plant(s) of inter-

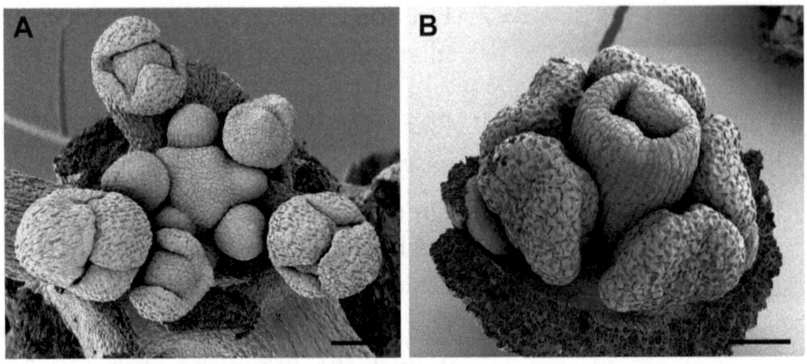

Fig. 9.2. **Scanning electron micrographs of an Arabidopsis inflorescence apex and a developing flower.** (**A**) Wild-type Landsberg *erecta* (L*er*) inflorescence meristem producing floral meristems in a spiral phyllotaxy. Stage 1 through stage 5 floral meristems are shown. (**B**) Wild-type L*er* developing flower with all four sepals removed to reveal the petal, stamen, and carpel morphology. Scale bars, 50 μm.

est have opened, and remove each open flower for analysis. Because organ number in unopened flower buds is difficult to accurately quantitate, when all the open flowers have been analyzed, stop and continue the analysis the next day, once more buds have opened.

4. Some genotypes may produce flowers with fused or mosaic organs consisting of two types of tissue. If an abnormal floral organ is observed, add a new category to the results table and note the frequency of its occurrence (8). The cellular composition of mosaic floral organs may be investigated using scanning electron microscopy.
5. To prevent damage to the gynoecium, use a sharp cutting motion rather than a sawing motion.
6. Formaldehyde and formaldehyde-containing solutions are toxic and should be handled wearing gloves under a fume hood. Treat all materials (pipets, tubes, etc.) that touch the solutions as hazardous waste.
7. Pipette the liquid off with a Pasteur pipet or a micropipette, trying not to touch the samples to avoid damaging them.
8. Propidium iodide is a nucleic acid intercalating agent and should always be handled wearing gloves and suitable protecting clothing. Treat all materials (pipets, tubes, etc.) that touch the solution as hazardous waste.
9. To remove the flower buds, use two sets of fine forceps. With one set hold the inflorescence by the stem. With the other set touch the top of the bud and apply gentle pressure down and outwards away from the inflorescence stem. When the flower pedicel is bent outwards the flower bud can be safely pulled off the stem without damage to the surrounding tissues.

10. It is important to let the inflorescence meristem touch the coverslip. The surface of the meristem should lay flat against the coverslip for best imaging of the flower meristems.
11. Sealing with nail polish is preferable to other methods because it can easily be dissolved with acetone in the event that samples are dropped or need to be better oriented for imaging.
12. Osmium tetroxide (OsO_4) may be used as a secondary fixative if necessary to add additional density and contrast to the tissue (9). Prepare 4 mL of a 1% OsO_4 solution: 1 mL 4% OsO_4, 1 mL 0.1 M PB, and 2 mL distilled water for each sample vial and add it to the vial using a Pasteur pipet. Note that OsO_4 is highly poisonous, even at low exposure levels, so it should always be handled wearing gloves under a fume hood. Samples are incubated from overnight to several days, and the tissue should turn black. Usually overnight is sufficient for both inflorescences and single flowers. If the samples are left too long in the solution, the OsO_4 may begin to sediment and leave a grainy black residue on the sample surfaces. After incubation, empty the 1% OsO_4 fixation solution into a hazardous waste bottle using a pasteur pipet and replace with 25 mM PB. Treat all materials (pipets, tubes, etc.) that touch the OsO_4 solution as hazardous waste.
13. For long periods, samples should be stored in 70% ethanol rather than in 100% ethanol.
14. Use of safety glasses is advised while operating the critical point dryer.
15. Directly stick the tissues to the stub using the adhesive sticker. The surface of the stub should be as smooth and free of structure as possible to avoid confusing background.
16. The dried tissues are brittle and prone to damage unless handled extremely carefully (9).
17. Place 3–4 inflorescences, or as many as five single flowers horizontally aligned, on a single mounting stub.
18. This step coats the samples with a conductive metal to prevent the buildup of high-voltage charges on the surface during the microscopy process (9). Generally a 15–40 nm coating thickness is adequate, but use the minimum coating thickness possible. Over-coating obscures the surface detail and prevents high-resolution imaging, whereas under-coating can lead to charge buildup in the electron microscope. Samples may be recoated if charging occurs. When first performing this protocol, use a few wild-type samples to empirically determine the optimal coating thickness needed to satisfactorily obtain the data.

Acknowledgments

We thank Elliot Meyerowitz, Beth Krizek, Joshua Levin, and Mark Running for sharing protocols and Cristel C. Carles, Chanman Ha, and JiHyung Jun for providing helpful suggestions concerning the techniques. This work is supported by USDA CRIS 5335-21000-016-00D and NSF IOS-0718843.

References

1. Simpson, G. G. and Dean, C. (2002) Arabidopsis, the Rosetta stone of flowering time. *Science* **296**, 285–289.
2. Krizek, B. A. and Fletcher, J. C. (2005) Molecular mechanisms of flower development: an armchair guide. *Nature Rev Genet* **6**, 688–698.
3. Smyth, D. R., Bowman, J. L., and Meyerowitz, E. M. (1990) Early flower development in Arabidopsis. *Plant Cell* **2**, 755–767.
4. Clark, S. E., Running, M. P., and Meyerowitz, E. M. (1995) CLAVATA3 is a specific regulator of shoot and floral meristem development affecting the same processes as CLAVATA1. *Development* **121**, 2057–2067.
5. Zhao, Y., Medrano, L., Ohashi, K., et al. (2004) HANABA TARANU is a GATA transcription factor that regulates shoot apical meristem and flower development in Arabidopsis. *Plant Cell* **16**, 2586–2600.
6. Running, M. P., Clark, S. E., and Meyerowitz, E. M. (1995) Confocal microscopy of the shoot apex. *Methods Cell Biol* **49**, 217–229.
7. Bowman, J. L., Smyth, D. R., and Meyerowitz, E. M. (1991) Genetic interactions among floral homeotic genes of Arabidopsis. *Development* **112**, 1–20.
8. Levin, J. Z. and Meyerowitz, E. M. (1995) UFO: an Arabidopsis gene involved in both floral meristem and floral organ development. *Plant Cell* **7**, 529–548.
9. Bozzola, J. J. and Russell, L. D. (1999) Electron Microscopy: Principles and Techniques for Biologists. 2nd ed. Jones and Bartlett, Sudbury, MA.

Chapter 10

Female Gametophytic Mutants: Diagnosis and Characterization

Ronny Völz and Rita Groß-Hardt

Abstract

In plants, gametes are formed in multicellular haploid structures, termed gametophytes. The female gametophyte of most higher plants comprises seven cells, which develop from a single haploid spore through nuclear proliferation and subsequent cellularization. The female gametophytic cells differentiate into four distinct cell types, which play specific roles during fertilization and seed formation thereby ensuring reproductive success. In recent years many new techniques and cell type-specific marker lines have been established, making the female gametophyte an attractive system to study mechanisms of reproduction as well as cell specification. The following chapter describes a basic protocol for, first of all, recognizing a female gametophytic mutant and subsequently analyzing the phenotype on a morphological, molecular, and functional level.

Key words: Female gametophyte, cell specification, segregation distortion, pollen tube attraction, fertilization.

1. Introduction

The female gametophyte of most higher plants develops from a single haploid spore through three incomplete mitotic division cycles (*see* **Fig. 10.1**). The resulting eight-nucleate syncytium subsequently cellularizes giving rise to seven cells, which differentiate into four distinct cell types. Synergids, egg cell, central cell, and antipodal cells differ morphologically, molecularly, and with respect to their function in the reproductive process (20).

In contrast to the plethora of well-described sporophytic mutants, defects that affect the female gametophytic life phase are

L. Hennig, C. Köhler (eds.), *Plant Developmental Biology*, Methods in Molecular Biology 655,
DOI 10.1007/978-1-60761-765-5_10, © Springer Science+Business Media, LLC 2010

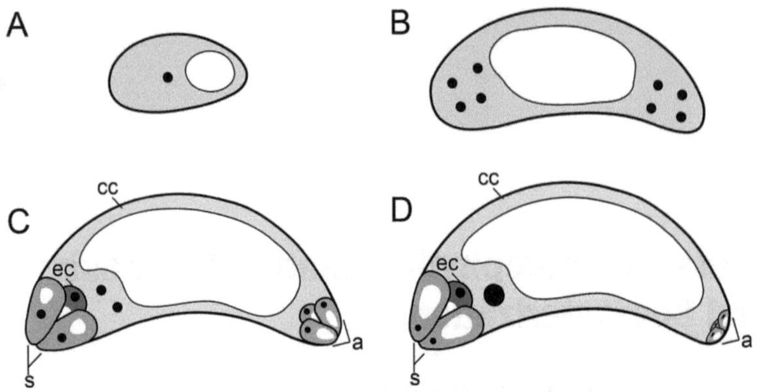

Fig. 10.1. **Development of the female gametophyte.** (**A**) Female gametophyte development is initiated with meiosis and the formation of a single haploid spore. (**B**) Three incomplete mitotic divisions result in the formation of an eight-nucleate syncytium. (**C**) Subsequent cellularization gives rise to seven cells that differentiate into four distinct cell types: Two synergids, one egg cell, one central cell, and three antipodal cells. (**D**) Prior to fertilization, the two polar nuclei of the central cell fuse and the antipodal cells degenerate. a, antipodal cells; cc, central cell; ec, egg cell; s, synergids.

often neglected. This is partially due to the fact that the female gametophyte is less accessible than most sporophytic structures. A concomitant lack of knowledge and tools has maintained the wallflower image of the female gametophyte. The past few years, however, have seen a tremendous progress in the field making the female gametophyte an attractive system to study mechanisms of reproduction and cell specification. One of the achievements is the recent generation of cell-specific marker lines that greatly facilitate the characterization of mutants affected in the haploid life phase (e.g. 4, 9). Together with morphological clues and a functional characterization of the distinct cell types, a comprehensive picture can be drawn for any new female gametophytic mutant.

2. Materials

1. Corney's solution: 9:1 ratio of ethanol and glacial acetic acid.
2. 80 and 70% ethanol.
3. Clearing solution I: 8:1:2 (w:v:v) ratio of chloral hydrate:glycerol:water (*see* **Note 1**).
4. Fixation buffer: 4% glutaraldehyde, 12.5 mM cacodylate buffer, pH 6.9.
5. Ethanol dilution series (10, 20, 40, 60, 80, 95, and 100%).

6. Clearing solution II: 2:1 ratio of benzyl benzoate:benzyl alcohol (v:v).
7. Immersion oil.
8. Nail polish.
9. 5% glycerol.
10. GUS staining solution: 10 mM EDTA, 0.1% Triton X-100, 2 mM $K_4Fe(CN)_6$, 2 mM $K_3Fe(CN)_6$, and 1 mg/mL 5-bromo-4-chloro-3-indolyl glucuronide (X-Gluc) in 50 mM sodium phosphate buffer, pH 7.2, 100 μg/mL chloramphenicol (*see* **Note 2**).
11. 80% glycerol solution.
12. 10% chloral hydrate.
13. 5 M NaOH.
14. 0.1 M K_3PO_4 buffer, pH 8.3.
15. Aniline blue staining buffer: 0.1% aniline blue in 0.1 M K_3PO_4 buffer, pH 8.3.
16. Clearing solution III: 8:1:2 (w:v:v) ratio of chloral hydrate:glycerol:water diluted with water to 10%.

3. Methods

The haploid life phase is initiated by meiosis and terminated by the fusion of gametes. These abrupt transitions between sporophytic and gametophytic life phase can obscure the nature of a given mutation. To test whether a defect is of female gametophytic origin, a few complementary experiments have to be performed. The phenotype can subsequently be analyzed in detail using various morphological and molecular tests (*see* **Note 3**). Finally, the functional relevance of a given defect can be assessed by several techniques. For many female gametophytic mutants, homozygous plants cannot be obtained. This can be either due to the lack of transmission of the mutation through female gametes or an additional sporophytic requirement of the respective gene. Therefore, the subsequent techniques will be exemplified for plants heterozygous for a given female gametophytic mutation (+/−).

3.1. Determining Penetrance and Inheritance Pattern of a Given Mutation

1. Female gametophytic mutants often carry unfertilized ovules (*see* **Fig. 10.2A** and **Note 4**), which in many cases can be distinguished morphologically from aborted seeds (*see* **Fig. 10.2B**). The latter are indicative of a developmental arrest postfertilization, which can be both of sporophytic

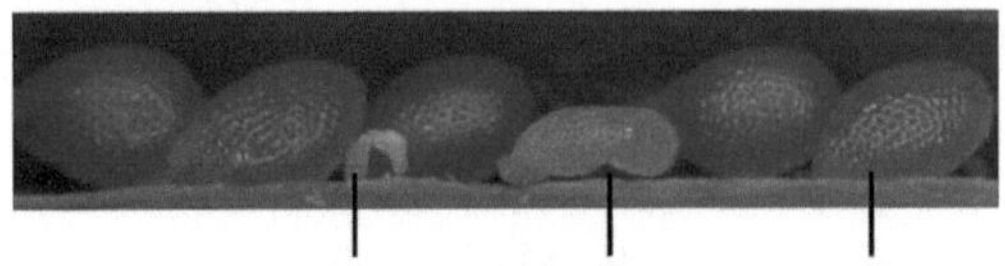

Fig. 10.2. **Silique showing ovule and seed defects.** (A) Sterile ovule. (B) Collapsed aborting seed, which will eventually adopt a brownish color. (C) Normal seed.

and gametophytic origins. Siliques are easily inspected after fixing them on an adhesive tape. Subsequently, a fine needle is used to slit open one carpel along its entire length (*see* **Note 5**). By bending the carpel to one side and fixing it on the tape the seeds become exposed.

2. To determine whether a given defect is of female gametophytic nature, the heterozygous female plant is fertilized by wild-type pollen. This allows bypassing zygotic embryo lethal effects that often obscure the analysis (*see* **Fig. 10.3A**). If the defect is of female gametophytic origin and impairs fertility, the progeny of this plant will more likely be derived from ovules containing a wild-type female gametophyte than a mutant female gametophyte. This is reflected by a concomitant shift of the expected segregation from 1:1 (wild-type:heterozygous mutant) up to 1:0 in case of a fully penetrant mutation (*see* **Fig. 10.3B**). The female transmission efficiency (TE_F) is defined as the percentage of heterozygous mutant progeny to the expected mutant progeny. The latter value equals the number of wild-type progeny as the normal segregation rate is 1:1(5) (*see* **Note 6**).

$$TE_F\ (\%) = \frac{\#\text{heterozygous mutant progeny}}{\#\text{wild-type progeny}} \times 100$$

Many gametophytic mutants are affected both in male and female gametogenesis (16). The reciprocal cross, ♀ +/+ × ♂ +/−, reveals whether and to what extent a male gametophytic defect contributes to the observed phenotype. Segregation analysis and transmission studies are greatly facilitated by the presence of easy traceable traits that cosegregate with the mutation, like the presence of T-DNAs conferring antibiotic resistance (*see* **Fig. 10.3**).

3. The determination whether a given mutant defect results from loss of gene function or gain of gene function is hampered by the haploid nature of the gametophytes. One possibility for tackling that problem is to generate tetraploid mutant plants, which segregate diploid heterozygous mutant gametophytes that either display the defect (dominant) or not (recessive) (*see* **Note 7**).

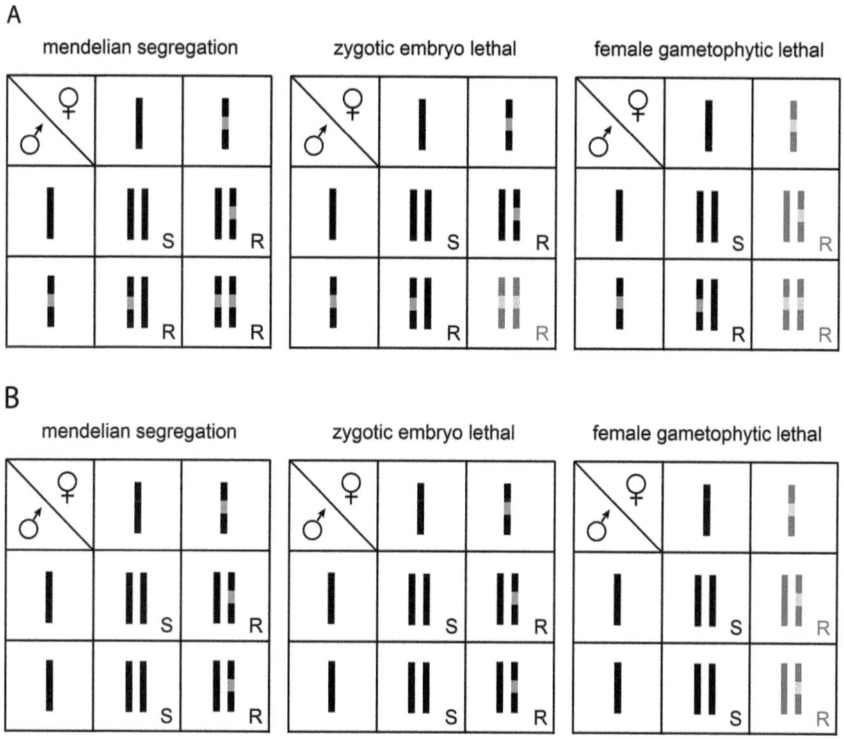

Fig. 10.3. **Segregation rates of various mutants.** (**A**) Segregation of a self-fertilized, heterozygous mutant. In comparison to the Mendelian segregation, zygotic embryo and female gametophytic lethal effects can result in distorted segregation. (**B**) Segregation of a heterozygous mutant fertilized with wild-type pollen. Segregation of recessive zygotic embryo lethal mutations is not affected. Female gametophytic lethal mutations are not transmitted to the next generation. The letters indicate resistance (R) and sensitivity (S) and apply to mutants that cosegregate an antibiotic resistance, for example, T-DNA insertion mutants conferring kanamycin resistance. Dimmed fields indicate lethality (modified after (14)).

3.2. Morphological Characterization of Female Gametophytic Mutants

The previously described experiments can determine whether a given defect is of female gametophytic nature and whether it affects the gametophytic or the sporophytic life phase. Many female gametophytic mutants are impaired at discrete levels of development, including nuclear divisions, cell formation, fusion of polar nuclei, or in reproductive processes such as pollen tube guidance and fertilization (16). To characterize the defect on a morphological level, cleared whole mounts can be generated and inspected using Nomarski optics.

Ovules in a given pistil develop slightly asynchronously and analysis is hence facilitated if plants are emasculated to synchronize the wild-type stages at maturity (*see* **Fig. 10.1D**) (*see* **Note 8**). If the defect is to be analyzed in more depth, close inspection using a confocal microscope is recommended (*see* **Note 9**). This technique is superior to cleared whole mounts in that it displays cytological structures like vacuoles and cell membranes in more detail. However, it is far more time consuming than cleared whole mounts.

3.2.1. Sample Preparation for Light Microscopy

1. Emasculate the oldest (largest) closed flower (4, 21).
2. Fix flowers in Corney's solution 2 days after emasculation (*see* **Note 10**).
3. Infiltrate flowers under vacuum for 30 min.
4. Incubate in Corney's solution at 4°C overnight.
5. Incubate flowers in 80 and 70% ethanol for 30 min each.
6. Remove pistils from flowers with a fine needle and clear in a drop of clearing solution I (*see* **Note 11**) for at least half an hour and observe by Nomarski optics.

3.2.2. Sample Preparation for Confocal Microscopy

1. Harvest pistils and fix on an adhesive tape (3, 18).
2. Slit open the pistil replum on both sides using, for example, an injection needle.
3. Transfer the pistils to fixation buffer and incubate for at least 4 h at room temperature.
4. Dehydrate pistils by an ethanol dilution series: 10, 20, 40, 60, 80 and 95%, 10 min each.
5. Fix pistils in absolute ethanol and apply vacuum for 30 min (~ 200 torr).
6. Keep pistils in absolute ethanol overnight or for a minimum of 4 h.
7. Incubate pistils in clearing solution II for 20 min.
8. Embed pistils in one drop of immersion oil on a glass slide and cover by a coverslip.
9. For stabilization and fixation, seal the sample with nail polish (*see* **Note 12**).

3.2.3. Molecular Characterization of a Female Gametophytic Mutant

After cellularization, the female gametophytic cells of wild-type plants adopt different cell fates, reflected not only by a distinct morphological profile (*see* **Note 13**) but also onset of various cell-specific marker genes (*see* **Table 10.1**). Marker genes can be used to assess various aspects of female gametophyte development, e.g., whether the micropylar–chalazar polarity is established correctly, or whether single cell types correctly differentiate. A very useful collection of marker genes was published by Steffen et al. (9). However, all existing cell-specific markers should be interpreted with care. Thus far, no single master regulator for a given cell type has been identified, therefore, the markers available reflect some, but not necessarily all the aspects of a given cell identity. Still, they are valuable tools to molecularly characterize a given cell. To analyze fluorescence markers, siliques are dissected and mounted in 5% glycerol (*see* **Note 14**). For markers

Table 10.1
Selection of marker lines for the female gametophyte

Expression	Name	AGI	Reporter gene	Ecotype	References
Synergid cells	*DD2*	At5g43510	*GFP*	Col	(9)
	DD31	At1g47470	*GFP*	Col	(9)
	DD35	At5g12380	*GFP*	Col	(9)
	ET2634	–	*GUS*	Ler	(4)
	ET884	–	*GUS*	Ler	(4)
Egg cell	*DD45*	At2g21740	*GFP*	Col	(9)
	ET1119	–	*GUS*	Ler	(4)
Central cell	*DD7*	At2g20595	*GFP*	Col	(9)
	DD9	At1g26795	*GFP*	Col	(9)
	DD22	At1g26795	*GFP*	Col	(9)
	DD65	At5g38330	*GFP*	Col	(9)
	AGL61	At2g24840	*GFP*	Col	(19)
	Medea	At1g02580	*GUS*	C24	(11)
	pMEA	At1g02580	*GUS*	Ler	(4)
	AGL80	At5g48670	*GFP*	Col	(17)
Antipodal cells	*DD1*	At1g36340	*GFP*	Col	(9)
	pAt1g36340	At1g36340	*GUS*	Ler	(22)
	DD6	At2g42930	*GFP*	Col	(9)
	DD13	At3g59260	*GFP*	Col	(9)
	GT3733		*GUS*	Ler	(4)
Entire female gametophyte	*DD33*	At2g20070	*GFP*	Col	(9)
	pAt5g40260	At5g40260	*GUS*	Ler	(22)

that contain the β-glucuronidase (GUS) reporter gene, it is crucial to remove carpel walls to obtain a specific staining.

3.2.3.1. GUS Staining Assay

1. To analyze mature ovules, emasculate flowers and harvest 48 h later (*see* **Note 10**).
2. Harvest pistils and fix them on tape. Slit open and remove carpel walls.
3. Transfer pistils to GUS staining solution and apply vacuum for half an hour to facilitate infiltration of the GUS staining solution.
4. Incubate pistils at 37°C. Incubation time greatly depends on promoter activity and ranges between a few hours to 3 days.

5. Dissect and mount pistils in one drop of 80% glycerol, cover by a coverslip, and inspect under the light microscope using Nomarski optics.

3.3. Functional Characterization of Female Gametophytic Mutants

For successful fertilization to occur several processes must be governed by the female gametophyte, like pollen tube attraction, synergid degeneration, sperm cell discharge, sperm cell guidance, gamete fusion, and the initiation of embryo or endosperm development. To specify a given fertility defect, any of these processes can be analyzed.

1. To analyze pollen tube attraction, mutant plants are pollinated 2 days after emasculation with pollen from a plant harboring a homozygous pollen tube GUS marker like *pAt5g40260::GUS* (22). Plants are harvested 2 days later and subjected to a GUS-staining assay. Successful pollen tube attraction is reflected by a blue staining in the remnants of the degenerating synergids (12).
2. Synergid degeneration is easily detected using the fixation technique by Christensen et al. (3) as described above (*see* **Fig. 10.4**).

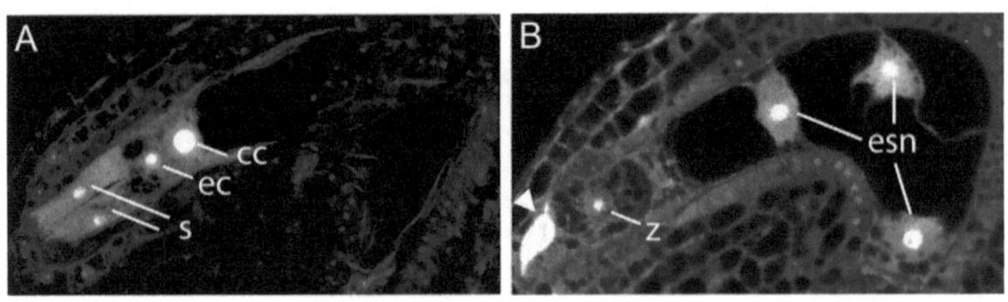

Fig. 10.4. **Analysis of female gametophytes before and after fertilization by confocal microscopy.** (**A**) Mature female gametophyte before fertilization. (**B**) After fertilization: the degenerated synergid appears white (*arrowhead*). s, synergid; ec, egg cell; cc, central cell; z, zygote; esn, endosperm nuclei.

3. Defects in sperm cell discharge as e.g. shown for *feronia/sirene* (6), *abstinence by mutual consent* (1), and *lorelei* mutants (2) can be easily viewed using aniline blue, which detects callose incorporated in the cell wall of pollen tubes (*see* **Section 3.3.1**).
4. The analysis of sperm cell guidance within the female gametophyte can be addressed by live imaging as performed by Ingouff et al. (7). To visualize sperm cells, the red fluorescent reporter construct *HTR10::mRFP1* was used that is specifically expressed in sperm cells. Live imaging is recommended, as the relevant time window is very narrow.
5. Cleared whole mounts, as described above, allow to determine whether and to what extent embryo and endosperm formation is initiated correctly (*see* **Section 3.2**). When

preparing fertilized seeds for light microscopy, it is recommended to use clearing solution III.

3.3.1. Aniline Blue Staining Assay

1. For pollen tube staining, clear pistils in 10% chloral hydrate at 65°C for 5 min (15).
2. Wash with water, softened with 5 M NaOH at 65°C for 5 min.
3. Wash with water and incubate pistils in aniline blue staining buffer for 3 h in the dark.
4. Wash pistils with 0.1 M K_3PO_4 buffer and mount them on a microscope slide using a drop of glycerol. With the ends of a forceps apply gentle pressure to the coverslip.
5. The samples are observed under UV light to visualize callose of pollen tubes and vascular bundles.

4. Notes

1. Clearing solution should be stored at 4°C and mixed thoroughly prior to use.
2. X-Gluc should be protected from light.
3. The application of molecular markers is limited in the case of mutants that arrest prior to cellularization, as most cell-specific marker lines known to date are only initiated after cellularization (10).
4. In most cases, unfertilized ovules are caused by a female rather than a male gametophytic defect. The reason is that a heterozygous mutant still generates 50% wild-type pollen, which often outcompete the mutant pollen resulting in full seed set. However, male gametophytic mutants are known that can compete with wild-type pollen with respect to pollen germination and growth rate, resulting in 50% sterile ovules (e.g. 13).
5. The ovules are attached to the false septum. Accordingly the slit should be only superficial so as not to destroy that connection. The carpel wall is not easily bent aside until the carpel is thoroughly cut along its entire length.
6. A contamination by self-pollination cannot always be excluded. It is thus advisable to use a paternal plant that is homozygous for an easily detectable dominant trait, e.g. a fluorescence marker, which allows the identification of progeny derived from the crossing.
7. Haploinsufficiency and incomplete penetrance of the mutation can obscure the result. As with sporophytic mutations, a distinction can possibly be made after complementing

the mutant with either a wild-type copy of the gene or introducing a mutant allele into a wild-type plant.

8. For mutants that arrest prior to the four-nucleate stage, further inspections should be adjusted to an earlier stage, according to the respective defect.
9. Christensen et al. (3) published a detailed study of different stages of wild-type megagametogenesis, which is very helpful in order to interpret wild-type and mutant specimens.
10. Development rate critically depends on the respective growth conditions and harvesting time might deviate by +/− 1 day.
11. Best results are obtained if as little as possible clearing solution is used.
12. Better results are achieved if a weight (40 g) is put on the coverslip 10 min before it is sealed by nail polish.
13. At maturity, egg cell and synergid cells differ with respect to nuclei size and cell polarity: The synergid cell nuclei are smaller than the egg cell nucleus and face the micropylar end of the female gametophyte, whereas the egg cell nucleus is directed towards the chalazal end of the female gametophyte. These differences are not evident directly after cellularization (10) (*see* **Fig. 10.1C**, D).
14. Crossing of different accessions can cause silencing of marker gene expression and should, if possible, be avoided.

Acknowledgments

The authors would like to thank F. de Courcy and members of the Gross-Hardt laboratory for critical reading of the manuscript. Work in the Gross-Hardt laboratory is supported by grants from the Deutsche Forschungsgemeinschaft (DFG).

References

1. Boisson-Dernier, A., Frietsch, S., Kim, T., Dizon, M. B., and Schroeder, J. I. (2008) The Peroxin Loss-of-Function Mutation abstinence by mutual consent Disrupts Male-Female Gametophyte Recognition. *Curr Biol* **18**, 63–68.
2. Capron, A., Gourgues, M., Neiva, L. S., Faure, J., Berger, F., Pagnussat, G., Krishnan, A., Alvarez-Mejia, C., Vielle-Calzada, J. P., Lee, Y. R., Liu, B., and Sundaresan, V. (2008) Maternal control of male-gamete delivery in Arabidopsis involves a putative GPI-anchored protein encoded by the *LORELEI* Gene. *Plant Cell* **20**, 3038–3049.
3. Christensen, C. A., King, E. J., Jordan, J. R., and Drews, G. N. (1997) Megagametogenesis in Arabidopsis wild type and the *gf* mutant. *Sex Plant Reprod* **10**, 49–64.
4. Gross-Hardt, R., Kägi, C., Baumann, N., Moore, J. M., Baskar, R., Gagliano, W. B., Jürgens, G., and Grossniklaus, U. (2007)

LACHESIS restricts gametic cell fate in the female gametophyte of *Arabidopsis*. *PLoS Biol* **5**, e47

5. Howden, R., Park, S. K., Moore, J. M., Orme, J., Grossniklaus, U., and Twell, D. (1998) Selection of T-DNA-tagged male and female gametophytic mutants by segregation distortion in Arabidopsis. *Genetics* **149**, 621–631.
6. Huck, N., Moore, J. M., Federer, M., and Grossniklaus, U. (2003) The Arabidopsis mutant feronia disrupts the female gametophytic control of pollen tube reception. *Development* **130**, 2149–2159.
7. Ingouff, M., Hamamura, Y., Gourgues, M., Higashiyama, T., and Berger, F. (2007) Distinct dynamics of HISTONE3 variants between the two fertilization products in plants. *Curr Biol* **17**, 1032–1037
8. Ingouff, M., Jullien, P. E., and Berger, F. (2006) The female gametophyte and the endosperm control cell proliferation and differentiation of the seed coat in Arabidopsis. *Plant Cell* **18**, 3491–3501.
9. Joshua, G., Steffen, J. G., Kang, I., Macfarlane, J., and Drews, G. N. (2007) Identification of genes expressed in the Arabidopsis female gametophyte. *Plant J* **51**, 281–292.
10. Kägi, C. and Groß-Hardt, R. (2007) How females become complex: Cell differentiation in the gametophyte. *Curr Opin Plant Biol* **10**, 633–638.
11. Luo, M., Bilodeau, P., Dennis, E. S., Peacock, W. J., and Chaudhury, A. (2000) Expression and parent-of-origin effects for FIS2, MEA, and FIE in the endosperm and embryo of developing Arabidopsis seeds. *Proc Natl Acad USA* **12**, 10637–10642.
12. Moll, C., von Lyncker, L., Zimmermann, S., Kägi, C., Baumann, N., Twell, Grossniklaus, U., and Gross-Hardt R. (2008) CLO/GFA1 and ATO are novel regulators of gametic cell fate in plants. *Plant J* **56**, 913–921.
13. Nowack, M. K., Grini, P. E., Jakoby, M. J., Lafos, M., Koncz, C., and Schnittger, A. (2006) A positive signal from the fertilization of the egg cell sets off endosperm proliferation in angiosperm embryogenesis. *Nat Genet* **38**, 63–67.
14. Page, D. R. and Grossniklaus, U. (2002) The art and design of genetic screens: *Arabidopsis thaliana*. *Nat Rev Genet* **3**,124–136.
15. Pagnussat, G. C., Yu, H., and Sundaresan, V. (2007) Cell-fate switch of synergid to egg cell in *Arabidopsis eostre* mutant embryo sacs arises from misexpression of the BEL1-like homeodomain gene *BLH1*. *Plant Cell* **19**, 3578–3592.
16. Pagnussat, G. C., Yu, H., Ngo, Q. A., Rajani, S., Mayalagu, S., Johnson, C. S., Capron, A., Xie, L. F., Ye, D., and Sundaresan, V. (2005) Genetic and molecular identification of genes required for female gametophyte development and function in Arabidopsis. *Development* **132**, 603–614.
17. Portereiko, M. F., Lloyd, A., Steffen, J. G., Punwani, J. A., Otsuga, D., and Drews G. N (2006) AGL80 is required for central cell and endosperm development in Arabidopsis. *Plant Cell* **18**, 1862–1872.
18. Sandaklie-Nikolova, L., Palanivelu, R., King, E. J., Copenhaver, G. P., and Drews, G. N. (2007) Synergid cell death in Arabidopsis is triggered following direct interaction with the pollen tube. *Plant Physiol* **144**, 1753–1762.
19. Steffen, J. G., Kang, I. H., Portereiko, M. F., Lloyd, A., and Drews, G. N. (2008) AGL61 interacts with AGL80 and is required for central cell development in Arabidopsis. *Plant Physiol* **148**, 259–268.
20. Yadegari, R. and Drews, G. N. (2004) Female gametophyte development. *Plant Cell* **16**, 133–141.
21. Yadegari, R., Paiva, G., Laux, T., Koltunow, A. M., Apuya, N., Zimmerman, J. L., Fischer, R. L, Harada, J. J., and Goldberg, R. B. (1994) Cell differentiation and morphogenesis are uncoupled in Arabidopsis raspberry embryos. *Plant Cell* **6**, 1713–1729.
22. Yu, H. J., Hogan, P., and Sundaresan, V. (2005) Analysis of the female gametophyte transcriptome of Arabidopsis by comparative expression profiling. *Plant Physiol* **139**, 1853–1869.

Chapter 11

Pollen Tube Development

Mark A. Johnson and Benedikt Kost

Abstract

Pollen tubes grow rapidly in a strictly polarized manner as they transport male reproductive cells through female flower tissues to bring about fertilization. Vegetative pollen tube cells are an excellent model system to investigate processes underlying directional cell expansion. In this chapter, we describe materials and methods required for (1) the identification of novel factors essential for polarized cell growth through the isolation and analysis of Arabidopsis mutants with defects in pollen tube growth and (2) the detailed functional characterization of pollen tube proteins based on transient transformation and microscopic analysis of cultured tobacco pollen tubes.

Key words: Arabidopsis, tobacco, pollen tube, fertilization, tip growth, mutant screening, transient transformation, live-cell microscopy.

1. Introduction

The formation of a pollen tube by germinating pollen grains represents the final stage of the development of the haploid male gametophyte. Pollen tubes consist of one large vegetative cell, which contains much smaller reproductive cells, either a single generative cell or two sperm cells, enclosed in its cytoplasm. Division of the generative cell into two sperm cells occurs either before pollen germination (in Arabidopsis and other species producing tricellular pollen), or during pollen tube elongation (in tobacco and other species producing bicellular pollen).

Growing pollen tubes mediate fertilization by transporting male reproductive cells through different female flower tissues from the stigma to egg cells within ovules. The elaborate signaling mechanisms that must be guiding pollen tubes along this path are

L. Hennig, C. Köhler (eds.), *Plant Developmental Biology*, Methods in Molecular Biology 655, DOI 10.1007/978-1-60761-765-5_11, © Springer Science+Business Media, LLC 2010

only poorly understood. After penetrating an ovule through the micropyle, pollen tubes burst at the tip. Their cytoplasm along with the sperm cells it contains is discharged into the extracellular space freed-up by the degeneration of one of the two synergid cells, which are flanking the egg cell adjacent to the large central cell. Subsequently, double fertilization characteristic of angiosperms occurs when one sperms cell fuses with the egg cell to form a zygotic embryo, whereas the other fuses with the central cell to initiate endosperm development (1).

Pollen tubes grow extremely rapidly in a strictly polarized manner exclusively at the tip. Pollen tube tip growth is widely used as model system to investigate cellular polarization, directional cell expansion and the control of these processes by extracellular cues. The characterization of Arabidopsis mutants with defects in pollen tube development is a very potent approach to identify novel genes with essential functions in these processes. Pollen tubes grow deeply buried within female flower tissues. Normal development of these cells is essential for the sexual transmission of mutations and generation of homozygous mutant lines. The effective identification and characterization of pollen tube mutants therefore require specifically developed tools and procedures (2).

The investigation of many processes underlying tip growth is greatly facilitated by growing pollen tubes in culture. Large numbers of pollen tubes can be grown in vitro, free from contaminating other cell types, for RNA isolation or biochemical analyses (3). Cultured pollen tubes are excellent material for live-cell microscopy as they are transparent, free of auto-fluorescence and have a diameter of only 10–20 μm. Efficient methods for the transient or stable expression of transgenes in pollen tubes under the control of specific promoters are available (4). RNAi-based techniques can be employed to down-regulate the expression of specific pollen tube genes (5; Cottier and Kost, unpublished). GFP (green fluorescent protein)-fusion proteins and staining procedures based on specific dyes have been developed, which enable noninvasive visualization of a variety of structures (e.g., organelles and cytoskeletal elements) and processes (e.g., membrane traffic, accumulation of signaling lipids, and Rac/Rop activation) in cultured pollen tubes (e.g. 4, 6). These tools and techniques are extensively employed to functionally characterize genes involved in pollen tube tip growth by analyzing, in detail, effects of increasing or decreasing their expression levels and monitoring distribution as well as dynamic behavior of the proteins they encode.

Although pollen tubes of most species can be cultured in simple media containing just carbohydrates, borate, and calcium, reproducing normal pollen germination and pollen tube growth in vitro is not trivial. Pollen tubes generally elongate at lower rates in vitro than in situ. Culturing pollen tubes of Arabidopsis and other species forming tricellular pollen is particularly challenging. In vitro germination rates of Arabidopsis pollen, as well as growth rates and morphology of cultured Arabidopsis pollen tubes, tend to display considerable variations even under optimized conditions (7). Transient transformation of Arabidopsis pollen tubes has not been reported, presumably because these cells do not survive long enough in culture to support detectable transgene expression. Because of the relatively small amount of pollen produced by individual plants, collecting enough Arabidopsis pollen for biochemical experiments requires major resources. As a consequence of these issues, tobacco pollen tubes are much more widely employed than Arabidopsis pollen tubes as an experimental system for the investigation of tip growth in vitro. Often, Arabidopsis proteins are transiently expressed in tobacco pollen tubes to contribute to their functional characterization. Culture conditions have been established under which tobacco pollen reproducibly germinates at high rates and forms morphologically normal pollen tubes that elongate at rates approaching those observed in situ (8, 9). All tools and techniques listed in the previous paragraph are readily applicable to tobacco pollen tubes cultured under these conditions.

In this chapter, we describe materials and methods required for the analysis of pollen tube growth and guidance in mutants of Arabidopsis, or for the culture, transient transformation and microscopic analysis of tobacco pollen tubes. These protocols take advantage of the individual strengths of the tobacco and Arabidopsis pollen experimental systems.

2. Materials

2.1. In Vitro Determination of Pollen Germination Rate and Pollen Tube Length of Heterozygous Arabidopsis 'Blue SAIL' Mutants (see Note 1)

1. Flowers with dehiscent anthers (stage 14; 10) from a plant that is heterozygous for a single-locus 'Blue SAIL' insertion in gene of interest.
2. Flowers with dehiscent anthers from *LAT52:GUS* positive control line (76224, ABRC stock # CS16336; 2).
3. Fresh Arabidopsis pollen germination medium (APGM): 0.01% boric acid, 5 mM $CaCl_2$, 5 mM KCl, 1 mM $MgSO_4$, 10% sucrose, pH 7.5 (7).
4. Glass microscope slide and coverslips (24 × 30 mm).
5. Dissecting microscope for sample preparation; compound microscope for analysis.
6. Hydrophobic barrier pen (1-mm Edge pen, http://www.vectorlabs.com, cat #H-4000).

7. Forceps (e.g., Fine Science Tools, http://www.finescience.com, Dumont Inox 5, cat #91150–20).
8. Humidity chamber (sealed plastic container with water in the bottom and rack to hold microscope slides).
9. 22°C incubator.
10. GUS staining solution: 5 mM potassium ferrocyanide, 5 mM potassium ferricyanide, 50 mM $NaPO_4$, pH 7, 0.5 mg/mL 5-bromo-4-chloro-3-indolyl-β-D-glucuronic acid (X-Gluc).
11. 60% glycerol.

2.2. The Blue Dot Assay: In Vivo Analysis of Pollen Tube Growth and Guidance in Arabidopsis 'Blue SAIL' Mutants (see Note 1)

1. Flowers with dehiscent anthers (stage 14; 10) from a plant that is heterozygous for a single-locus 'Blue SAIL' insertion in gene of interest.
2. Flowers with dehiscent anthers from *LAT52:GUS* positive control line (76224, ABRC stock # CS16336; 2).
3. *male sterile 1* (*ms1*, ABRC stock number CS75) plants at stage 14 (10).
4. Dissecting microscope for sample preparation; compound microscope for analysis.
5. 27.5 gauge needle (Becton Dickinson, Franklin Lakes, NJ) and 1-mL syringe.
6. Double-sided tape and Petri dish.
7. Glass slide and coverslips (18 mm).
8. Forceps (e.g., Fine Science Tools, http://www.finescience.com, Dumont Inox 5, cat #91150–20).
9. GUS staining solution (*see* **Section 2.1**)
10. 96-Well microtiter dish with lid.
11. 60% glycerol.

2.3. Tobacco Pollen Tube Culture

1. 10× PTNT salts: Dissolve 1470 mg of $CaCl_2 \times 2H_2O$, 989 mg of H_3BO_3, 746 mg of KCL, 1972 mg of $MgSO_4 \times 7H_2O$, and 75 mg of $CuSO_4 \times 5H_2O$ in 1 L of water, keep 50 mL aliquots at −20°C (4, 8, 9).
2. 1.5% Casein hydrolysate: Dissolve 4.5 g of casein acid-hydrolysate (Sigma 'amicase,' A-2427) in 300 mL of water, keep 10 mL aliquots at −20°C.
3. 20 mg/mL Rifampicin: Dissolve 250 mg of rifampicin (Sigma, R-7282) in 12.5 mL of dry methanol, keep 600 mL aliquots at −20°C.
4. 2× PTNT: Dissolve 50 g of sucrose and 3 g of MES in 150 mL of water. Add 100 mL of 10× PTNT salts, 20 mL

of 1.5% casein hydrolysate, and 500 μL of 20 mg/mL rifampicin. Adjust pH to 5.8 with KOH, before adding 125 g of PEG-6000 (BDH, 4427). Stir for 30 min to dissolve PEG, add water to total volume of 500 mL, filter sterilize and keep at 4°C.

5. 1× PTNT: 1 mM $CaCl_2$, 1.6 mM H_3BO_3, 1 mM KCL, 0.8 mM $MgSO_4$, 30 μM $CuSO_4$, 5% (w/v) sucrose, 0.03% (w/v) casein hydrolysate, 12.5% (w/v) PEG, 10 mg/L rifampicin, 0.3% (w/v) MES, pH 5.8. Mix 2× PTNT 1:1 (v/v) with sterile water.
6. 0.5% (w/v) Phytagel: Dissolve 150 mg of phytagel (Sigma, P-8169) in 30 mL water, autoclave.

2.4. Gene Transfer by Particle Bombardment to Tobacco Pollen on Solid Medium

1. Particle suspension: Suspend 60 mg of gold particles (1.6 μm diameter; Bio-Rad, 165–2264) in 1 mL of absolute ethanol by vortexing. Centrifuge for 10 s at 13,000×*g* and aspirate supernatant. Wash twice with 1 mL of sterile water. Resuspend in 1 mL of sterile 50% glycerol and store at room temperature (4)
2. 2.5 M $CaCl_2$: Dissolve 3.675 g of $CaCl_2 \times H_2O$ in 10 mL of water and filter sterilize. Keep 0.5 mL aliquots at −20°C.
3. 0.1 M spermidine: Dissolve 5 g of spermidine (Sigma, S-2626) in 344 mL of water and filter sterilize. Keep 10 mL aliquots at −70°C for long-term storage, and 0.5 mL aliquots at −20°C for short-term storage.

2.5. Microscopic Analysis of Pollen Tubes Cultured on Solid Medium

1. 1 M Sodium phosphate at pH 7.0 (~100 mL): Titrate 1 M Na_2HPO_4 (~60 mL) with 1 M NaH_2PO_4 (~40 mL) to pH 7.0
2. 50 mM Potassium ferricyanide: Dissolve 1.65 g of potassium ferricyanide (Sigma, P-3667) in 100 mL of water, filter sterilize, and freeze 5 mL aliquots at −20°C.
3. 50 mM Potassium ferrocyanide: Dissolve 2.112 g of potassium ferrocyanide (Sigma, P-3289) in 100 mL of water, filter sterilize, and freeze 5 mL aliquots at −20°C.
4. GUS substrate solution composition: 0.2% (w/v) X-Gluc, 0.1% Triton X-100, 5% mannitol, 5 mM potassium ferricyanide, 5 mM potassium ferrocyanide, 0.1 M sodium phosphate, pH 7.0. Dissolve 2.5 g of mannitol in 31.5 mL of water. Add 5 mL of 1 M sodium phosphate at pH 7.0, 0.5 mL of 10% Triton X-100, 5 mL of 50 mM potassium ferricyanide, 5 mL of 50 mM potassium ferrocyanide, and 100 mg of X-GLUC dissolved in 3 mL of dimethyl formamide.

3. Methods

3.1. In Vitro Determination of Pollen Germination Rate and Pollen Tube Length of Heterozygous Arabidopsis 'Blue SAIL' Mutants (see Note 1)

1. Prepare media and slides. APGM should be made freshly and the pH should be adjusted to 7.5 immediately before beginning an in vitro pollen germination assay (*see* **Note 2**). Pollen tubes are grown in upside-down drops of APGM (11). We draw a 9 mm × 9 mm square on the surface of a microscope slide using a hydrophobic barrier pen (*see* **Fig. 11.1**).
2. Apply pollen to medium. Pipette 50 μL of APGM into the square so that a dome forms (*see* **Fig. 11.1A**). Dust pollen from two flowers onto the surface of the APGM, invert the slide, and immediately place the slide into a humidity chamber. Remove petals from the pistil using forceps and then hold the flower with forceps by the pedicel and touch the anthers to the surface of APGM. Do this under the dissecting microscope, to ensure that pollen is applied to the APGM. Flowers should be at stage 14 (10); petals are obvious and either still closed around the stigma, or they have just opened. Pollen should be abundant and appear flaky on anthers. Slides should be kept humid at all times – evaporation of APGM diminishes germination.

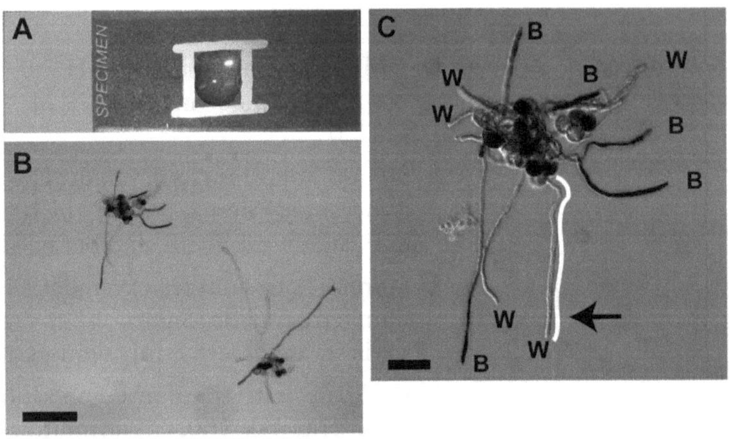

Fig. 11.1. **Measuring pollen tube growth in vitro after staining for GUS activity in Arabidopsis pollen tubes.** (**A**) Set up for pollen tube growth on a microscope slide. A 9 cm × 9 cm hydrophobic barrier has been drawn on the slide and pollen growth media was pipetted into the center. (**B**) A representative image after 6 h of pollen tube growth of a control line hemizygous for *LAT52:GUS*. This image was obtained using a 10× objective. Scale bar: 200 μm. (**C**) A higher magnification image of a group of pollen tubes. GUS+ (B) pollen tubes and GUS− (W) pollen tubes are shown. A freehand line was drawn next to one GUS− pollen tube (*arrow*) and measured at 210.65 μm using ImageJ software. Scale bar: 50 μm.

3. Pollen tube growth. Allow 6 h for pollen tube germination and growth (*see* **Fig. 11.1B, C**). This time is sufficient for maximal germination, and wild-type pollen tubes will be long enough to measure significant differences from mutants affecting pollen tube growth. Longer incubation times are possible, but pollen tubes become difficult to measure when they are too long and pollen tube viability decreases quickly after longer periods of growth.
4. Differential staining of mutant and wild-type pollen tubes. Remove slides from humidity chamber and carefully invert them so they are medium-side-up. Under a dissecting scope, carefully remove APGM using a pipette. Add 100 μL of 80% acetone and let incubate for 20 min. Do not let the acetone evaporate completely. Remove residual 80% acetone using a pipette and add 50 μL of GUS staining solution. Incubate the slide (right-side-up) in the humidity chamber at 37°C for 16 h (overnight). Length of staining can be reduced significantly for most transgenic lines (as little as 3 h); however, lines with weak GUS expression will require 16 h or more of staining.
5. Imaging and analysis. When staining is complete, carefully remove GUS staining solution under the dissecting scope using a pipette. Add 20 μL of 60% glycerol and place a 24 × 30-mm coverslip on the sample. The coverslip will flatten the sample and liquid will spread beyond the hydrophobic barrier square; however, the pollen and pollen tubes will generally be found inside this square. We use the 10× objective on a Zeiss Axiovert 200 M fluorescence microscope (Carl Zeiss, Germany) equipped with optics for differential interference contrast (DIC) microscopy to obtain images using a Zeiss AxioCam MRc5 (Carl Zeiss, Germany) color digital camera. However, any compound microscope equipped with a color digital camera should work. We take 2584 × 1936 pixel images, using ~70 ms exposure time. We use axiovision software (v. 4.2, Carl Zeiss, Germany) to adjust the numerical aperture of the condenser lens to 0.16 – this increases the depth of field so that entire pollen tubes are in focus in a single image (*see* **Fig. 11.1B, C**). Image the entire area marked by the hydrophobic barrier pen taking a picture of every frame with at least one tetrad in it.
6. Determine relative percent germination. Pollen tube germination rates are determined by counting the number of tetrads, the number of blue pollen tubes, and the number of white pollen tubes on each image. One can also count pollen germination rates while at the scope; this may be useful because it allows one to change the focal plane while scoring germination. A germinated pollen tube is defined

as a projection from the pollen grain that is at least half the width of a pollen grain. The counts for each image are entered into a spreadsheet and the percent germination is calculated for white and blue pollen grains. A typical experiment will result in 30 images containing ~250 tetrads or 1000 pollen grains. Data are reported as relative germination of blue versus white pollen. The value of this method is that in each experiment the wild-type pollen (GUS−, white) are growing alongside the mutant pollen (GUS+, blue); germination rates vary between experiments for both wild-type and mutant pollen, however, we have found that relative germination rate is consistent across experiments.

7. Determine relative pollen tube length. Pollen tube lengths are determined using ImageJ software (http://rsbweb.nih.gov/ij/). Images (tiff) are opened in ImageJ and a freehand line (*see* **Fig. 11.1C**) is drawn along the length of each pollen tube in the image. The length is obtained using the measure feature of the software. This reports the length of the freehand line, so curving pollen tubes can be accurately measured. A scale in μm can be set by obtaining an image of a stage micrometer and using the set scale feature. We record the length (μm) and color (white or blue) in a spreadsheet and calculate average tube lengths for blue and white pollen tubes. There is often a wide range in tube lengths for mutant and wild-type pollen tubes; it is useful to report the average tube length along with the range in tube lengths. A typical experiment will routinely yield 50 wild-type and 50 mutant pollen tubes. In cases where mutant pollen have severe germination defects, the number of mutant pollen tubes may be reduced relative to wild-type and multiple experiments are required to measure an adequate number of mutant pollen tubes.

3.2. The Blue Dot Assay: In Vivo Analysis of Pollen Tube Growth and Guidance in Arabidopsis 'Blue SAIL' Mutants (see Note 1)

1. Pollinate *male sterile 1* (*ms1*) pistil with mutant pollen. Pollen tube growth and guidance are affected by the developmental stage of the pistil and female gametophyte (12), so care must be taken to consistently choose flowers at the same appropriate developmental stage. We use *ms1* as pollen acceptor in our experiments. *ms1* is a recessive, sporophytic mutation in the Landsberg *erecta* background that completely blocks pollen development. We use a single, just-opened *ms1* flower per inflorescence. We do not use older pistils and have noted significant differences in pollen tube growth patterns between pistils of different stages. Pollen donors are used that are just about to open; at this stage pollen is abundant, flaky, and still on the anther surface. For manual pollination, we use forceps to remove sepals,

petals, and the pistil from pollen donor flowers and then apply pollen to the *ms1* stigma by holding onto the pedicel of the pollen donor and using the flower like a paint brush. The stigma must be saturated with pollen; more than one pollen donor flower may be required.

2. Pollen tube growth in the pistil. *ms1* plants are returned to growth chambers to allow pollen tube growth. A time-course of growth can be performed by stopping the assay at desired times. The apical ovules begin to be targeted after as little as 3 h and basal ovules will be reached by 12 h (*see* **Fig. 11.2**; 13); pollen tube growth is complete by 20 h after pollination. 16 h of pollen tube growth is a good end-point that allows one to assess how long mutant pollen tubes will grow and how likely they are to target ovules. Furthermore, we have found that GUS activity is abundant and easily detected in pollen tubes at this stage; GUS activity begins to decrease with longer periods of time after pollination.
3. Remove ovary walls from pistil. When pollen tube growth is complete, it is important to remove the ovary walls from the pistil so that pollen tube growth can be imaged. Using forceps, remove the pistil from the plant where the pedicel meets the inflorescence (*see* **Fig. 11.2**). Lay the pistil down on a piece of double-sided tape that has been placed on top of a Petri dish; position the dish and the pistil under a dissecting scope. The pistil should be oriented such that the replum is facing up and the two carpels are on either side (*see* **Fig. 11.2**). Use a 27.5-gauge needle attached to a 1-mL syringe as a scalpel to make a shallow incision on both sides of the replum (*see* **Note 3**). Make incisions at the top and bottom of each carpel and push the ovary wall to either side of the pistil, securing it to the surface of the tape. Finally cut the ovary wall away by running the needle under the ovules, cutting the ovary wall away from lower surface of the pistil. Gently lift the sample by the pedicel and place it in 80% acetone immediately. We use 96-well microtiter plates to handle samples. Incubate the pistil in 80% acetone for at least 1 h; samples can be left in acetone for up to 4 h as long as they do not dry out. This minimum incubation time is required to clear the pistil tissue, which facilitates imaging of GUS+ pollen tubes. Remove acetone using pipette and replace with 50 μL of GUS staining solution; incubate pistils overnight at 37°C in a humid chamber.
4. Mount pistils on microscope slides. Pipette 30 μl of 50% glycerol onto the center of a microscope slide and transfer the pistil into the center of the glycerol; carefully place an 18-mm coverslip over the pistil. The glycerol should form a seal around the pistil and between the slide and the coverslip.

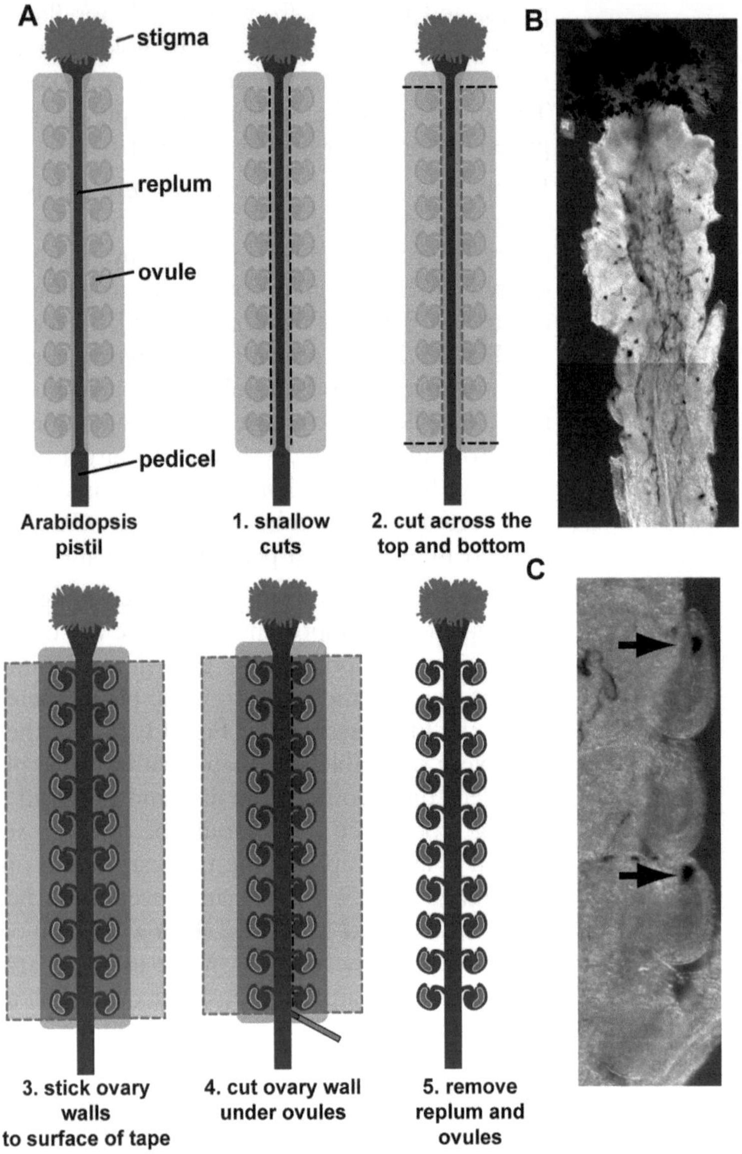

Fig. 11.2. **The blue dot assay.** (**A**) How to remove the ovary walls from the Arabidopsis pistil. *Dotted lines* represent incision sites. (**B**) An *ms1* pistil pollinated with homozygous *LAT52:GUS* control pollen. The *blue dots* appear as dark spots on each ovule. (**C**) A series of four ovules in an *ms1* pistil pollinated with hemizygous *LAT52:GUS* pollen. *Arrows* represent ovules that have been targeted by GUS+ pollen tubes.

5. Obtain images. To image the pistil, we use DIC optics on a 10X objective a Zeiss Axiovert 200 M fluorescence microscope (as above) and obtain images using Zeiss AxioCam MRc5 (*see* **Fig. 11.2**). This low magnification view will reveal obvious mutant phenotypes such as failure of mutant pollen tubes to grow in the pistil (*see* Fig. 4 in reference (2)). Around 2–3 10× images are required to document the entire length of the pistil. We use Adobe Photoshop to

generate composite images of the entire pistil length. ImageJ can be used to measure the length of the longest GUS+ pollen tube in these images. Comparison of these values to those of the control line allows one to make conclusions about the pollen tube growth potential of the mutant. We use a 20× magnification objective to obtain images of pollen tubes growing on the funiculus and into the micropyle (*see* **Fig. 11.2**, *see* also Fig. 5 in reference (2)). These images are useful for documenting abnormal growth behavior as pollen tubes approach the micropyle.

6. Quantification of ovule targeting. Apply gentle pressure to the coverslip using a pipette tip. This will squash the sample slightly and move ovules away from each other so they are easier to count. By scanning down the length of the pistil with the 20× objective, one can count the total number of observed ovules and the number that have a characteristic blue dot of GUS activity in the synergid cell (*see* **Fig. 11.2**). This blue dot represents GUS activity that has been injected from the pollen tube into the synergid cell as the pollen tube bursts. When counting, be careful to ensure that the blue dot is in the correct position at the micropylar end of the ovule. This positional information is important to ensure that the blue dot being counted is a *bona fide* ovule-targeting event. To obtain ovule-targeting rates, we simply divide the number of ovules with blue dots by the number of total ovules observed (*see* **Note 4**).

3.3. Tobacco Pollen Tube Culture

3.3.1. Tobacco Pollen Tube Culture in Liquid Medium: Large Scale, e.g., for the Isolation of About 0.5 mg RNA (see **Note 5**)

1. Collect 100 mg of pollen from dehiscent anthers of about 70 freshly opened flowers using a vacuum-powered collection device constructed as outlined in **Fig. 11.3** (*see* **Note 6**).
2. Transfer pollen to 25 mL of 1× PTNT in a 50-mL screw-cap tube and suspend by vortexing.
3. Transfer pollen suspension through sieve (1-mm pore size: to remove stamen and anther tissue) to a 15-cm Petri dish and incubate for 3 h in the dark at 25°C.
4. Transfer pollen tube culture back to a 50-mL screw-cap tube and centrifuge for 1 min at 700×*g*.
5. Remove supernatant and wash pollen tube pellet with 0.4 M mannitol and 50 mM Tris-HCL, pH 6.0.
6. Add appropriate extraction buffer to pollen tube pellet (*see* **Note 7**).

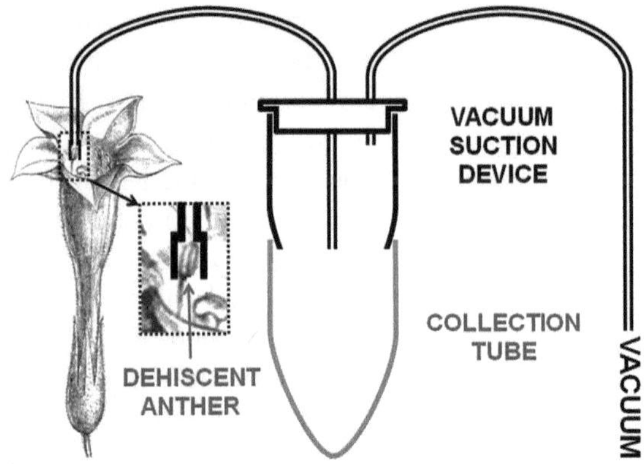

Fig. 11.3. **Collection of tobacco pollen using vacuum suction device.** A vacuum suction device assembled as outlined from standard 1.5-mL reaction tubes and plastic tubing greatly facilitates large-scale collection of tobacco pollen.

3.3.2. Tobacco Pollen Tube Culture on Solid Medium: 16 Culture Plates, for Gene Transfer by Particle Bombardment or Microscopic Analysis of Living Pollen Tubes (see ***Note 5****)*

1. Transfer 25 mL of 2× PTNT medium to a 50-mL screw-cap tube and place in hot water (about 95°C) to heat up.
2. Boil 30 mL of 0.5% (w/v) phytagel in microwave.
3. Add 25 mL of boiling 0.5% (w/v) phytagel to preheated 2×PTNT in screw-cap tube, mix well (vortex), and keep in hot water.
4. Transfer 3 mL of 1× PTNT + 0.25% phytagel (w/v) to a 5-mL Petri dish and spread by gently shaking (*see* **Note 8**).
5. Deposit pollen grains on solidified 1× PTNT + 0.25% phytagel (w/v) either by dipping dehiscent anthers from freshly opened flowers onto the surface of the medium (for microscopic analysis of pollen tube growth) or using vacuum filtration as described below (for gene transfer by particle bombardment).
6. Seal plates with parafilm and incubate in the dark at 25°C.

3.4. Gene Transfer by Particle Bombardment to Tobacco Pollen on Solid Medium (see Notes 5 and 9)

3.4.1. Particle Coating with Plasmid DNA: For One Plasmid Mix and Two Bombardments (see ***Note 10****)*

1. Vortex 25 μL of particle suspension in a 1.5-mL reaction tube (hold with forceps at the lid hinge).
2. Continue vortexing and add in the indicated order: 6 μL or less of a plasmid mix containing a total of 2–6 μg of DNA (*see* **Note 11**), 25 μL of 2.5 M $CaCl_2$, and 10 μL of 0.1 M spermidine.
3. Keep vortexing for 2 min.

4. Centrifuge for 10 s (13,000×*g*) and aspirate supernatant.
5. Add 200 μL of absolute ethanol and resuspend particles (pipette up and down, vortex).
6. Centrifuge for 10 s (13,000×*g*) and aspirate supernatant.
7. Add 15 μL of absolute ethanol and resuspend particles (pipette up and down, vortex).
6. Place about half the volume of the particle suspension (changes rapidly due to ethanol evaporation) in the center of each of two macro carriers (Bio-Rad, 165–2335) mounted in a macro carrier holder (Bio-Rad, 165–2322).
7. Let ethanol evaporate in a vibration-free environment.

*3.4.2. Plating of Pollen Grains on Solid Medium for Particle Bombardment: For One Plasmid Mix and Two Bombardments (see **Note 10**)*

1. Transfer 10 stamen with dehiscent anthers from two freshly opened flowers to 10 mL of 1× PTNT in a 50-mL screw-cap tube and vortex.
2. Pour pollen suspension through sieve (1-mm pore size) into a fresh tube to remove stamen and anther tissue.
3. For each of the two plates to be prepared, collect the pollen contained in 5 mL of suspension on a circular membrane filter (4.5-cm diameter, Millipore HAWP 047 00) by vacuum filtration. To ensure even spreading of pollen grains, prewet filter before use by slowly submerging it in sterile water and vortex pollen suspension immediately before pouring it onto the filter.
4. Place membrane filter upside-down on 3 mL of solid PTNT in a 5-cm Petri dish, such that pollen grains are in contact with the medium surface. Remove air bubbles trapped between filter and medium by gently applying pressure from above. Lift filter off (can be reused), leaving pollen grains on the surface of the medium.

*3.4.3. Particle Bombardment (see **Note 12**).*

1. Bombard pollen tube cultures as soon as possible after pollen plating (within 5–10 min).
2. Set-up Bio-Rad PDS-1000/He particle gun (Bio-Rad 165–2257) according to the instruction manual with standard settings for solenoid valve, vacuum flow rate, as well as distance between macro carrier and stopping screen.
3. Load particle gun with an 1100-psi rupture disc, and with a lunch assembly containing coated particles dried down on a macro carrier facing a stopping screen underneath.
4. Place culture plate containing pollen tube cultures without lid on the target plate shelf inserted into the particle gun at position L2 (3rd slot from the bottom), 6 cm below the stopping screen.

5. Evacuate the sample chamber to 28 in. of mercury and trigger particle delivery (*see* **Note 13**).
6. Release vacuum and remove culture plate. Close plate with lid, seal with parafilm, and incubate in the dark at 25°C.

3.5. Microscopic Analysis of Pollen Tubes Cultured on Solid Medium (see Note 14)

3.5.1. Pollen Tubes in Culture Plates: Monitor Quality of Cultures, Particle Delivery, and Transformation Efficiency (see Note 15)

1. Place culture plates on the stage of an inverted microscope to observe pollen tubes growing on the surface of solid medium at low or intermediate magnification (4×–40× objectives designed for working distances of several mm).
2. Use epifluorescence illumination or confocal laser scanning to noninvasively visualize pollen tubes expressing GFP (*see* **Note 16**).
3. To identify pollen tubes transformed with GUS expression constructs, add 1 mL of GUS substrate solution to culture plates and incubate at 37°C for 3–12 h before microscopic analysis (*see* **Note 17**).

3.5.2. Pollen Tubes on a Coverslip: Analysis of Length, Morphology, Growth Rate, and Intracellular Distribution of Fluorescent Fusion Proteins

1. Use scalpel to cut two ca. 2×3-cm rectangular sections out of the solid medium in a pollen tube culture plate.
2. Transfer a single section upside-down on a coverslip (standard thickness: 170 μm), such that pollen tubes are in direct contact with the glass surface.
3. Place coverslip on the stage of an inverted (or upright) microscope with the glass surface facing the objective and the culture medium facing the condenser (*see* **Note 18**).
4. Statistical analysis of pollen tube length: 5–6 h after gene transfer, take digital images of at least 50 different GFP- or GUS-expressing pollen tubes per plate at 4× or 5× magnification (*see* **Note 19**). Measure the length of individual pollen tubes using ImageJ (*see* **Note 20**).
5. Analysis of cellular morphology: Image the tips of GFP-labeled, living pollen tubes (*see* **Note 21**) using 10× or 20× objectives, and epifluorescence or 3-dimensional confocal microscopy. Additional information may be obtained by taking transmitted light differential interference contrast (DIC, Nomarski) reference images.
6. Analysis of growth rate: At an interval of 2 min, take two sequential fluorescence or DIC high-magnification (63× or 100× objectives) images of the tip of an individual growing

pollen tube using a digital camera or a confocal microscope. To determine growth rate, measure the distance the extreme apex has traveled between the two images. Confocal microscopes generally are equipped with software functions for this purpose. Alternatively, the distance between the pollen tube apex on sequential images can be measured using ImageJ, after the images have been converted to a stack (*see* **Note 22**).

7. Analysis of the intracellular distribution of fluorescent fusion proteins: Use the epifluorescence equipment of a confocal microscope and intermediate magnification (25× or 40× objectives) to search samples for pollen tubes suitable for confocal imaging. Transiently transformed pollen tubes display a range of transgene expression levels and fluorescence brightness. Select pollen tubes with tips lying flat on the coverslip surface, which display a normal morphology and weak fluorescence just bright enough for confocal imaging (*see* **Note 23**). At an interval of 2 min, take two sequential confocal images of such pollen tubes at high magnification (63× or 100× objectives; *see* **Note 24**). Determine the growth rate of the imaged pollen tube as described above (*see* **Section 3.5.2** and step 6). A growth rate in the range of several μm/min (*see* **Note 5**), together with normal pollen tube morphology on the second image, establish that the first image displays the intracellular distribution of the analyzed fusion protein in a normally elongating pollen tube (*see* **Fig. 11.4**).

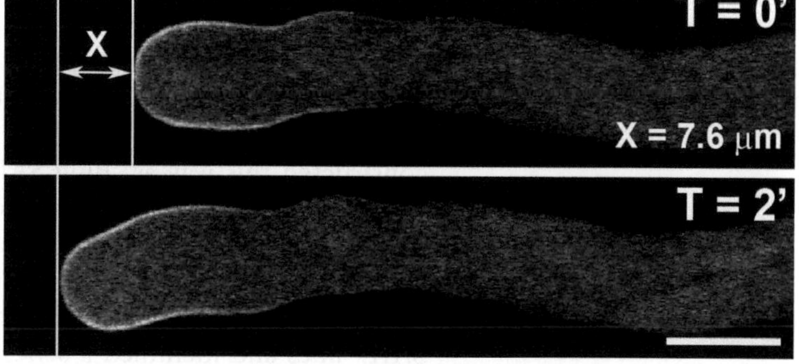

Fig. 11.4. **Analysis of the intracellular distribution of fluorescent fusion proteins.** Single medial confocal sections through a pollen tube expressing a Cys1:YFP fusion protein that serves as a marker for the membrane lipid diacyl glycerol (6) imaged at an interval of 2 min. The analyzed pollen tube was growing at a rate of 3.8 μm/min between the two images and showed a normal morphology on the second image (*T*=2′). The first image (*T*=0′) therefore displays Cys1:YFP distribution in a normally elongating pollen tube. Scale bar: 10 μm.

4. Notes

1. Mutations that completely block pollen tube growth and/or guidance, or the ability of sperm to fertilize the egg and central cell cannot be transmitted to progeny by pollen and have to be maintained in the heterozygous state. Analysis of pollen mutant phenotypes in heterozygous plants is challenging because half of the pollen are wild-type and it is difficult or impossible to differentiate these from mutant pollen. Fortunately, a large collection of SAIL (syngenta arabidopsis insertion library) T-DNA insertion mutants are available, which have been mutagenized with a T-DNA that carries the *LAT52:GUS* reporter gene (14). Therefore, in mutants that carry the T-DNA at a single insertion site, mutant pollen tubes are marked by GUS activity and can be clearly differentiated from their wild-type meiotic siblings that do not express GUS. SAIL inserts can be identified within your gene of interest and ordered through the Salk Institute's T-DNA express website (http://signal.salk.edu/cgi-bin/tdnaexpress) (15). All SAIL lines with numbers beginning from 1 to 456, 1052 to 1057, 1142 to 1205, or 1206 (A to D) were generated in the *qrt1-2* mutant background and carry the *LAT52:GUS* T-DNA (14). The *qrt1-2* mutant, which maintains the four male meiotic products in a tetrad (16), is very useful for studying gametophytic mutations that disrupt pollen development or function because these mutations will generate tetrads with two wild-type and two mutant pollen grains (2, 17, 18). *LAT52:GUS* provides a visible marker that is only expressed in mutant pollen grains, enabling one to efficiently associate phenotype with genotype in pollen tetrads; tetrads generated by plants with a single insertion site will have two mutant (GUS+) and two wild-type (GUS−) pollen grains. We refer to the insertion mutants constructed this way as 'Blue SAIL' lines and found that we could identify at least one 'Blue SAIL' insertion in 55% of 400 Arabidopsis genes we surveyed. We previously performed a forward genetic screen on a collection of T-DNA mutants that was constructed in the same way as the 'Blue SAIL' lines and identified a series of *hapless* mutants with defects in pollen tube growth, guidance, and fertilization (2, 18). We have also found this collection to be advantageous for reverse genetic analysis of pollen-expressed genes and here, we provide protocols that enable one to take advantage of the 'Blue SAIL' lines to carefully analyze pollen tube growth and guidance in vitro and in the Arabidopsis pistil.

2. Arabidopsis pollen has notoriously low, and variable, germination rates in vitro and it is critical to standardize and optimize all variables in the assay. Pollen growth medium has recently been optimized for Arabidopsis (APGM) (7).
3. Incisions should just cut the ovary wall, no deeper; deeper cutting will remove ovules. Hold the syringe so that the point of the needle cuts the tissue and the angled face of the needle points up; this will help you make straight incisions. The needle must be sharp; we use a fresh needle for each pistil.
4. This calculation is based on the assumption that all ovules are targeted and that those without blue dots have been targeted by wild-type (GUS−) pollen tubes. This assumption has proven to be valid when the stigma is at the appropriate developmental stage and saturated with pollen. To assess whether ovule targeting is complete in your assay conditions, perform the assay using control pollen that is homozygous for *LAT52:GUS* (all four members of each tetrad are GUS+). Ovule targeting should approach 100%, and the value obtained with this control experiment can be used to normalize values obtained from experiments done in parallel.
5. All methods described in **Sections 3.3, 3.4, 3.5** have been developed for pollen produced by *Nicotiana tabacum* cultivar Petit Havana SR1 plants grown in a green house. Except for excessive summer heat, which may reduce the efficiency of gene transfer by particle bombardment, seasonal changes in plant growth conditions do not noticeably affect pollen tube culture or transformation. Pollen tube cultures can be established and maintained under semisterile conditions, as the culture medium PTNT contains an antibiotic that effectively prevents the growth of most contaminating microorganisms. Most pollen grains germinate within the first hour of culture in liquid or on solid PTNT. Emerging pollen tubes deposit callose plugs at regular intervals, contain a cytoplasmic generative cell that divides into two sperm cells (8, 9), and grow to a total length of about 15 mm within 48 h. Highest growth rates (up to 10 μm/min; average 5 μm/min) are reached a few hours after germination.
6. Although pollen can be stored at −70°C, germination rates are highest with fresh pollen.
7. Osmotic pressure will cause pollen tubes to burst in extraction buffer; freezing and grinding are not required for RNA isolation. Different protocols for RNA isolation will work including the one described in (3).

8. Work rapidly, PTNT + 0.25% phytagel (w/v) rapidly solidifies even at high temperature above 50°C (high Ca^{2+} concentrations strongly promote phytagel polymerization).
9. This protocol routinely results in successful gene transfer to at least 50 pollen tubes per bombarded plate (*see* **Fig. 11.5A**).

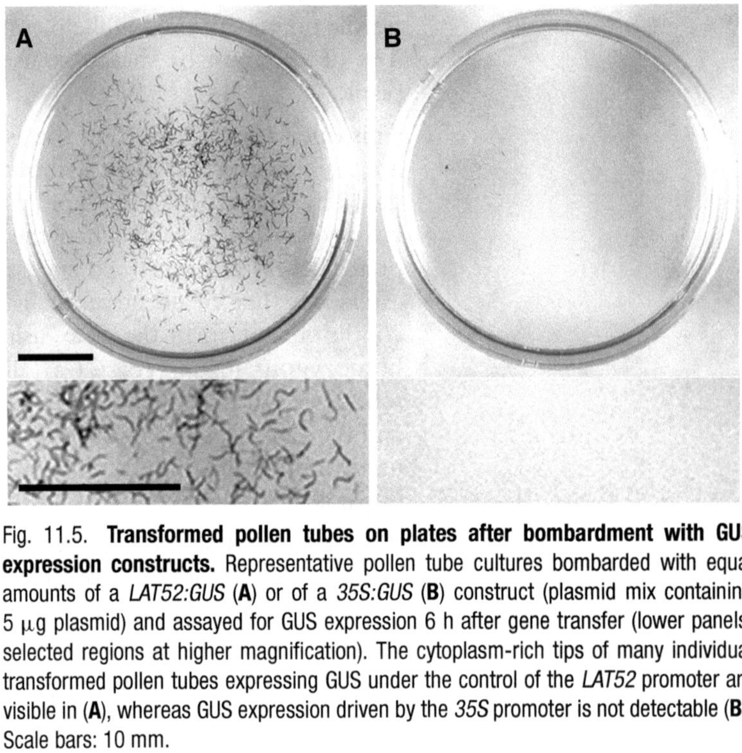

Fig. 11.5. **Transformed pollen tubes on plates after bombardment with GUS expression constructs.** Representative pollen tube cultures bombarded with equal amounts of a *LAT52:GUS* (**A**) or of a *35S:GUS* (**B**) construct (plasmid mix containing 5 μg plasmid) and assayed for GUS expression 6 h after gene transfer (lower panels: selected regions at higher magnification). The cytoplasm-rich tips of many individual transformed pollen tubes expressing GUS under the control of the *LAT52* promoter are visible in (**A**), whereas GUS expression driven by the *35S* promoter is not detectable (**B**). Scale bars: 10 mm.

10. It is advisable to prepare enough material for the bombardment of at least two plates with each plasmid mix. Occasional failure of the particle gun can cause single bombardments to result in inefficient gene transfer (*see* **Note 13**).
11. Plasmid solutions containing at least 0.5 μg/μL DNA prepared using alkaline-lysis spin-column miniprep kits, or with cleaner methods, generally work well. Impurities in plasmid solutions may cause excessive particle clumping (some clumping is always observed) and may reduce gene transfer efficiency. To allow identification of successfully targeted cells after gene transfer, particles can be coated with plasmids conferring cytoplasmic expression of GFP (noninvasively detectable; *see* **Section 3.5.1., Fig. 11.6A** and **11.7B**) or GUS (detectable using a destructive assay; *see* **Section 3.5.1., Fig. 11.5A** and **11.6B**) under the control of the *LAT52* promoter (19), which is highly active in tobacco pollen tubes (*see* **Fig. 11.5A**). The *CaMV 35S* promoter, which drives strong gene expression in most

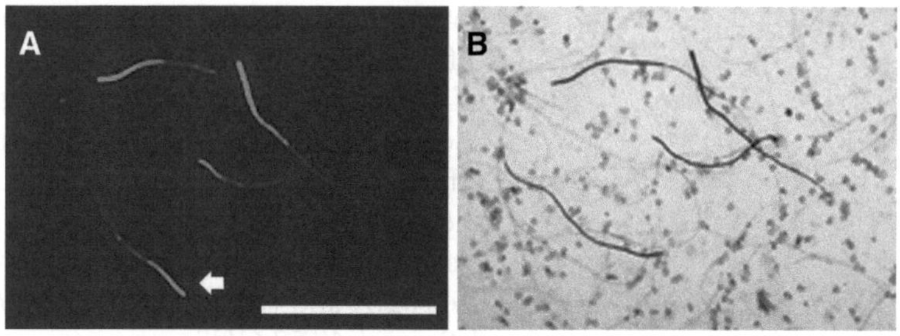

Fig. 11.6. **Transformed pollen tubes after cobombardment with GFP and GUS expression constructs.** Pollen tubes bombarded with equal amounts of a *LAT52:GFP* and a *LAT52:GUS* construct (plasmid mix containing 2 μg of each plasmid) were analyzed by epifluorescence microscopy 6 h after gene transfer. Noninvasive imaging of GFP fluorescence (**A**), followed by destructive analysis of GUS expression (**B**), showed that transformed pollen tubes expressed both marker genes. Note that a pollen tube visible in the lower left corner in (A, *arrow*) slightly shifted position on the surface of the culture medium upon addition of the GUS substrate solution. Scale bar: 500 μm.

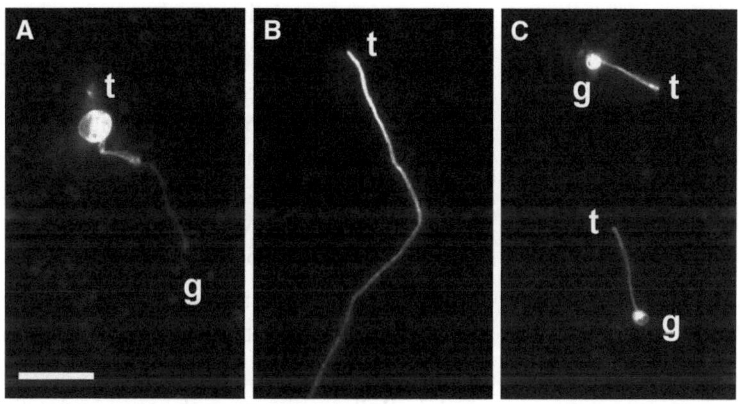

Fig. 11.7. **Transformed pollen tubes expressing Nt-Rac5 and Nt-RhoGDI2 at different levels.** GFP-labeled pollen tubes analyzed by epifluorescence microscopy 6 h after bombardment with plasmid mixes containing 2 μg of a *LAT52:GFP* construct, along with different amounts of *LAT52:Nt-Rac5* (*L:R5*) and *LAT52:Nt-RhoGDI2* (*L:G2*) constructs: (**A**) 3.2 μg *L:R5* and 0.8 μg *L:G2*, (**B**) 2 μg *L:R5* and 2 μg *L:G2*, and (**C**) 0.8 μg *L:R5* and 3.2 μg *L:G2*. Excess *Nt-Rac5* activity depolarizes pollen tube tip growth (**A**: ballooning tips), whereas high-level *Nt-RhoGDI2* expression inhibits this process (**C**: short tubes). Tip growth is not affected when both proteins are cooverexpressed at similar levels (**B**) (3). Scale bar: 200 μm.

types of plant cells, only confers marker gene expression at barely detectable levels in tobacco pollen tubes (*see* **Fig. 11.5B**). Simultaneous coating of particles with a marker gene construct, and one or more additional plasmids containing different *LAT52* expression cassettes (e.g., 3 μg each of two plasmids or 2 μg each of three plasmids), will result in expression of all cassettes in pollen tubes (*see* **Fig. 11.6**). To titrate levels of gene expression, particles can be coated with varying amounts of an expression construct (*see* **Fig. 11.7**). In such experiments, it is advisable to compensate unequal amounts of an expression construct in different plasmid mixes by adding corresponding amounts of a marker gene construct that is different from the one used to identify successfully targeted cells: e.g., to compare effects of the expression of gene X at different levels on the growth of GFP-labeled pollen tubes, the following two plasmid mixes can be used: (a) 2 mg *LAT52:GFP* + 2 mg *LAT52:X* and (b) 2 mg *LAT52:GFP* + 0.2 mg *LAT52:X* + 1.8 mg *LAT52:GUS*.

12. Because the culture medium PTNT contains the antibiotic rifampicin (*see* **Note 5**), gene transfer by particle bombardment can be performed under semisterile conditions.
13. Rupture disks occasionally burst or are dislodged before the target pressure for particle delivery (1100 PSI) is reached. This generally drastically reduces gene transfer efficiency.
14. In liquid medium, it is difficult to obtain high-quality microscopic images because most pollen tubes are floating.
15. Microscopic analysis of pollen tubes in culture plates allows rapid, nondestructive observation of cultured pollen tubes, particle (visible as small black dots) distribution after bombardment, and transient marker gene expression. However, Petri dish material and solid culture medium in the light path, along with a long working distance, prevent high-quality microscopic imaging.
16. GFP expression starts to become visible 3 h after bombardment with particles coated with 2 μg of a *LAT52:GFP* construct.
17. Pollen tube growth stops immediately after the addition of GUS substrate solution. This is an advantage when comparing the lengths of pollen tubes in many plates based on digital imaging (*see* **Section 3.5.2** and step 4). Continuing pollen tube growth during the time it takes to collect digital images can reduce data accuracy.
18. With only a coverslip between pollen tubes and the surface of objectives, conditions are optimal for high-quality microscopic imaging at all available magnifications. Pollen tubes continue to grow normally for 20–30 min on the coverslip, before the osmolarity in the drying medium rises too high.
19. An epifluorescence microscope equipped with a sensitive digital camera, or a confocal microscope, can be used for this purpose. More than 6 h after gene transfer, normally growing pollen tubes are too long to fit on a single image from grain to tip.
20. ImageJ is platform independent, freely available software (http://rsbweb.nih.gov/ij/). Calibrate ImageJ by taking

a digital image of a stage micrometer at the same magnification as used for pollen tube imaging and use the 'calibrate' function in the 'analyze' menu. Draw a freehand line along the image of a pollen tube to be measured and determine its length by applying the 'measure' function in the 'analyze' menu. Export data to spread sheet software (e.g., Excel) for statistical analysis.

21. Destructive GUS assays (*see* **Section 3.5.1**) cause pollen tubes to collapse and are therefore not suitable for the analysis of cellular morphology.
22. Open sequential images in ImageJ and convert to a stack by applying the 'images to stack' function in the 'stacks' sub-menu of the 'image' menu. Draw a straight line from the apex on the first image to the apex on the second image and determine its length as described above (*see* **Note 20**).
23. Most fluorescent fusion proteins affect pollen tube growth when expressed at high levels. Weakly fluorescent pollen tubes are therefore most likely to display normal growth. Most fluorescent fusion proteins can be expressed at levels sufficient for high-quality confocal imaging without interfering with normal growth. The complete absence of autofluorescence at the tip of pollen tubes strongly facilitates imaging of weakly fluorescent structure.
24. Pairing 20–25× and 63× water immersion objectives, or 40× and 100× oil immersion objectives, allows rapid switching between searching samples at intermediate magnification and performing high-magnification confocal imaging.

Acknowledgments

B.K. was supported by DFG, BBSRC, FORMAS, and VR funding. M.J. was supported by NSF grant IOS-0644623 and would like to thank Rebecca Macri for the illustration in **Fig. 2** and Alexander R. Leydon for micrographs.

References

1. Bedinger, P. A., Hardeman, K. J., and Loukides, C. A. (1994) Travelling in style: The cell biology of pollen. *Trends Cell Biol* **4**, 132–138.
2. Johnson, M. A., von Besser, K., Zhou, Q., Smith, E., Aux, G., Patton, D., Levin, J. Z., and Preuss, D. (2004) Arabidopsis hapless mutations define essential gametophytic functions. *Genetics* **168**, 971–982.
3. Klahre, U., Becker, C., Schmitt, A. C., and Kost, B. (2006) Nt-RhoGDI2 regulates Rac/Rop signaling and polar cell growth in tobacco pollen tubes. *Plant J* **46**, 1018–1031.

4. Kost, B., Spielhofer, P., and Chua, N.-H. (1998) A GFP-mouse talin fusion protein labels plant actin filaments *in vivo* and visualizes the actin cytoskeleton in growing pollen tubes. *Plant J* **16**, 393–401.
5. Gu, Y., Fu, Y., Dowd, P., Li, S., Vernoud, V., Gilroy, S., and Yang, Z. (2005) A Rho family GTPase controls actin dynamics and tip growth via two counteracting downstream pathways in pollen tubes. *J Cell Biol* **169**, 127–138.
6. Helling, D., Possart, A., Cottier, S., Klahre, U., and Kost, B. (2006) Pollen tube tip growth depends on plasma membrane polarization mediated by Tobacco PLC3 activity and endocytic membrane recycling. *Plant Cell* **18**, 3519–3534.
7. Boavida, L. C. and McCormick, S. (2007) Temperature as a determinant factor for increased and reproducible in vitro pollen germination in *Arabidopsis thaliana*. *Plant J* **52**, 570–582.
8. Read, S. M., Clarke, A. E., and Bacic, A. (1993) Stimulation of growth of cultured *Nicotiana tabacum* W38 pollen tubes by poly(ethylene glycol) and $Cu_{(II)}$ salts. *Protoplasma* **177**, 1–14.
9. Read, S. M., Clarke, A. E., and Bacic, A. (1993) Requirements for division of the generative nucleus in cultured pollen tubes of *Nicotiana*. *Protoplasma* **174**, 101–115.
10. Smyth, D. R., Bowman, J. L., and Meyerowitz, E. M. (1990) Early flower development in Arabidopsis. *Plant Cell* **2**, 755–767.
11. Hicks, G. R., Rojo, E., Hong, S., Carter, D. G., and Raikhel, N. V. (2004) Geminating pollen has tubular vacuoles, displays highly dynamic vacuole biogenesis, and requires VACUOLESS1 for proper function. *Plant Physiol* **134**, 1227–1239.
12. Palanivelu, R. and Preuss, D. (2006) Distinct short-range ovule signals attract or repel *Arabidopsis thaliana* pollen tubes in vitro. *BMC Plant Biol* **6**, 7.
13. Faure, J. E., Rotman, N., Fortune, P., and Dumas, C. (2002) Fertilization in *Arabidopsis thaliana* wild type: Developmental stages and time course. *Plant J* **30**, 481–488.
14. Sessions, A., Burke, E., Presting, G., Aux, G., McElver, J., Patton, D., Dietrich, B., Ho, P., Bacwaden, J., Ko, C., Clarke, J. D., Cotton, D., Bullis, D., Snell, J., Miguel, T., Hutchison, D., Kimmerly, B., Mitzel, T., Katagiri, F., Glazebrook, J., Law, M., and Goff, S. A. (2002) A high-throughput Arabidopsis reverse genetics system. *Plant Cell* **14**, 2985–2994.
15. Alonso, J. M., Stepanova, A. N., Leisse, T. J., Kim, C. J., Chen, H., Shinn, P., Stevenson, D. K., Zimmerman, J., Barajas, P., Cheuk, R., Gadrinab, C., Heller, C., Jeske, A., Koesema, E., Meyers, C. C., Parker, H., Prednis, L., Ansari, Y., Choy, N., Deen, H., Geralt, M., Hazari, N., Hom, E., Karnes, M., Mulholland, C., Ndubaku, R., Schmidt, I., Guzman, P., Aguilar-Henonin, L., Schmid, M., Weigel, D., Carter, D. E., Marchand, T., Risseeuw, E., Brogden, D., Zeko, A., Crosby, W.L., Berry, C. C., and Ecker, J. R. (2003) Genome-wide insertional mutagenesis of *Arabidopsis thaliana*. *Science* **301**, 653–657.
16. Preuss, D., Rhee, S. Y., and Davis, R. W. (1994) Tetrad analysis possible in Arabidopsis with mutation of the *QUARTET (QRT)* genes. *Science* **264**, 1458–1460.
17. Johnson-Brousseau, S. A. and McCormick, S. (2004) A compendium of methods useful for characterizing Arabidopsis pollen mutants and gametophytically-expressed genes. *Plant J* **39**, 761–775.
18. von Besser, K., Frank, A. C., Johnson, M. A., and Preuss, D. (2006) Arabidopsis *HAP2* (*GCS1*) is a sperm-specific gene required for pollen tube guidance and fertilization. *Development* **133**, 4761–4769.
19. Twell, D., Yamaguchi, J., Wing, R. A., Ushiba, J., and McCormick, S. (1991) Promoter analysis of genes that are coordinately expressed during pollen development reveals pollen-specific enhancer sequences and shared regulatory elements. *Genes Dev* **5**, 496–507.

Chapter 12

Analysis of Root Meristem Size Development

Serena Perilli and Sabrina Sabatini

Abstract

Plant post-embryonic development takes place in the meristems. In the root of the model plant *Arabidopsis thaliana*, stem cells organized in a stem-cell niche in the apex of the root meristem generate transit-amplifying cells, which undergo additional division in the proximal meristem and differentiate in the elongation/differentiation zone. For meristem maintenance, and therefore continuous root growth, the rate of cell differentiation must equal the rate of generation of new cells: how this balance is achieved is a central question in plant development. We have shown that maintenance of the *Arabidopsis* root meristem size is established by a balance between the antagonistic effects of cytokinin, which promotes cell differentiation, and auxin, which promotes cell division. Cytokinin antagonizes auxin in a specific developmental domain (the vascular tissue transition zone) from where it controls the differentiation rate of all the other root tissues. Here, we describe protocols to analyze development of root meristems.

Key words: Meristem, cell division, cell differentiation, auxin, cytokinin.

1. Introduction

Root growth and development are sustained by the root meristem, where multipotent stem cells for all root tissue types surround a small group of mitotically inactive, organizing cells, the quiescent centre (QC). Together they form a stem-cell niche (1, 2). The QC maintains stem cells providing short-range cell-nonautonomous signals that inhibit differentiation (1, 3, 4, 5). Each stem cell undergoes asymmetric cell division, giving rise to a self-renewing cell and a daughter cell that is allowed to differentiate in the upper part of the root meristem (3, 6). Owing to a stereotyped division pattern, columns or files of cells develop in which the spatial relationship of cells in a file reflects their age: younger cells lie near the root tip; older cells are higher up in the

L. Hennig, C. Köhler (eds.), *Plant Developmental Biology*, Methods in Molecular Biology 655,
DOI 10.1007/978-1-60761-765-5_12, © Springer Science+Business Media, LLC 2010

root. Therefore, all developmental stages are present in every root and anatomy reflects ontogeny (3, 7).

Along the longitudinal axis, the root meristem can be divided into three different developmental zones (*see* **Fig. 12.1**). In the stem-cell niche, stem cells continuously produce transit amplifying cells, which undergo a finite number of cell divisions in the proximal meristem (the division zone) until they leave the meristem, rapidly expand and differentiate to reach maturity (the elongation/differentiation zone, *see* **Fig. 12.1**). Cell differentiation is initiated at the transition zone encompassing the boundaries between dividing and expanding cells in the different files (*see* **Fig. 12.1**, inset). The transition zone is different for each cell type, giving a jagged shape to the boundary between dividing and expanding cells (*see* **Fig. 12.1**).

The balance between the rate of cell proliferation in the meristem and the extent of cell differentiation at the transition zone determines the overall rate of root growth and root meristem size (3, 6). How this balance is achieved is a central question in plant developmental biology.

Analysis of simple systems is useful to understand the molecular mechanisms underlying a specific developmental process. Because of the simplicity of its structural and functional organization, the *Arabidopsis thaliana* root meristem has become one of the best-studied model systems in plant biology (7). Indeed, *Arabidopsis* roots display a radial and symmetric organization, a small diameter and a transparent structure, which permit an easy observation at the microscope. For this reason, the *Arabidopsis* root meristem has been used to identify the molecular mechanisms controlling the balance between cell division and cell differentiation necessary to ensure meristem maintenance and continuous root growth.

2. Materials

2.1. Seed Sterilization

1. Hypochlorite solution: Prepare a solution with 1% final concentration of active Cl in water. Store at room temperature (*see* **Note 1**).
2. Sterile distilled water.
3. 0.1% Agarose. Prepare in water, autoclave and store at room temperature (*see* **Note 2**).

2.2. Plant Growth Conditions

1. MS Medium: 0.5× Murashige and Skoog (MS) salt mixture, 1% sucrose, 0.5 g/L 2-(*N*-morpholino)ethanesulfonic acid (MES) at pH 5.8, 0.8% agar. Autoclave before use.
2. Square Petri dishes, 120×120×17 mm. For square Petri dishes of this dimension, use 50 mL of medium.

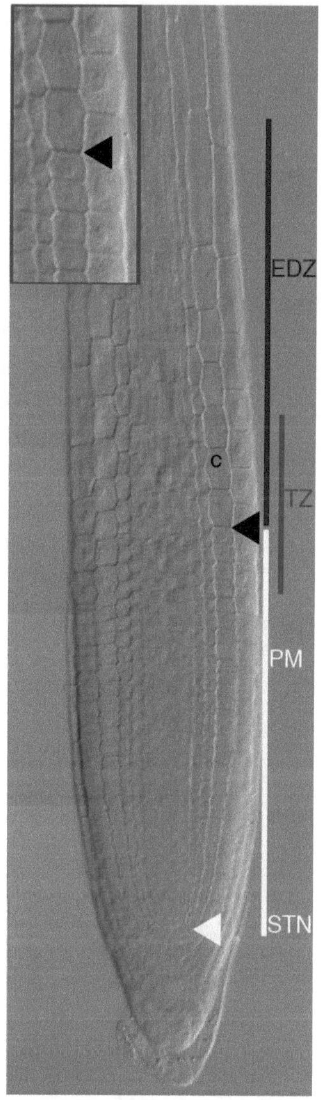

Fig. 12.1. **Structure of the *Arabidopsis* root meristem.** Along the longitudinal axis, the *Arabidopsis* root meristem can be divided into three developmental zones: the stem-cell niche (STN), the proximal meristem (PM) or division zone and the elongation/differentiation zone (EDZ). At the transition zone (TZ), cells leave the meristem and enter the EDZ. Note that the transition zone is different for each cell type, giving a jagged shape to the boundary between dividing and expanding cells. Root meristem size is expressed as the number of cortex cells in a file (c) extending from the quiescent center (*white arrowheads*) to the first elongated cortex cell (*black arrowheads* and *inset*).

2.3. Meristem Size Analysis

1. Chloral-hydrate: Prepare an 8:3:1 mixture of chloral hydrate:distilled water:glycerol (8). Store at room temperature.
2. Glass slides (25.4 × 76.2 mm, 1.0–1.2 mm thick).
3. Cover glasses (24 × 50 mm).

4. Optical microscope with Nomarski optics, connected to a digital camera (for our analysis we use a Nikon DX1200).

2.4. Root Meristem Size Analysis over Time

Same materials as under **Sections 2.1, 2.2, 2.3.**

2.5. Root Meristem Size Analysis After Hormone Treatment

1. Same materials as under **Sections 2.1, 2.2, 2.3, 2.4.**
2. Auxin treatment: Prepare a 1 μM stock of indole-3-acetic acid in ethanol. Store at –20°C.
3. Cytokinin treatment: Prepare a 30 mM stock of zeatin in 1 N NaOH. Store at –20°C.

2.6. Cell-Division Rate Index

1. Same materials as in **Section 2.3.**
2. X-gluc solution: 100 mg/ml X-gluc (5-bromo-4-chloro-3-indolyl glucuronide) dissolved in *N*-*N*-dimethyl-formamide. Prepare it fresh for each experiment.
3. X-gluc solution: 100 mM Na_2HPO_4, 100 mM NaH_2PO_4, 0.5 mM $K_3Fe(CN)_6$, 0.5 mM $K_4Fe(CN)_6$, 0.1% Triton X-100 and 0.5 mg/ml X-gluc. Store at +4°C or −20°C (*see* **Note 3**).

3. Methods

The *Arabidopsis* root can be viewed as a set of concentric cylinders: epidermis, cortex, endodermis and pericycle surrounding the vascular tissue in the middle of the root. The epidermis is made up of two different cell types, hair and non-hair, organized in contiguous cell files. The inner tissues are all composed of a single cell type (7).

Root meristem size can be measured as the number of meristematic cortex cells in a file extending from the QC to the first elongated cell excluded (from white to black arrowheads in **Fig. 12.1**). The cortex is the best suitable tissue to count meristematic cells because it is composed of a single cell type and its cell number is constant between different roots. The number of epidermal cells, instead, is largely variable, because cells giving rise to root hairs divide more frequently than cells giving rise to non-hair cells (9). Since the two different cell types are indistinguishable in the meristem, the number of cells can drastically change depending on the epidermis cell file which has been counted. Inner tissues are difficult to count owing to their smaller size.

At the microscope, meristematic cells are easily distinguishable from differentiating cells because of their size and morphology. Meristematic cells, indeed, have a smaller size and a large central vacuole so that the cytoplasm appears denser than for differentiating cells. Moreover, differentiating cells undergo cell expansion along the longitudinal axis; thus they appear elongated (*see* **Fig. 12.1, inset**).

For each experiment, a minimum of 90 plants should be analyzed, and three independent analyses should be performed to ensure statistical significance.

3.1. Seed Sterilization

1. Place the seeds in separate collecting tubes. Use at least 90 seeds for each experiment (*see* **Note 4**).
2. Add 500 μL to 1 mL of hypochlorite solution to each tube and gently shake for 10 min.
3. Shortly centrifuge the tubes to allow seeds deposition. From this step onwards, use sterile solutions and work under a laminar hood.
4. Remove the supernatant and wash the seeds three times for 10 min each with distilled water (*see* **Note 5**).
6. Suspend the seeds in 300–500 μL of 0.1% agarose. The volume can be varied depending on the amount of seeds.
7. Place the tubes at 4°C in the dark for 2–5 days (*see* **Note 6**).

3.2. Growth Conditions

1. Plate seeds on solid MS medium under a laminar hood. Seeds have to lie on the medium at a 2–3 mm distance from each other to avoid contact between seedlings (*see* **Note 7**).
2. Incubate plates in a near vertical position at 22°C with 16 h light and 8 h dark cycle.

3.3. Root Meristem Size Analysis

1. For meristem size analysis, prepare a glass slide with chloral-hydrate.
2. Place the seedlings on the glass slide. When necessary, use a surgical blade to cut away the shoot and place only the root on the glass slide. Put on the cover glass.
3. Place the slides under the microscope and use the manual focusing drive to find the right longitudinal plane of the root, i.e. until all tissue layers and the QC are visible. To determine root meristem size, count the number of cortex cells in a file extending from the QC to the first elongated cell excluded (from white to black arrowheads in **Fig. 12.1**; *see* **Note 8**).
4. Calculate mean and standard deviation.

3.4. Root Meristem Size Analysis over Time

Sometimes it can be useful to measure root meristem size development over time (*see* **Fig. 12.2**). To this aim, follow the steps described under **Sections 3.1, 3.2, 3.3** but be aware that in this case as many plates are needed as the number of the points of the

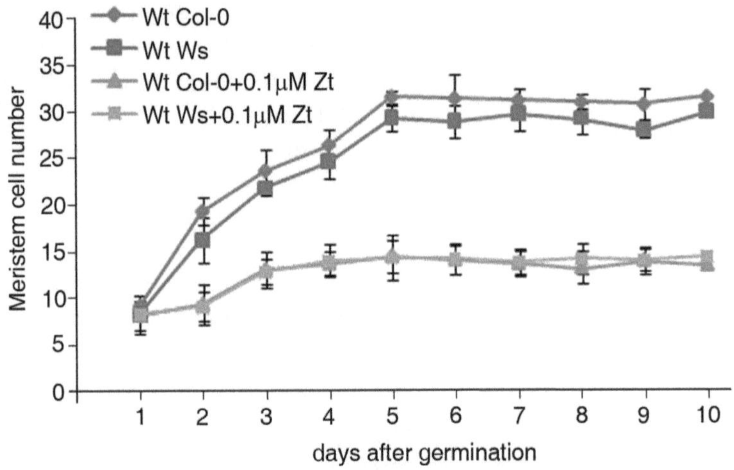

Fig. 12.2. **Time course of root meristem size development over time in untreated and cytokinin-treated plants.** Root meristem cell number of wild-type control plants and wild-type plants grown on 0.1 μM zeatin measured over time. Note that root meristem size increases until 5 days after germination, when a stable number of approximately 30 cells is established in the meristem and maintains constant in time. Exogenous cytokinin application causes a decrease in meristem size because of a progressive decrease in the number of meristematic cells. For each experiment at least 90 plants were analyzed. Col-0: Columbia; Ws: Wassilewskija; Wt: wild-type; Zt: zeatin.

time course (i.e. to perform an analysis at 3, 5 and 7 days after germination, you need at least three plates, one for each day). In this way, during plant growth you can avoid to open plates, which may easily get contaminated. However, if necessary, open the plate under a laminar hood, taking care to work with sterile materials to avoid contamination (*see* **Note 9**).

3.5. Root Meristem Size Analysis After Hormone Treatment

In order to determine if a hormone could be involved in the control of root development, it is useful to assess whether exogenous application of that hormone interferes with root meristem activity. For example, it has been shown that exogenous application of cytokinins alters the dynamic equilibrium between cell division and cell differentiation, leading to a progressive decrease in the number of meristematic cells due to an increase of the rate of cell differentiation at the transition zone; on the other hand, exogenous application of auxin to wild-type roots during growth causes an increase in meristem size due to an increase of the rate of cell division (10) (*see* **Fig. 12.3**). In fact, in the root meristem these two hormones act on the same molecules in a synergistic, coordinated and antagonistic way to balance cell differentiation with cell division, thus determining root meristem size and the overall rate of root growth (11).

1. Take 5-day old seedlings, grown as described under **Section 3.2** (*see* **Note 10**).

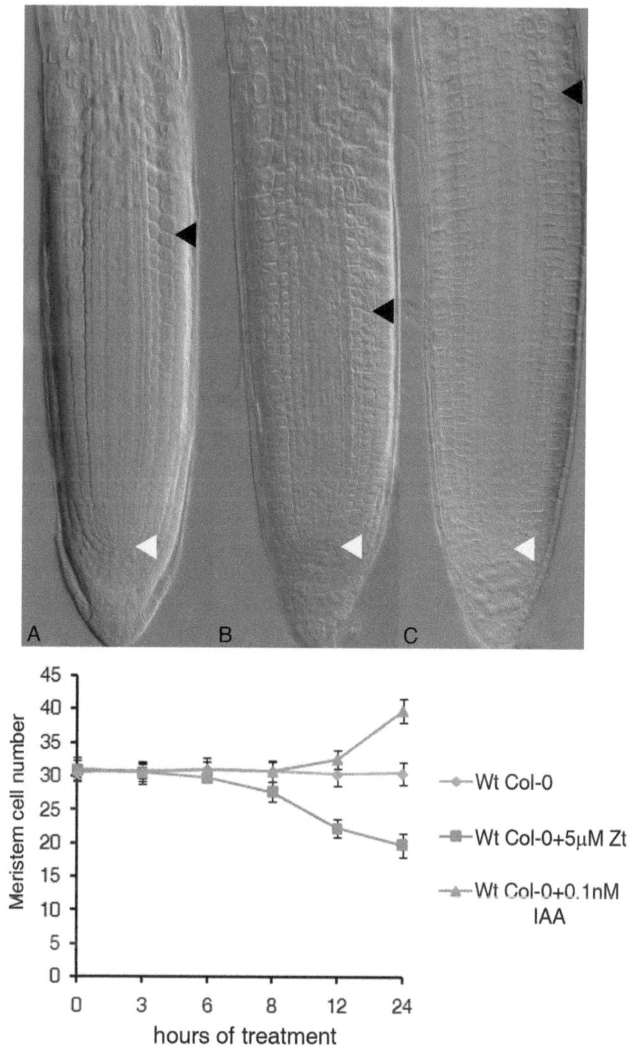

D

Fig. 12.3. **Effect of hormone treatment on root meristem size. A–C** Root meristems of wild-type plants (**A**), wild-type plants treated for 12 h with 5 μM zeatin (**B**) and wild type plants treated for 24 h with 0.1 nM IAA. Roots were analyzed 5 days after germination. *White* and *black arrowheads* indicate, respectively, the QC and the cortex transition zone. (**D**) Root meristem cell number of plants depicted in **A–C** detected after different hours of hormone treatment. For each experiment, a minimum of 90 plants were analyzed. Col-0: Columbia; Ws: Wassilewskija; Wt: wild-type; IAA: indole-3-acetic acid; Zt: zeatin.

2. Transfer seedlings (at least 90 plants for each experiment) to solid MS medium containing mock conditions or a suitable concentration of hormone (*see* **Note 11**).
3. Check root meristem size after several times of treatment, even just several hours (*see* **Note 11**).

3.6. Cell-Division Rate Index Calculation

As described above, root meristem maintenance depends on the coordinated activity of its three developmental zones (stem-cell niche, proximal meristem and elongation/differentiation zone).

To assess whether a variation in root meristem size can be caused by alteration of meristematic cell division potential in the division zone, or by a change in the rate of elongation-differentiation of the meristematic cells at the transition zone, it is useful to calculate the cell-division rate index. To this aim, use plants harbouring a cell division marker, for example the *D-Box CYCB1::GUS* construct, which allows visualization of cells in the G2-M phase of the cell cycle (12). These plants can be used to test either the effect of a specific substance (i.e. hormones) or the effect of a mutation on root meristem development (*see* **Table 12.1**). In the latter case, the construct has to be transferred by genetic cross to the mutant that has to be analyzed.

Table 12.1
Cell-division rate in root meristems of cytokinin-treated wild-type plants and in cytokinin-signalling mutants

Genotype	GUS-stained cells (*X*)	Meristem cell number (*Y*)	Cell-division rate index (*X*/*Y*)
WT *CycB:GUS*	25.22±1.63	30.9±1.43	0.82±0.09
WT *CycB:GUS* + 5 μM Zt	17.57±1.36	21.6±1.05	0.81±0.09
ahk3-3, *CycB:GUS*	33.1±1.95	40.68±1.23	0.81±0.07
arr1-4, *CycB:GUS*	32.89±1,58	39.57±1.76	0.83±0.06

Cell-division rate index has been calculated for untreated roots, roots treated for 12 h with 5 μM Zeatin and roots carrying the *CycB:GUS* construct in a *ahk3-3* or *arr1-4* mutant background. Notice that the *X*/*Y* value is the same in treated, mutant and control plants, indicating that in this case meristem size variation is independent of cell division. The root-meristem cell number is expressed as the number of cortex cells in the cortex file extending from the QC to the first elongated cell. For each experiment, a minimum of 90 plants were analyzed.

1. Grow plants (wild-type or/and a mutant) carrying the *CYCB1::GUS* construct as described under **Section 3.2** (*see* **Note 12**).
2. Transfer plantlets into X-gluc solution. For each experiment use at least 90 plants.
3. Apply 10 min of vacuum treatment to facilitate substrate entry into cells.
4. Incubate for 1 h at 37°C in the dark to visualize β-glucuronidase (GUS) activity.
5. Prepare glass slides as described under **Section 3.3**.

6. For each root, count the number of GUS-stained cells (X) and meristematic cortex cells (Υ).
7. Calculate average and standard deviation for X and Υ.
8. Calculate the cell-division rate index, which is equal to $X/\Upsilon \pm \sqrt{(\partial X/X)^2 + (\partial \Upsilon/\Upsilon)^2}$, where X and Υ are the mean, and ∂X and $\partial \Upsilon$ are their standard deviations (*see* **Note 13** and **Table 12.1**).

4. Notes

1. One can use commercial bleach (containing 5% of active Cl) and dilute it at 1:5 in distilled water.
2. To avoid contamination of 0.1% agarose, prepare it fresh for each experiment. This solution tends to form clumps, so include a magnet in the bottle and, after sterilization in autoclave, let it cool at room temperature under continuous stirring.
3. X-gluc is light sensitive. This solution can form precipitates: make sure that no crystals are in the solution before use.
4. On average, 1 g of dry seed material contains about 50,000 seeds.
5. To sterilize a large amount of samples, one can use this alternative method: open the tubes and place them in a desiccator (250 mm diameter) next to a beaker with 100 mL of commercial household bleach containing 5% Cl and add 3 mL of 37% HCl. This mixture produces a high concentration of toxic fumes of Cl_2 (so use a fume hood). Cover the desiccator to allow its saturation with Cl_2 fumes and wait for 3–4 h. Do not apply vacuum! Open the desiccator under the fume hood and rapidly place the open tubes under a laminar hood for 1 h, to allow complete volatilization of toxic fumes. When fumes in the tubes are completely volatilized, continue at step 6 of **Section 3.1**.
6. This step, called stratification, is important to synchronize germination. Since old seeds might have problems to germinate, we recommend using seeds not older than 1 year. For dry seeds (stored at room temperature with drierite), 2 days of stratification are usually sufficient. In contrast, fresh seeds (for example, collected from a not yet completely dried plant) need 5 days of stratification. We suggest keeping fresh seeds with drierite for at least 4 days before sterilization.

7. Use a Gilson pipette to plate the seeds. We suggest cutting the tip to facilitate seeds passing. Place the seeds at about 1 cm from the top of the dish and take care that the roots have enough space to elongate depending on the number of days they have to grow. Before closing the plate and sealing with Parafilm™, make sure that the agarose solution in which the seeds were embedded is dried.
8. Sometimes it may be difficult to track the boundary between the last meristematic cell and the first elongating one. If you are in doubt, observe the other cortex file.
9. For example, one can address the question at which day after germination the dynamic equilibrium between cell division and cell differentiation in the meristem is achieved. To this aim, follow the development of wild-type plants over time and count the number of meristematic cells at different days (from 1 to 10) after germination. As shown in **Fig. 12.2**, root meristem size progressively increases during early developmental stages, because the rate of cell division is higher than the rate of cell differentiation; this condition persists until 5 days after germination, when the two rates become equal and a stable number of approximately 30 cells is established in the meristem and maintains constant in time (10).
10. We suggest using seedlings grown for 5 days because at this stage root meristem size is already established. However, this analysis can be performed at every stage of development to determine the developmental stage of hormone action.
11. When treating a sample with a hormone for the first time, it is important to set up the experimental conditions, i.e. concentration to be used and time of incubation. Apply various concentrations of the hormone and follow the effect on root meristem size after several hours of treatment. For example, we noticed that 24 h on 0.1 nM indole-3-acetic acid (IAA, the most abundant naturally occurring auxin) are needed to induce an increase in root meristem size, while 12 h on 5 μM zeatin (the most abundant natural cytokinin) are sufficient to induce root meristem size decrease (10) (*see* **Fig. 12.3**).
12. We suggest performing this analysis using cell-division markers specific for different stages of the cell cycle.
13. For example, from the analysis, the following average ± standard deviation values were obtained: $X=25.15\pm1.82$ and $Y=31.4\pm1.32$. Then, the cell-division rate index will be $25.15/31.4\pm\sqrt{(1.82/25.15)^2+(1.32/31.4)^2} = 0.80\pm0.08$.

Acknowledgments

The authors would like to thank Raffaele Dello Ioio, Laila Moubayidin, Riccardo Di Mambro, Lorenzo Mariotti and Francesco Spinelli for helpful advice and encouragement.

References

1. Van den Berg, C., Willemsen, V., Hendriks, C., Weisbeek, P., and Scheres, B. (1997) Short-range control of cell differentiation in the Arabidopsis root meristem. *Nature* **390**, 287–289.
2. Sabatini, S., Heidstra, R., Wildwater, M., and Scheres, B. (2003) SCARECROW is involved in positioning the stem cell niche in the Arabidopsis root meristem. *Genes Dev* **17**, 354–358.
3. Dolan, L., Janmaat, K., Willemsen, V., Linstead, P., Poethig, S., Roberts, R., and Scheres, B. (1993). Cellular organisation of the *Arabidopsis thaliana* root. *Development* **119**, 71–84.
4. Leyser, O. and Day, D. (2005). *Mechanism in Plant Development.* Blackwell Publishing, Malden, MA.
5. Scheres, B. (2007) Stem-cell niches: nursery rhymes across kingdoms. *Nat Rev Mol Cell Biol* **8,** 345–354.
6. Scheres, B., Wolkenfelt, H., Willemsen, V., Terlouw, M., Lawson, E., Dean, C., and Weisbeek, P. (1994). Embryonic origin of the *Arabidopsis* primary root and root meristem initials. *Development* **120**, 2475–2487.
7. Benfey, P. N. and Scheres, B. (2000) Root development. *Curr Biol* **10**, R813–R815.
8. Mayer, U., Torres Ruis, R., Berleth, T., Misera, S., and Jürgens, G. (1991) Mutations affecting body organization in the *Arabidopsis* embryo. *Nature* **353**, 402–407.
9. Berger, F., Hung, C.Y., Dolan, L., and Schiefelbein, J. (1998). Control of cell division in the root epidermis of *Arabidopsis thaliana*. *Dev Biol* **194**, 235–245.
10. Dello Ioio, R., Linhares, F.S., Scacchi, E., Casamitjana-Martinez, E., Heidstra, R., Costantino, P., and Sabatini, S. (2007) Cytokinins determine Arabidopsis root-meristem size by controlling cell differentiation. *Curr Biol* **17**, 678–682.
11. Dello Ioio, R., Nakamura, K., Moubayidin, L., Perilli, S., Taniguchi, M., Morita, M.T., Aoyama, T., Costantino, P., and Sabatini, S. (2008) A genetic framework for the auxin/cytokinin control of cell division and differentiation in the root meristem. *Science* **322**, 1380–1384.
12. Colon Carmona, A., You, R., Haimovitch-Gal, T., and Doerner, P. (1999) Technical advance: Spatio-temporal analysis of mitotic activity with a labile cyclin-GUS fusion protein. *Plant J* **20**, 503–508.

Chapter 13

Phenotypic Characterization of Photomorphogenic Responses During Plant Development

Thomas Kretsch

Abstract

Light is one of the most important exogenous factors regulating plant development throughout the entire life cycle. Light is involved in the breaking of seed dormancy, the regulation of photomorphogenic seedling development, the adaptation of plant morphology toward spectral composition of incident light, and the transition to flowering. Plants have evolved with several photoreceptor families that sense UV-A, blue, red, and far-red light. Here, basal methods to measure light-regulated changes in plant morphology and pigment accumulation will be described. The methods include the determination of apical hook angle and cotyledon opening, the measurement of stem elongation, the determination of leaf surface area, the measurements that characterize light-controlled transition to flowering, and the determination of anthocyanin and chlorophyll accumulation. Furthermore, different light programs are listed that can be used to test for the functional involvement of separate light response modes controlling photomorphogenic plant development.

Key words: Light regulation, photomorphogenesis, germination induction, flowering, low-fluence response, very-low-fluence response, high-irradiance response, end-of-day treatment, shade avoidance response, night break.

1. Introduction

Light is an important exogenous factor that substantially influences plant development at all times during the life cycle. In many plant species, seed germination is induced by light, with red being the most efficient waveband. After seed imbibition and dark incubation, even extremely low amounts of red light are often sufficient to break seed dormancy (1). In darkness, seedlings of higher plants undergo a special kind of development called

L. Hennig, C. Köhler (eds.), *Plant Developmental Biology*, Methods in Molecular Biology 655,
DOI 10.1007/978-1-60761-765-5_13, © Springer Science+Business Media, LLC 2010

skotomorphogenesis or etiolation, which is most often characterized by the presence of folded cotyledons, an apical hook, elongated stems, short roots, and the lack of chlorophyll and anthocyanin pigments. Transition to photomorphogenesis or de-etiolation is induced by UV-A, blue, red and far-red light. Several developmental changes occur upon induction of photomorphogenesis: the opening of the cotyledons and the apical hook, the decrease in the rate of stem elongation, the stimulation of root growth, and the initiation of pigment synthesis (2, 3).

Light-grown plants respond to a reduced ratio of red:far-red light that is caused by the absorption of red light from chlorophyll present in the leaves of the above canopy, in other words, shade caused by green plants. Reduced red to far-red ratios trigger shade avoidance responses (SARs) that include the stimulation of petiole and stem elongation, the reduction of chlorophyll content per leaf surface area, the increase of chlorophyll b/a ratios, and the acceleration of flowering (2–4). The duration of the daily dark period is another important factor that controls plant development, including stem elongation and transition to flowering. Night breaks are often used to study light-dependent induction of flowering. Normally, continuous blue, red and far-red light are efficient in flower induction; however, even single red light pulses are sufficient for controlling transition to flowering in many plant species. End-of-day red and far-red light treatments (EOD-R/-F) trigger light responses similar to SARs, but signaling is most probably related to the measurement of day-length and the spectral light composition at sunset (3, 4).

To sense light quality, intensity, and direction, plants have evolved several classes of photoreceptors, including cryptochromes, phototropins, and phytochromes. Phototropins and cryptochromes function in the UV-A and blue light spectra, whereas phytochromes mainly sense red and far-red light (2, 3). Plant phytochromes are synthesized during dark periods in the inactive Pr conformation, which absorbs red light. Upon absorption of red light, the Pr conformation converts to the active far-red-absorbing Pfr form and activates photomorphogenic responses. Subsequent absorption of far-red light reconverts Pfr back to Pr. Because Pr and Pfr have distinct but overlapping absorption spectra, the Pr:Pfr ratio is wavelength dependent and reaches maximum levels under red light and minimum levels under far-red light (3–5).

Phytochromes can be subdivided into light labile and light stable types. Light labile phytochromes accumulate to very high levels in darkness and are rapidly degraded in red light (3, 4, 6). At high levels in the dark, light labile phytochromes are able to sense extremely low amounts of light and regulate the so-called very low-fluence responses (VLFR). Furthermore, light labile phytochromes trigger far-red-light-dependent high-irradiance responses (HIR), which are activated by continuous irradiation with light of high photon-fluence rates (2, 3). Light stable phytochromes exhibit the classical red/far-red photoreversible mode of action and trigger low-fluence responses (LFR) toward intermediate amounts of red light. In addition, light stable phytochromes predominantly regulate HIRs toward strong continuous red light. In light-grown plants, light stable phytochromes control SARs and EOD responses (2–4). Blue light HIRs are mainly controlled by cryptochromes and light labile phytochromes (2, 3).

2. Material

2.1. Equipment for Light Treatments

1. Blue (410–450 nm), red (650–660 nm), far-red (720 nm), and extreme far-red (740–750 nm) light can be obtained from light-emitting-diode (LED) panels (for example: Roithner Laser, Vienna, Austria, www.roithner-laser.com/LED_diverse.htm) that produce less heat then conventional fluorescence tubes or light bulbs. Light sources should be kept in an air-conditioned growth chamber that is ideally localized in a dark room, in order to minimize contamination by other light sources.
2. Neutral glasses are used to reduce light intensities (Schott, Mainz, Germany; or Optics Balzers, Balzers, Liechtenstein). Useful filter sets include 30% (OD 0.5), 10% (OD 1), 1% (OD 2), and 0.1% (OD 3) relative transmittance (optical density) glasses that can be combined to reach lower light intensities. Samples should be kept in black boxes below neutral glasses to avoid light scattering from the surroundings.
3. Black boxes made from metal, wood, thick cardboard, or plastic can be used together with black drapery to protect samples from any light. Alternatively, samples can be packed into aluminum foil for dark incubations.
4. Green safelight for manipulation of etiolated plants can be obtained from 525-nm LEDs (for example: Roithner Laser, Vienna, Austria, www.roithner-laser.com/LED_diverse.htm).
5. Commonly, light meters can only sense a certain range of light spectra (*see* **Note 1**). Therefore, usually two light meters with two different sensor heads are necessary to measure visible light and light in the far-red range of the spectra (for example: LI-COR LI250A Light Meter together with the LI 190SA measuring head for visible light

(400–700 nm). www.licor.com/env/; Gigahertz-Optik X11 Optometer together with the PS-3703 measuring head for far-red light (650–850 nm), www.gigahertz-optik.com

2.2. Surface Sterilization of Seeds

1. 70% ethanol (v/v).
2. 100% ethanol.
3. Sterile filter paper circles: Wrap filter paper circles (MN 615; 80-mm diameter; Macherey-Nagel; www.mn-net.com) in aluminum foil, autoclave, and store under dessication or in a drier cabinet.

2.3. Determination of Germination Rates

1. Most seeds germinate well in plastic boxes on four layers of sterile filter paper (*see* **Section 2.1**, Step 3) supplemented with sterile distilled water (4.5–5 mL).
2. Magnifying glass or binocular microscope.

2.4. Measurement of Growth Parameters During Seedling Development

1. Growth on paper: Prepare four layers of filter paper (MN 615; 80-mm diameter; Macherey-Nagel; www.mn-net.com) supplemented with 4.5 mL of sterile distilled water in Petri dishes or plastic containers. Growth on agar plates: Autoclave distilled water supplemented with 1.2% (w/v) agar or distilled water containing Murashige and Skoog basal salt mixture and 1.2% agar. Cast into Petri dishes or sterile plastic containers (*see* **Note 2**).
2. Agar plates for seedling preparation: Melt 1.5% (w/v) Bacto agar (Difco Laboratories, Sparks, USA) in a microwave oven and cast into large quadratic plastic culture plates. Alternatively, use transparent adhesive tape to hold samples for image preparation.
3. To take pictures: Digital camera, binocular with digital camera, or flatbat scanner.
4. ImageJ: Image Processing and Analysis in Java; open source software (rsbweb.nih.gov/ij/).

2.5. Measurement of Anthocyanin Accumulation

1. Extraction buffer: 18% (v/v) 1-Propanol (flammable, irritant) supplemented with 1% (v/v) concentrated hydrochloric acid (corrosive, causes burns, irritant to respiratory system).
2. Glass cuvettes (0.6 or 1 mL) together with a photometer (wavelengths: 535 nm, 650 nm).

2.6. Determination of Chlorophyll Content

1. Grinding mill: Plastic tubes (1.5 mL) supplemented with seven glass beads (diameter: 1.7–2 mm; Roth, Karlsruhe, Germany, www.carl-roth.de/website/de-de/carl-roth_index.jsp) together with a Silamat S5 shaker (ivoclar vivadent, Ellwangen, Germany, www.ivoclarvivadent.de/worldwide.aspx).

2. *N*, *N*-Dimethylformamide (harmful by inhalation and in contact with skin, irritating to eyes).
3. Glass cuvettes (0.6 or 1 mL) together with a photometer (wavelengths: 664, 647, 625, and 603 nm).

2.7. Determination of Leaf Blade Area and Petiole Length

1. Planting pots with soil.
2. Transparent adhesive tape to spread leaves.
3. Flatbed scanner.
4. ImageJ: Image Processing and Analysis in Java; open source software (rsbweb.nih.gov/ij/).

2.8. Determination of Flowering Time

1. Planting pots with soil.
2. Growth chamber with short-day (8 h white light: 16 h darkness) and long-day photoperiods (16 h white light: 8 h darkness).

3. Methods

To determine growth parameters, plant material is spread on agar plates or on transparent adhesive tape, and pictures are taken using camera systems or a flatbed scanner. Measurements are done using the ImageJ software (7) together with size standards. Anthocyanin content is measured spectroscopically after heat inactivation of samples and extraction of the pigment using an acidic propanol solution (8). Chlorophyll content is determined after disruption of plant material and extraction with *N,N*-dimethylformamide (9, 10).

3.1. Surface Sterilization of Seeds

1. Transfer seeds into a sterile tube and add 70% ethanol. The volume of 70% ethanol (in mL) should be at least 5- to 10-times the fresh weight of seed material (in g).
2. Shake in an end-over-end tumbler for 10 min.
3. Spin down seed material ($<1000 \times g$, 10 s) and carefully remove the supernatant.
4. Add 100% ethanol (same volume as for 70% ethanol) and shake seeds in an end-over-end tumbler for 5 min.
5. Spin down seed material and remove most of the supernatant.
6. Pour the suspension onto a stack of sterilized filter papers in a sterile culture dish and allow to air dry in a laminar flow hood until seeds become completely dry.
7. Seeds can either be stored in the culture dish or can immediately be used for sowing.

3.2. Determination of Germination Rates

1. Sow seeds on growth medium. Try to reach an equidistant distribution. If necessary, separate seeds with surface-sterilized forceps. Use about 50–100 seeds for each experiment.
2. Stratification (if necessary): Transfer samples into a photoresistent box or wrap with aluminum foil. Incubate plant material at 4–6°C in darkness for 3 days.
3. Incubate samples at the appropriate growth temperature (species dependent) in darkness for 24 h.
4. Germination can be induced under VLF (*see* **Note 3**) and LF (*see* **Note 4**) conditions. To check HIR, apply 4 h of continuous far-red, red or blue light of variable photon-fluence rates (*see* **Note 5**).
5. Incubate samples in complete darkness at the appropriate growth temperature for an additional 3–7 days to complete germination.
6. Count the number of germinated seeds (g) and nongerminated seeds (n). Seeds are regarded as being germinated when the radicle has penetrated the testa. Use a magnifying glass or binocular microscope for small seeds (such as Arabidopsis or tobacco).
7. Calculate the germination rate as $g/(n + g)$ for each experiment.
8. Repeat the experiment at least 4–5 times.

3.3. Measurement of Growth Parameters During Seedling Development

1. Sow seeds on growth medium. Try to create an equidistant distribution of seeds. If necessary, separate seeds with surface-sterilized forceps.
2. Stratification (if necessary): Wrap plant material with aluminum foil or transfer it into a photoresistent box. Incubate at 4°C in darkness for 3 days.
3. Germination induction (if necessary): Incubate samples at a temperature appropriate for the species in darkness for 2 h to adjust temperature. Transfer plant material to continuous red or white light for 3 h to induce germination.
4. Store samples at the appropriate growth temperature in darkness for 1 day to complete germination.
5. Grow samples under the respective light conditions (*see* **Notes 6, 7**, and **8**) or incubate in darkness at the appropriate growth temperature for an additional 3–5 days.
6. Transfer seedlings to plates with 1.5% water agar and try to spread the roots and hypocotyls without disturbing hypocotyl hooks and the angle of cotyledon opening. Alternatively, fix the seedlings on transparent adhesive tape

and cover them with an additional tape before drying (*see* **Note 9**). Use at least 10 seedlings for each experiment.

7. Take a photo of samples together with a size standard using a digital camera or a binocular with digital camera. Alternatively, scan in with a flatbed scanner. Resolution should be 300 dpi or higher.
8. ImageJ is utilized to perform measurements. The scaling for measurements is adjusted using the picture of the size standard. Mark a line of known distance with the straight line selection tool → Analyze → Set Scale → adjust parameters, select global, and click OK (*see* **Note 10**).
9. Root and hypocotyl lengths are measured using the segmented line selection tool of ImageJ (*see* **Fig. 13.1**A). The segmented line should follow the central axis of the respective organ → double click after selection → press 'control + M.' Repeat measurements for all seedlings.

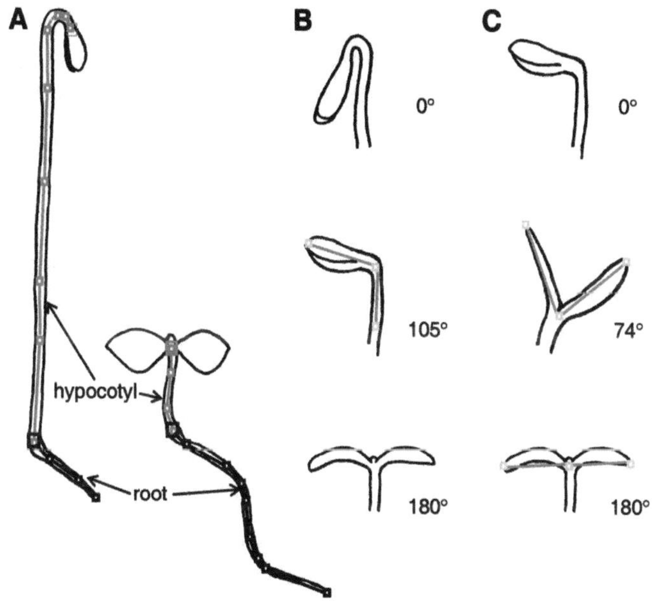

Fig. 13.1. **Measurement of growth parameters during seedling development.** (**A**) Determination of hypocotyl elongation and root growth. (**B**) Determination of the angle of light-induced hook opening. (**C**) Determination of the angle of light-induced cotyledon opening.

10. Angles for hypocotyl hook opening and cotyledon opening are measured using the angle tool of ImageJ (*see* **Fig. 13.1B, C**). Mark 3 points → double click after selec tion → press 'control + M' (*see* **Note 11**).
11. Repeat the experiment at least 2 times for each light treatment.

3.4. Measurement of Anthocyanin Content

1. Heat water to 100°C. Prepare plastic tubes with extraction buffer for each sample.
2. Harvest a defined amount of tissues or seedlings (*see* **Notes 12** and **13**) and transfer into a fixed volume of extraction buffer (about 5–10 volumes of the fresh weight of plant material). Store samples on ice until the next step.
3. Heat the samples in a boiling water bath for 1 min with extraction buffer volumes up to 1 mL or for 2 min with higher volumes.
4. Transfer samples to ice immediately after boiling to cool them down.
5. Shake samples in a refrigerated room (4–8°C) in darkness overnight.
6. Centrifuge the samples at 4°C for 15 min at not less than 10,000 g.
7. Transfer the clear supernatant to a new tube. Avoid contamination with any scattering material.
8. Measure absorbance at 535 nm (A_{535}) and 650 nm (A_{650}) against extraction buffer control. The relative anthocyanin content A is calculated as:

$$A = A_{535} - 2.2 \times A_{650}.$$

9. To compare different samples, the value for the relative anthocyanin concentration must be normalized to a definite amount of tissues (for example: $A/50$ seedlings, $A/$cotyledon pairs, etc.).
10. In total, at least five independent measurements should be done for each light treatment in at least two independent experiments.

3.5. Determination of Chlorophyll Content

1. All manipulations are done under green safelight to avoid photodestruction of extracted chlorophyll.
2. Harvest a defined amount of tissues or seedlings (*see* **Notes 14** and **15**) and transfer into 1.5-mL plastic tubes supplemented with seven glass beads. Freeze immediately in liquid nitrogen. Samples can be stored in darkness at –80°C.
3. Shake frozen material for 10 s in a disrupter to grind the material.
4. Add 1 mL of *N,N*-dimethylformamide and let the samples thaw with open lids.
5. Shake samples at 4–8°C in darkness overnight to extract chlorophyll.

6. Centrifuge the samples at 4°C for 15 min at not less than 10,000×*g*.
7. Transfer the clear supernatant to a new tube. Avoid contamination with any scattering material.
8. Measure absorbance at 664 nm (A_{664}), 647 nm (A_{647}), 625 nm (A_{625}), and 603 nm (A_{603}) against *N,N*-dimethylformamide. During deetiolation, protochlorophyllid (PChl), chlorophyll a (Chla), and chlorophyll b (Chlb) concentrations can be calculated according to the formulae (μg/mL):

$$c\,(\text{Chla}) = 12.81A_{664} - 2.16A_{647} + 1.44A_{625} - 4.91A_{603}$$
$$c\,(\text{Chlb}) = -4.93A_{664} + 26.01A_{647} + 3.74A_{625} - 15.55A_{603}$$
$$c\,(\text{Chlt}) = -2.52A_{664} - 0.79A_{647} + 36.55A_{625} - 27.08A_{603}$$

   In the absence of PChl (after prolonged irradiation with red or blue light), concentrations of Chla, Chlb, and total chlorophyll (Chlt) can be calculated according to the formulae (μg/mL):

$$c\,(\text{Chla}) = 12.91A_{664} - 2.12A_{647} - 3.85A_{603}$$
$$c\,(\text{Chlb}) = -4.67A_{664} + 26.09A_{647} - 12.79A_{603}$$
$$c\,(\text{Chlt}) = 8.24A_{664} + 23.97A_{647} - 16.64A_{603}$$

9. To compare different samples, the value for the relative chlorophyll concentration must be normalized to a defined amount of tissues or constant leaf area.
10. In total, at least five independent measurements should be done for each light treatment in at least two independent experiments.

3.6. Determination of Flowering Time

1. Sow seeds on soil in a pot appropriately sized for your plant material. Grow a single plant per pot.
2. Stratification (if necessary): Transfer pots to a photoresistent box and incubate at 4°C in darkness for 3 days.
3. Grow plants under short-day (8 h white light: 16 h darkness) or long-day (16 h white light: 8 h darkness) conditions at an appropriate temperature. For night break experiments, plants are kept under short-day conditions and light treatments are given in the middle of the dark phase (*see* **Note 16**).
4. Start of flowering is normally defined as the stage when the first anther opens to release its pollen. Note the number of days between transfer to light and onset of flowering. Count number of leaves that are formed until the onset of flowering.

5. Repeat the experiment at least twice using at least ten plants for each experiment.

3.7. Determination of Leaf Blade Area and Petiole Length

1. Grow seedlings as described above (*see* **Section 3.3** and steps 1–5). To obtain adult plants *see* **Section 3.6** and steps 1–3.
2. Fix cotyledon pairs or adult plant leaves (*see* **Note 17**) on a transparent adhesive tape and cover them with an additional tape before drying out. A transparent sheet is used for larger leaves.
3. Scan with a flatbed scanner in the black and white mode and save pictures as bitmap files (8 bit). Include scanning of a size standard (millimeter paper). Resolution should be 300 dpi or higher.
4. ImageJ is utilized to perform measurements. The scaling for measurements is adjusted using the picture of the size standard. Mark a line of known distance with the straight line selection tool → Analyze → Set Scale → Adjust Parameters, Select Global, and click OK (*see* **Note 18**).
5. Chose Analyze → Set Measurement → Mark Area and click OK.
6. Open scanned pictures with ImageJ and choose the wand (tracing) tool → select leaf area with a double click and check selected area → press 'control + M.' Repeat measurements for all samples.
7. Repeat the experiment at least twice using at least ten leaves or cotyledon pairs for each experiment.

4. Notes

1. Light meters normally exhibit some variability in their sensitivity to different wavelengths. Therefore, correction factors have to be estimated for the operation characteristics of the equipment and the actual light irradiance at the respective wavelength. Wavelength of maximum light emission of narrow-banded filters and LED light can be used to determine the appropriate correction factor.
2. Do not add sugar, because this compound interferes with light signaling and reduces light responsiveness (11).
3. VLFR: VLFRs depend on the presence of high levels of light-labile phytochrome (phytochrome A). Therefore, plants should be grown in darkness for at least 1 d to enable accumulation of the photoreceptor molecule. Photon fluences of red light (650–670 nm) are adjusted to between 10^{-5} and 10^{-2} μmol photons m^{-2} by varying

photon-fluence rates of light fields and/or irradiation time. Duration of light pulses must not exceed 5 min. For fast testing, 2.5 min strong far-red light pulses (λ ~720 nm; ~10 μmol photons/m^2/s) can be used, which produce sufficient levels of Pfr-A for induction of VLFR, but do not trigger LFR. Sometimes multiple weak red or far-red light pulses are applied with intermitting dark phases of 1–4 h to induce VLFR (12).

4. LFR: Red light photon fluences are adjusted between 10^{-2} and 10^{+2} μmol photons m^{-2} by varying photon-fluence rates of light fields and/or irradiation time. Duration of light pulses must not exceed 5 min. Classical red/far-red reversible phytochrome responses also fall into this class of light responses. The rule of thumb for photoreversion experiments: Maximal levels of Pfr are formed by red light photon-fluence rates of 5 μmol photons/m^2/s (~1 W/m^2) after 1 min of irradiation, whereas photoreversion with 740 nm (720–750 nm) light takes about 3 min at the same light intensity.
5. HIR: Full induction of HIRs needs prolonged irradiation under high photon-fluence rates of blue (410–450 nm), red (650–670 nm), or far-red (~720 nm) light. Photon-fluence rates should be varied between 10^{-3} μmol photons/m^2/s and 50 μmol photons/m^2/s to test fluence rate dependencies from threshold to saturation. Normally, at least 4 h of irradiation are necessary for full establishment of an HIR. To measure hypocotyl elongation and root growth, seedlings are normally grown under continuous light for several days. If germination induction depends on red light treatments, far-red light irradiation should be started 24 h after the onset of germination induction to avoid photoreversion (e.g., Arabidopsis: 3 h of red or white light → 21 h darkness → 3–4 days of irradiation).
6. Root growth and hypocotyl elongation in seedlings normally show characteristics of HIR (*see* **Note 5**), and thus, experiments are done under continuous irradiation for at least 2 days. Hook opening and cotyledon opening are often more sensitive and can also be used to check responses under VLF (*see* **Note 3**) and LF (*see* **Note 4**) conditions. Hypocotyl elongation also shows SAR (*see* **Note 6**) and responds toward EOD (*see* **Note 7**) treatments. Thus, this parameter can easily be used to test for these response modes of light action.
7. SAR: Shade avoidance responses are induced by low levels of Pfr during the light phase. Plants are grown under an array of white-light fluorescent tubes supplemented with red-light fluorescent tubes to enable formation of high Pfr

levels. Low Pfr levels can be adjusted by supplementary far-red light from lateral light sources (740 or 750 nm LED panels). The Pfr level can be varied using different intensities of supplementary far-red light. Fast responses are studied after the onset of supplementary far-red light. In principle, plants can be grown under continuous light, but it might be more favorable to use 12 h light: 12 h dark cycles. There are several operating errors that should be avoided: (1) Far-red light must be shut down together with the other light sources to avoid EOD effects, (2) Control plants must be protected from far-red light by effectual shielding, and (3) Far-red light sources must not reduce light intensity from additional light sources by shading effects. In some cases, it might be helpful to grow control plants under similarly arranged light panels that are switched off.

8. EOD: For end-of-day treatments, plants are irradiated with 5 min of strong saturating red or far-red light to establish maximum (red) or minimum (far-red) levels of Pfr at the beginning of the dark phase. Plants are best kept under short-day conditions (8 h light: 16 h darkness) to reach maximum responses.
9. Agar plates keep seedlings humid and normally do not cause severe troubles by changing the angles of hook and cotyledon opening, especially with tiny and sensitive Arabidopsis seedlings. Drying of seedlings and alterations in hook and cotyledon angles are a more severe problem with transparent adhesive tapes, but work is much faster. Therefore, this technique is preferred for measurements of hypocotyl elongation or root growth.
10. Always test the accuracy of ImageJ scaling by measuring a size standard.
11. Angles for cotyledon opening can sometimes exceed 180° due to curling of these leaf organs. ImageJ only measures angles between 0 and 180° and thus, correct angles have to be noted during measurements to correct data for calculations with spreadsheet programs.
12. Examples for typical extraction procedures: 50 Arabidopsis seedlings + 0.6 mL extraction buffer and 1 Arabidopsis rosette leaf + 1 mL extraction buffer. Fresh weight of plant material can also be used for normalization of anthocyanin content in samples. The use of fresh weight is not appropriate with plant material that exhibits strong differences in morphology, e.g., with etiolated and deetiolated seedlings.
13. Induction of anthocyanin accumulation is normally rather insensitive toward light. In most plant species, anthocyanin

accumulation belongs to the class of HIR and thus, continuous strong far-red, red, blue, and UV-A light should first be tested in the experiments.

14. In a typical experiment, 50 Arabidopsis seedlings are extracted in 0.8 mL of DMFA. A cork borer with defined diameter is very helpful to harvest samples with identical area. Leaf area can also be determined according to the described protocol (*see* **Section 3.7**).
15. Chlorophyll accumulation in etiolated seedlings needs blue, red, or white light. Light intensities should be in the range of 1–10 $\mu mol/m^2/s^1$ to avoid photobleaching and block of greening, especially with etiolated seedlings. Far-red light is not able to trigger protochlorophyllide to chlorophyll photoconversion. Preirradiation of seedlings with far-red light increases protochlorophyllide content and thus, can accelerate greening, but also makes the plant more sensitive to photobleaching. Alterations in chlorophyll a/b ratios are characteristic for SAR.
16. Night break experiments: Plants are usually grown under 8 h light: 16 h dark phases and light treatments are done around the middle of the dark phase. Light responses belong either to the class of LFR (*see* **Note 4**) or HIR (*see* **Note 5**). Species and even ecotypes respond variably to night break experiments, and thus, optimal conditions have to be established for each species. Several species also need some time to establish competence toward night breaks. For the Landsberg *erecta* and Wassilevskija ecotypes of Arabidopsis, it is sufficient to apply night breaks in 3 consecutive dark periods with 14-day-old seedlings grown under short-day conditions.
17. Cotyledon expansion is normally very sensitive to all light treatments and exhibits light effects under all response modes, including VLFR (*see* **Note 3**), LFR (*see* **Note 4**), HIR (*see* **Note 5**), SAR (*see* **Note 7**), and toward EOD treatments (*see* **Note 8**). At later stages of development, leaf size is strongly altered under shade-avoidance conditions.
18. Always test the accuracy of ImageJ scaling be measuring standards with known surface area.

Acknowledgments

This work was supported by the DFG grant 'Analysis of phytochrome A-dependent light signalling in *Arabidopsis thaliana*' (KR2020/2-4).

References

1. Casal, J. J. and Sanchez, R. A. (1998) Phytochromes and seed germination. *Seed Sci Res* **8**, 317–329.
2. Chen, M., Chory, J., and Fankhauser, C. (2004) Light signal transduction in higher plants. *Annu Rev Genet* **38**, 81–117.
3. Franklin, K. A., Larner, V. S., and Whitelam, G. C. (2005) The signal transducing photoreceptors of plants. *Int J Dev Biol* **49**, 653–664.
4. Yanosky, M. J. and Kay, S. A. (2003) Living by the calendar: How plants know when to flower. *Nat Rev Mol Cell Biol* **4**, 265–275.
5. Rockwell, N. C., Su, Y. S., and Lagarias, J. C. (2006) Phytochrome structure and signaling mechanisms. *Annu Rev Plant Biol* **57**, 837–858.
6. Mancinelli, A. L. (1994) The physiology of phytochrome action. In: Photomorphogenesis in Plants, 2nd Edition, pp. 51–59. Kendrick, R. E. and Kronenberg, G. H. M., eds. Kluwer, Dordrecht.
7. Rasband, W. S., Image, J., U. S. National Institutes of Health, Bethesda, Maryland, USA, http://rsb.info.nih.gov/ij/, 1997–2008.
8. Schmidt, R. and Mohr, H. (1981) Time-dependent changes in the responsiveness to light of phytochrome-mediated anthocyanin synthesis. *Plant Cell Environ* **4**, 433–437.
9. Moran, R. (1982) Formulae for determination of chlorophyllus pigments extracted with *N,N*-dimethylformamid. *Plant Physiol* **69**, 1376–1381.
10. Moran, R. and Porath, D. (1980) Chlorophyll determination in intact tissues using *N,N*-dimethylformamid. *Plant Physiol* **65**, 478–479.
11. Smeekens, S. (2000) Sugar-induced signal transduction in plants. *Annu Rev Plant Physiol Plant Mol Biol* **51**, 49–81.
12. Zhou, Y.-C., Dieterle, M., and Kretsch, T. (2002) The negatively acting factors EID1 and SPA1 have distinct functions in phytochrome A-specific light signaling. *Plant Physiol* **128**, 1098–1108.

Chapter 14

Kinematic Analysis of Cell Division and Expansion

Bart Rymen, Frederik Coppens, Stijn Dhondt, Fabio Fiorani, and Gerrit T.S. Beemster

Abstract

Plant growth is readily analysed at the macroscopic level by measuring size and/or mass. Although it is commonly known that the rate of growth is determined by cell division and subsequent cell expansion, relatively few studies describing growth phenotypes include studies of the dynamics of these processes. Kinematic analyses provide a powerful and rigorous framework to perform such studies and have been adapted to the specific characteristics of various plant organs. Here we describe in detail how to perform these analyses in root tips and leaves of the model species *Arabidopsis thaliana* and in the leaves of the monocotyledonous crop species, *Zea mays*. These methods can be readily used and adapted to suit other species in most laboratories.

Key words: Cell division, cell expansion, kinematic analysis, image analysis, epidermal cells, Arabidopsis, leaf growth, root growth, maize.

1. Introduction

In plant sciences, growth is a key characteristic that is widely used to evaluate genotypes and responses to environmental conditions. Towards commercial application, growth rate under optimal or limiting conditions is the primary trait that ultimately determines crop yield. Measurements of growth at the whole plant or organ level are relatively straightforward: they involve measuring size or weight at multiple times and calculation of the rate of increase. However, in many cases the purpose of growth experiments is also to learn about the mechanisms that drive the observed differences in growth rate. A number of strategies have been developed

L. Hennig, C. Köhler (eds.), *Plant Developmental Biology*, Methods in Molecular Biology 655,
DOI 10.1007/978-1-60761-765-5_14,

to this end, each focussing on different aspects of growth regulation. Firstly, classical growth analysis focuses on the evolution of plant (dry) weight and its utilisation in roots, stems and leaves. Here, the main emphasis is on understanding relative growth rates (RGR), which expresses on how efficiently biomass is used to generate more biomass through photosynthesis (1).

A second strategy focuses on how the initiation and growth of individual leaves contribute to the development of the shoot as a whole. This approach is described in detail in **Chapter 7** of this volume.

The aim of the kinematic approaches that are the subject of this chapter is to understand how processes that operate at the cellular level, division and expansion, contribute to differences in rates of growth at the whole organ level. These methods have been pioneered halfway last century (2, 3) and a rigorous mathematical framework was developed based on laws of fluid dynamics a few decades later (4). Since then there has been a gradual increase in the experimental use of these methods, largely supported by the increasing power of computers and the availability of powerful, easy-to-use public-domain image-analysis software like ImageJ (http://rsbweb.nih.gov/ij/).

In this contribution, we describe the methods for analysis of cell division and expansion parameters in three different experimental systems: the first leaf pair of *Arabidopsis thaliana*, the primary root of the same species and the leaf of the monocotyledonous species maize. We focus here in detail on the practical implementation of such measurements. More extended reviews about its conceptual basis and the derivation of the formulae used have been published before (4, 5). Together these three experimental systems provide a comprehensive overview of the possibilities that this approach currently offers and the protocols can easily be adapted to suit other species.

2. Materials

2.1. Analysis of *Arabidopsis* Leaves

1. Round Petri dishes (12 cm) preferably with a grid on the bottom (*see* **Note 1**) and porous tape for sealing.
2. Murashige and Skoog (MS) medium: Mix 0.5 × MS salts, 10 g/L sucrose, 0.5 g/L 2-(*N*-morpholino)ethanesulfonic acid (MES) and 0.8 g/L plant tissue culture agar in nanopure water. Adjust pH to 5.8 before adding agar. Autoclave medium at 1 bar over-pressure for 20 min. In a flow bench, pour 50 mL of medium into each plate when the agar is still

hot. Allow the agar to set with the lid opened and close only when the medium is at room temperature to prevent excess condensation. Plates can be stored in a plastic bag at 4°C for about 2 weeks.

3. 3.5% bleach solution.
4. Sterile water.
5. 70% ethanol.
6. Lactic acid.
7. Hoyer medium: 80 g Chloral hydrate, 20 mL glycerol and 10 mL water (6).
8. Mounting material: Object slides and cover slips.
9. Binocular microscope equipped with a camera.
10. Microscope with 20 × and 40 × Plan Differential Interference Contrast (DIC) lenses and a drawing tube (*see* **Note 2** on the principle of a drawing tube).
11. Fladbed scanner.
12. Computer running image-analysis software (e.g. ImageJ; Public domain image-analysis software, freely available from http://rsbweb.nih.gov/ij/).
13. Spreadsheet (e.g. MS Excel or OpenOffice (freely available from http://www.openoffice.org/)).

2.2. Analysis of Arabidopsis Primary Roots

1. For preparation of MS medium and plates, *see* **Section 2.1**.
2. Square Petri dishes (12 cm) and porous tape for sealing.
3. 3.5% bleach solution.
4. Sterile water.
5. Mounting material: Object slides, cover slips and sticky tape (clear).
6. Tooth-pick with short hair (ca. 2 cm, e.g. eyelash) glued on it.
7. Toner powder in Petri dish.
8. Microscopy: For velocity measurements, a simple microscope with a Plan 5 × and 10 × long working distance lens. Ideally this microscope is capable of working in horizontal orientation, so that stage is vertical allowing the roots to grow unperturbed with regards to the gravistimulus. For cell-length measurements, a microscope with 20 × and 40 × plan differential interference contrast (DIC) lenses (*see* **Note 3** on importance of plan lenses). Digital camera for acquiring images on both microscopes.
9. Flatbed scanner.

10. Computer running image-analysis software (e.g. ImageJ; Public domain image-analysis software, freely available from http://rsbweb.nih.gov/ij/).
11. Spreadsheet (e.g. MS Excel or OpenOffice (freely available from http://www.openoffice.org/)).
12. R (Public domain statistical software, freely available from http://www.r-project.org/).

2.3. Analysis of Maize Leaves

1. Ruler.
2. Lactic acid.
3. 3:1 (v/v) absolute ethanol:acetic acid.
4. Mounting material: Object slides and coverslips.
5. 4′,6-Diamidino-2-phenylindole (DAPI) solution.
6. Buffer solution: 50 mM NaCl, 5 mM EDTA and 10 mM TRIS-HCl, pH 7.0.
7. Microscopy: For cell-length measurements: Microscope with 20 × and 40 × plan differential interference contrast (DIC) lenses. For meristem length measurements: An epi-fluorescence microscope with an excitation filter at wavelength 350 nm and emission filter at wavelength 420 nm, equipped with 20 × and 40 × lenses. Digital camera connected to a personal computer for acquiring images on both microscopes.
8. Computer running image-analysis software (e.g. ImageJ; Public domain image-analysis software, freely available from http://rsbweb.nih.gov/ij/).
9. Spreadsheet (e.g. MS Excel or OpenOffice; freely available from http://www.openoffice.org/).
10. R (Public domain statistical software, freely available from http://www.r-project.org/).

3. Methods

3.1. Analysis of *Arabidopsis* Leaves

After initiation at the shoot apex, leaves of dicotyledonous plants go through subsequent stages of cell division and expansion before they reach maturity. Although the transitions between these stages occur in a tip to base gradient (7), the approach outlined here calculates average rates of division and expansion across the entire leaf. The analysis is based on two sets of primary data: leaf blade area and cell area. For the first leaf pair of *A. thaliana* ecotype Columbia 0, we measure these variables from 2 days after germination (DAG) until 22 DAG. This period spans from leaf emergence until maturity under our environmental conditions (growth chamber: 21°C, fluorescent (cool white) light 80 μE/m^2/s, 16/8 h day/night and shelves cooled at 19°C to prevent condensation against the lids of the Petri dishes).

3.1.1. Preparation of Plant Material

1. Sterilise seeds for 15 min in 3.5% bleach solution in 1.5-mL microcentrifuge tubes and wash three times with sterile water. After the last wash leave the water in the tubes.
2. Sow seeds using a 20-μl pipette with tips of which the top is cut 2 mm with a razor blade to increase the opening. Aspirate about a dozen seeds with enough water to dispense them. Plate the seeds individually on the agar. In this step, it is important to gently touch the medium with the tip in order to break the surface of the agar layer, generating a small crack that allows the root to penetrate the medium (*see* **Note 4**).
3. Leave the plates open in the flow for a few minutes to let excess water evaporate.
4. Close the Petri dishes using porous tape that allows gas exchange.

3.1.2. Mounting on Slides

1. Harvest whole plants for the earliest stages of development. Dissect leaves once the first true leaves have formed a petiole (ca. 5 DAG).
2. Place the material in 70% ethanol to remove chlorophyll overnight.
3. Transfer material to lactic acid and incubate overnight for clearing.
4. Mount material on a microscopic slide using the lactic acid as mounting medium (*see* **Note 5**). For cell analysis, the abaxial epidermis is used; therefore, it is important to place the leaf on the microscopic slide with the adaxial-trichome-containing side down. For the earliest stages, expose the first leaf pair by gently pulling apart the cotyledons before placing the coverslip.

3.1.3. Leaf Area Measurements

1. Acquire images for about 10 leaves per genotype/treatment and time point using a binocular microscope equipped with a camera. Adjust the magnification to the size of the leaves under examination (×1.25–6.3) and make sure to photograph a ruler at the same magnification for image calibration (*see* **Note 6**).
2. Measure the leaf area using an image-analysis program, e.g. ImageJ. Calibrate the program using the ruler image, which allows converting pixels to corresponding distances in mm.

3. Use the 'polygon selection' tool to outline the leaf blade and measure the area. Because the leaves of Arabidopsis often have a curved surface, the leaf edge is sometimes folded double. To accommodate for this, it is necessary to measure the area of the folded parts of the leaf that can be easily recognised in addition to the outline of the whole leaf area.
4. Copy the measurements into a spreadsheet program and calculate the total leaf blade area. For the kinematic analysis, select at least 5 median-sized leaves and calculate average area and standard errors.

3.1.4. Measurements of Cell Area and Stomata Number

1. Visualise the cells in the cleared samples with a DIC microscope. Depending on the size of the cells, use a 20, 40 or 63 × magnification. Use preferably the abaxial epidermis of the first leaf pair for the analysis (*see* **Note 7** for a discussion on cell types). As there is a gradient in cell development from leaf tip to bottom, analyse the cells at two positions of the leaf, at about a quarter from the tip and bottom of the leaf and halfway between the leaf margin and the mid-vein. Avoid regions directly above the vasculature, because the epidermal cells can be more elongated and harder to visualise due to the optical disturbance of the underlying dense vascular strands.
2. Outline the leaf epidermal cells using a drawing tube (for an alternative *see* **Note 8**). To gather sufficient data, draw about 50 cells for each epidermal area examined. To minimise a bias in cell sizes, avoid the edge of the paper and avoid drawing a disproportionately high number of small cells. This problem may arise because small cells more easily fit on the paper compared to bigger ones. Draw a calibration grid for each magnification to allow the correct scaling.
3. The number of stomata per drawing is counted manually and recorded.

3.1.5. Processing of the Images

1. When the drawings are finished, they need to be digitised. Correct potential artefacts by closing cell walls that show gaps and erasing cell walls at the edge of the drawing, which do not delineate full cells. Also, avoid cell walls that touch the edges of the paper. Record the origin of the drawing by annotating next to the drawn area: genotype/treatment, time point, position in the leaf (tip or bottom) and magnification used for visualisation (for a typical example of the epidermis and a drawing, *see* **Fig. 14.1**).
2. Scan the drawings. Depending on the available scanner, adjusting the settings might take some trial and error. Scanning at 300 dpi and saving the scanned images as jpeg file are generally sufficient.

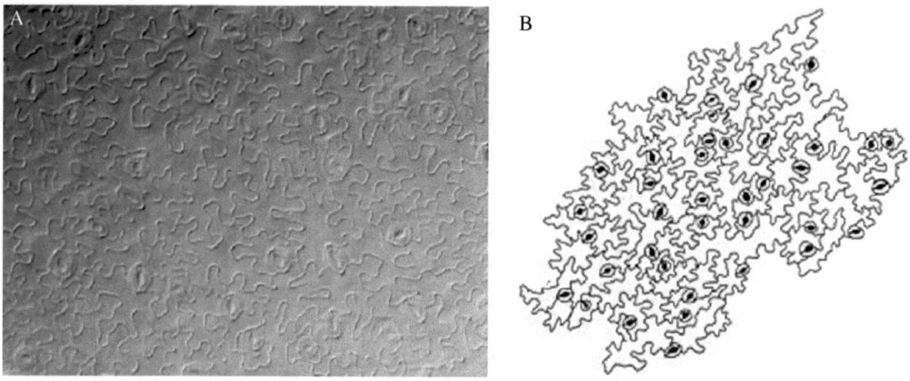

Fig. 14.1. **Image processing of the abaxialf epidermis of *Arabidopsis thaliana* leaves.** A. DIC image of the abaxial epidermis of cleared leaves. B. Typical drawing of cell outlines resulting from a similar specimen.

3. Process the images using image-analysis software (here described for ImageJ). Set the calibration by opening the appropriate calibration image and use the straight line tool to select a length of the grid. Use the 'Analysis/Set Scale' command and fill in the known distance and the units. Apply these settings for all images that are subsequently opened. Select the option 'global' under 'Analysis/Set Measurements' to verify that 'Area' and 'Display label' are selected.
4. Open an image and convert it to 8-bit greyscale under 'Image/Type.'
5. Apply a threshold using 'Image/Adjust/Threshold' that allows the cell walls to be best visualised.
6. Apply a closing step 'Process/Binary/Close' to close small gaps in the drawing possibly created by the thresholding.
7. Use the 'Magic Wand' tool and click to the left of the drawn area. This selects the outline of the entire area of the cell drawing. If the selection includes the interior of some cells not all cell walls are closed. Close gaps by selecting a line across the opening and clicking 'Edit/Draw' or 'Ctrl-D.' As this is tedious, check drawings beforehand for these artefacts.
8. Measure the drawn area by clicking 'Analyze/Measure' or 'Ctrl-M.'
9. Count all cells in the drawing. To easily keep track of which cells were counted, select every cell with the 'Magic Wand' tool and click 'Edit/Fill' or 'Ctrl-F' to fill it.
10. Copy the obtained data from the Result Window to a spreadsheet program (e.g. MS Excel).

3.1.6. Calculations

Following the above procedures, the primary data for each leaf consist of the leaf blade area and for two drawings (tip and base of leaf): number of cells, total area of the drawn cells and number of stomata. From these derived parameters can be calculated as follows:

1. Determine average leaf area as the average (and standard error) of all leaf areas per genotype/treatment at each time point (Units: mm^2). Due to the exponential nature of the growth process, present this parameter on a log-scale.
2. Divide the total area of the drawn cells by the number of cells in it (pavement cells + guard cells (2 per stoma)); this yields the average cell area in that drawing. Calculate an average (and standard error) of all the cell areas for all the leaves per genotype and time point.
3. Calculate for each drawing the Stomatal Index (*SI*) by dividing the number of guard cells (number of stomata (*S*) multiplied by two to correct for the presence of two guard cells) by the total number of cells (pavement cells (*PC*) + guard cells):

$$\mathrm{SI} = \frac{2S}{(\mathrm{PC} + 2S)}.$$

3. Divide the leaf area by the average cell area from the same leaf to obtain the number of cells per leaf. Calculate averages (and standard error) of the number of cells per leaf for all the leaves per genotype/treatment and time point. Due to the exponential nature of the division process, represent leaf numbers in log-scale with base 2.
4. The leaf expansion rate (LER) is the derivative of leaf area over time (on a $\log_n$ scale). Calculate this derivative by using the LocPoly algorithm or use a spreadsheet such as MS Excel. An R-script for the LocPoly or an example of an Excel sheet with those calculations (*see* **Note 9**) can be obtained from the corresponding author of this chapter.
5. Average cell division rate is the derivative of the cell number data with respect to time. The calculation is similar to the LER, using the $\log_2$ of the number of cells (cell/cell/h). Calculate cell-cycle duration as the inverse of cell-division rate. For additional information on effects on cell-cycle phase duration, these data can be combined with flow-cytometry measurements (*see* **Note 10**).

3.2. Analysis of Arabidopsis Primary Roots

The analysis consists of two parts, measurement of the velocity profile and cell-length distribution, and these need to be combined in order to calculate cell-division rates. In order for this to work, it is imperative to work relative to a common reference point. In the root typically the quiescent centre (QC) is a convenient point as it forms the origin of all cell files. Unfortunately, the QC is not recognisable in surface view of the root, which is needed for the velocity-profile measurements. Therefore, these measurements are made relative to the tip of the root and later corrected with the distance between the QC and tip of the root as measured on a median view using DIC optics.

3.2.1. Plant Growth and Measurements of Whole Root Growth Rates

To allow for roots to grow and be accessible for observations, grow plants in Petri dishes with agar-solidified growth medium. Although different nutrient mixes can be used, according to our experience full-strength MS salts are a convenient choice. First, this mix is available as ready-made salts so that no mixing of stock solutions is required. Second, the characteristics of the plants are good. The shoots look vigorous and green and although the roots grow slower than on, for example, Hoagland media, most ecotypes we tested grow at a steady rate from early after germination rather than accelerating (8), which simplifies the analysis significantly.

1. Sterilise seeds for 15 min in 3.5% bleach solution in 1.5-mL microcentrifuge tubes and wash three times with sterile water. After the last wash, leave the water in the tubes.
2. Sow seeds using a 20-μl pipette with tips from which the top 2 mm is cut with a razor blade to increase the opening. Aspirate about a dozen seeds with enough water to dispense them. Place the seeds individually on the agar. Pick up around 10 seeds and sufficient water and distribute 8–10 seeds at equal distance at about 1 cm from the top of the plate. Make sure not to touch the agar with the tip as breaking the agar surface will result in roots entering the agar rather than growing along the surface. When accidentally multiple seeds are sown at the same position, remove the additional ones with the pipette tip. Adding a bit of additional water often simplifies this task. Use about 10 roots for each line or treatment and grow another 20 or so in parallel plates for independent measurements of root growth rates.
3. Seal plates with porous tape to allow air-exchange and place under an angle of 80–85° in a growth chamber. As steady-state conditions are assumed, it is best to use continuous light (about 60–80 $\mu E/m^2/s$ photosynthetic active radiation) and temperature (ca 21°C) conditions.
4. Start measurements of root growth rates as soon as most seeds have germinated. Make each day a small scratch at the back of the plate perpendicular to the growth direction of the root marking the position of the tip. Keep track of the time when these marks are placed. Continue these measurements for at least 2 weeks.

5. Scan the plates using a flatbed scanner (*see* **Note 11**). Make sure always to use the same scanner resolution and also scan in a ruler as a scale for calibration.
6. Open the scanned images of plates with image-analysis software such as ImageJ. Calibrate the scale of all images using the image of the ruler.
7. Use the 'Freehand' tool to measure the distance between the marks.
8. Transfer these data for each root into a spreadsheet program. Calculate average root-elongation rates for each root at a particular day by dividing the measured distance between marks by the time difference between when the marks were made using a time scale of hours.
9. Calculate average velocity and standard error for each time interval across all roots. Use the growth rates calculated this way as a first screen to decide on which lines or treatments to concentrate on in more detail. Typically, preliminary experiments are done for this purpose in which also mature cortical cell length is measured (about 20 cells per root).
10. Divide root-elongation rate by mature cell length to obtain cell-production rates. As a general rule, differences for root-elongation rate, mature cell length or cell-production rate need to exceed 10% for kinematic analysis to be feasible, because too many replicates need to be measured to obtain significant differences for individual kinematic parameters. For kinematic analysis, measure ca 20 undisturbed roots growing on separate plates to control for effects of the experimental manipulations.

3.2.2. Velocity Measurements

The aim of these measurements is to determine the velocity at which cells are moving in function of position along the root. As a basis for this, it is necessary to make several images of the root over a specified time interval. Although the analysis can be done semi-automatically (*see* **Note 12**), we will describe here how to do this manually. This approach requires least setting up, is probably still the most reliable and increases understanding of the principles involved. The automation could be a valuable next step for those implementing the analysis for more routine usage.

1. Sprinkle the surface of the root with contrasting and recognisable marks. A convenient substance for this is toner powder that is used in copiers and laser printers. Pour some of this powder from a fresh cartridge into a small Petri dish. Open the plate(s) with roots to be analysed under a binocular and dip an eyelash or other fine hair glued to a toothpick in this powder. Then hold the hair above the root and

gently tap it to release the powder evenly, but thinly on the surface from the root tip to the region where mature root hairs appear. It is no problem to spill on the agar, but do not apply too much on the root as this will interfere with the cell-length measurements in the second stage. A good amount is shown in **Fig. 14.2**. After all the roots on a plate have been marked, quickly reseal the plate and bring it back to the growth room, leaving it to recuperate for an hour.

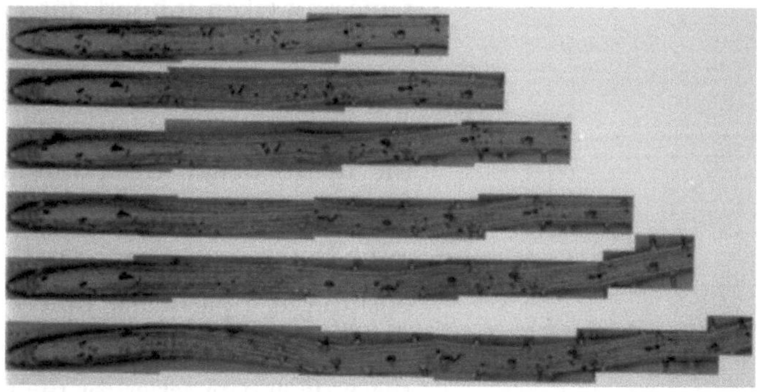

Fig. 14.2. **Time-lapse observations for a single Arabidopsis root tip.** Successive images were taken at intervals of 1 h and composed of a set of overlapping frames. The dark spots are larger aggregates of toner powder, showing the appropriate density at which they need to be applied. They also illustrate the rate of displacement in different parts of the root.

2. Mount the whole plate, preferentially on a horizontally oriented microscope fitted with a camera so as to keep the orientation of the plate vertical (*see* **Note 13**). Using a 5 or 10 × magnification make a series of overlapping images starting from the tip of the root till well into the region of fully developed root hairs, focussing on the root surface and toner particles. It is convenient to orient the root horizontally in the resulting images, so adjust the camera angle to achieve this.
3. Repeat this for all roots on the plate and make at least three sets of observations at hourly intervals. Also make sure that an image is made of a calibration grid using the same magnification.
4. Use the obtained images to make a composite image by stitching the individual images together (*see* **Note 14**). The individual particles form small aggregates with recognisable shapes that can be found back between two images of the same root taken at different times (*see* **Fig. 14.2**).
5. Draw a series of vertical lines through specific particles in a pair of images that are open at the same time. The distance between these lines should ideally be less than half of the root diameter. At some point along the root, cell walls and

root hairs become convenient natural landmarks that can (also) be used.

6. Open the resultant composite images in the image-analysis software, which is calibrated using the image of the calibration scale. Measure the distance from the tip of the root to the first mark along the mid-line of the root, followed by the distance between subsequent marks (*see* **Note 15**).
7. Repeat this procedure for each of the three images of the same root taken at hourly intervals so that the displacement of the reference marks can be determined.
8. Copy the distances from the tip to each particle into the spreadsheet and calculate the cumulative distances from the tip to define the position of each particle in every image.
9. Calculate the average position, $X_{avg} = \frac{X_{t1} + X_{t2}}{2}$, with X_t the position at a particular time t as well as velocity $V(x) = \frac{X_{t2} - X_{t2}}{t_2 - t_1}$, for all particles for each pair of images. Combine the x, $V(x)$ data series for the subsequent time intervals to increase the number of observations.
10. To obtain X values relative to the QC, subtract the distance between QC and root tip. These data will be needed in combination with cell-length data to calculate division rates.
11. Determine the velocity at specific, equally spaced positions along the root to allow averaging between roots and further calculations by combining these measurements with the cell-length data. A local polynomial smoothing procedure has been developed (9) and implemented as an R script that is available upon request from the corresponding author. This script gives a series of data with increasing smoothing. Smoothing should just remove the local noise, but not affect the overall curve.
12. Use the script to calculate the local derivative of the velocity curve, which equals the relative rate of cell expansion RLER or $R(x)$. Average the data for $V(x)$ and $R(x)$ (at x) between roots and calculate standard errors.

3.2.3. Cell-Length Measurements

While the velocity measurements give a detailed insight into the spatial distribution of cell expansion along the root tip as a whole, it lacks a direct link to the underlying cellular components. The same results would be obtained if a single giant cell was analysed instead of a root of same proportions consisting of thousands of cells. In order to get this cellular perspective, we need to measure the length of cells in function of position along the axis. Although typically this is done for a single cell type, like we will describe here, it is possible to compare multiple cell types (9).

The Arabidopsis root tip has the advantage that it has a small diameter (ca. 140 μm) because only a single layer of each tissue (10) is present. This allows for high-quality microscopy images using a whole-mount procedure that saves time and precludes potential problems such as shrinkage during an embedding procedure.

Although the mounting procedure is relatively simple, care needs to be taken to avoid the collapse of cells, particularly those in the meristem, which can be caused by both physical and osmotic pressures. Therefore, we recommend the use of spacers (a strip of sticky tape on both sides of the object slide to rest the coverslip on) and a solution that is isotonic to the growth medium of the root. The easiest for the latter is to make the same solution as used for the agar medium without adding the agar.

1. Cut the roots that were used for the velocity observations described above from the shoot with a scalpel or razor blade (make sure to keep track of individual plants to be able to combine velocity and cell-length data for each plant). Cut at least a centimetre of root and handle close to the cut to avoid damaging the tip where the measurements will be made. Place a droplet of the mounting solution on the surface of the object slide in the middle between the two spacers. Place the root in the droplet and lower the cover slide gently onto it.

2. Place the specimen under a microscope fitted with DIC optics. Find the tip of the root at low magnification (5 or 10 ×) and switch to 40 × for DIC imaging. Orient the root horizontally in the image; we will assume the tip to be pointing to the left. Adjust the microscope settings to optimise the image focussing on a median section in the region of the tip where the quiescent centre (QC) cells can be found. From the QC, files of cells can be seen radiating out.

3. Take a photo focused on the QC and make sure that the tip of the root cap is in the same frame. This image is required for determining the offset between the tip of the root and the QC in the velocity data (as described above). Capture and save the image for analysis at a later stage.

4. Typically, the cortical cells are easiest to analyse because they are the biggest in diameter; therefore, it is advisable to focus on those although other types can also be used. For the next frame, move the microscope so that the QC is close to the left edge of the frame. Focus on the cortical files so that a clear row of cells can be seen on at least one side of the root. Capture and save this image. In addition to this medium plane, it is possible to image additional cells by zooming out (without changing *X*,*Y* position on the stage) to the tangential plane through the cortical cell layer.

5. Make a series of images by moving along the root from the tip to the region where the root hairs are fully mature, capturing both the median and tangential plane whenever possible. Make sure that these images overlap by 10–20% to allow them to be aligned and stitched for further analysis.
6. Before making cell-length measurements, combine/stitch the images from each plane of view into a single image so that each file can be followed over the length of the root. Use the same approach as outlined for the images of the marks on the surface. When measuring cell sizes, it is important to bear in mind that they need to be positioned relative to the QC. For the median section, this is straightforward: position the line tool to select the length of the first cell adjacent to the QC and measure it. Then select the next cell in the file (in ImageJ by moving only one side of the line tool, leaving the side that marks the wall between the first and second cell in place) and measure its length. Continue doing this until the end of the file is reached (*see* **Note 16**).
7. Because this QC is only visual on the median section, use this image to measure the distance from the base of the QC to the left border of the image. On the tangential images, measure the distance from the first unambiguous cell to the left border of the image and subsequently measure all cells in this file.
8. Copy all cell lengths and distances that do not refer to a cell length for each file separately into a spreadsheet program.
9. Calculate the distance from the QC to the midpoint of each cell. For median files, add the total length of all preceding cells and sections that are not cells + 0.5 × the length of the cell itself.
10. Change the calculations of these positions from formulae into values (In MS Excel by 'Edit/Copy,' 'Edit/Paste Special: Values') and then remove the data for the sections that do not refer to cell lengths.
11. Combine the data from all files into a single pair of columns and sort for ascending position (x).
12. Use these data for interpolation and generation of equally spaced data using the same Locpoly routine in R that was used for the Velocity data (*see* **Section 3.1.6** and step 4). Calculate average cell length and standard errors between replicate roots from the interpolated data.

3.2.4. Calculation of Cell-Division Rates

1. For the calculation of cell-division rates, combine the equally spaced data for velocity and cell length and calculate flux ($F(x)$), the number of cells that are passing at a particular

position, by dividing local velocity by cell length: $F(x) = \frac{V(x)}{l(x)}$. In steady-state conditions (for non-steady state *see* **Note 17**), the derivative of the flux function $\left(\frac{dF(x)}{dx}\right)$ denotes the local rate of cell production per unit of length ($P(x)$). Calculate this derivative using 5-point equations (11).

2. Calculate local cell-division rates ($D(x)$) from the cell-production rates by dividing by local density, which equates to multiplying them with the local cell length: $D(x) = l(x) \times P(x)$.

3.2.5. Calculation of Kinematic Parameters

Based on the calculation of spatial data for cell size, velocity, cell expansion and division rates, it is relatively straightforward to calculate organ scale data for each replicate root.

1. Determine the size of the growth zone (L_{gz}) as the position where $V(x)$ becomes maximal and thus its derivative, $R(x)$, goes to 0.
2. Determine the size of the meristem (L_{mer}) as the position where $F(x)$ becomes constant, hence where $D(x)$ becomes 0.
3. Determine the size of the elongation zone (L_{el}) as the difference between L_{gz} and L_{mer}.
4. To determine the number of cells in each zone, calculate the number of cells for each interval of the interpolated cell-length data by dividing the size of the interval by the average size of the cells in it (by averaging the size of cells at the beginning and endpoint).
5. The number of cells in the meristem (N_{mer}) equals the cumulative number of cells in all the intervals located within the meristem.
6. The number of cells in the elongation zone (N_{el}) equals the cumulative number of cells in all the intervals located within the elongation zone.
7. The number of cells in the whole of the growth zone (N_{gz}) equals $N_{mer} + N_{el}$.
8. Determine the root-elongation rate (E) as the velocity in the mature part of the root by averaging the values obtained in this region.
9. Calculate mature cell size l_{mat} as the average cell length in the mature region.
10. Calculate cell production in the whole of the meristem as E/l_{mat}.
11. The average cell-division rate (D_{avg}) equals cell-production rate divided by N_{mer} and the average cell-cycle duration $T_c = \frac{\ln(2)}{D_{avg}}$.

12. Calculate all these parameters for each individual root and calculate averages and perform statistics.

3.3. Analysis of Maize Leaves

Similar to roots, the cells in the epidermis of monocotyledonous leaves are arranged in linear files, allowing the same basic approach as in the root system to analyse cell division (*see* above). The main difference between roots and monocotyledonous leaves, however, is that it is not possible to determine the velocity profile by direct observation of the epidermal cells in the growth zone, because the older leaves encapsulate the growing younger ones. Therefore, an indirect method based on cell-length profiling only has to be used. This method assumes that during steady-state growth, the cell-length profiles are constant. It entails leaf-elongation rates measurements, measurements of the cell-length profile along the axis of the leaf and estimation of the size of the leaf basal meristem. Unfortunately, it is not possible to perform all these measurements on the same individual leaves. Therefore, it is unavoidable to combine measurements of several distinct plants grown under the same conditions.

The method will be explained and discussed based on our experience with maize, but can be adapted to other monocotyledonous species.

3.3.1. Plant Growth and Measurements of Leaf-Elongation Rates

1. To perform a kinematic analysis of leaf growth in maize, a batch of at least 15 plants per condition is necessary. Record the leaf-elongation rate (LER) in function of time for a first subset of these plants by measuring with a ruler the length of the leaf (soil surface to leaf tip) under study at regular time intervals, preferably daily (*see* **Note 18**) for automation and a higher resolution approach. Straighten the leaf by hand and take caution not to break or damage the leaf as touching may influence growth rates. Start leaf measurement from leaf appearance (emergence from the sheath at the base of the shoot) until its complete extension.
2. Calculate LER from the recorded data as the difference in leaf length on two successive time points divided by the time interval between them (in mm/h). For a typical monocotyledonous leaf, leaf elongation is generally linear during the first days after appearance, followed by a period of progressive decline depending on leaf position and treatment (environmental conditions). The first days of linear increase can be considered as a situation of steady-state growth. At the cell level, it is assumed that also cell production and cell-length profiles in the growth zone are constant during the same time period (12, *see* **Note 19**).

3.3.2. Cell-Length Measurements

1. For profiling the cell length, harvest the whole growth zone during steady-state growth. The size of the growth zone depends on the environmental conditions, the species, the genotype and the developmental stage examined. Therefore, it is important to make an estimate of this size to make sure that the samples encompass the full extent of the growth zone (*see* **Note 20**).
2. To sample the growth zone, remove the older leaves that surround the growing leaf. Take special care not to damage the basal part of the growing leaf because this is where the leaf basal meristem is located. Some practise beforehand to optimise the dissection technique is required.
3. Segment the growth zone into smaller pieces (for example, segments of 10 mm) if the growth zone is too large to mount as a whole on microscope slides. Place the samples in absolute ethanol for 48 h for chlorophyll removal and fixation. To obtain better and faster clearing, renew the ethanol after 6 h. For further clearing and storage, place the samples in lactic acid.
4. For the cell-length measurements, mount the leaf (segments) on microscope slides. Unroll the samples on the object slide and remove the mid-vein with a scalpel. Mount the samples in lactic acid, so that the abaxial epidermis, which generally contains fewer stomata compared to the adaxial side, is placed face up.
5. Analyse the specimens under a microscope fitted with differential interference contrast (DIC) optics. Find the orientation of the samples and start from the most basal part. In maize, use the trichomes at the edge of the leaf, which point towards the leaf tip, as a reference to distinguish base from tip directions. Adjust the microscope to optimise the image focussing on the epidermal cell walls. Moving in distal direction from the base of the leaf sample, measure all the cells belonging to one epidermal cell file. Choose a cell file adjacent to stomatal rows on the abaxial side because these cell files are convenient to recognise and consist of a single cell type (*see* **Note 21**).
6. To take into account the variation in cell size at different positions across the leaf, repeat these measurements for a few equivalent cell files (e.g. cell files adjacent to stomatal rows) for each leaf.
7. Calculate the distance from the base to the mid-point of each cell in a spreadsheet. This is done by determining the cumulative lengths of all cells in more basal positions in the same file + 0.5 × the length of the cell itself.

8. Combine all data for a leaf by copying the data for the different cell files into the same position and size columns and sorting them for ascending position.
9. Use these data for interpolation and generation of equally spaced data using the same Locpoly routine in R that was used for the root data (*see* **Section 3.1.6** and step 4).
10. Calculate averages and standard errors between replicate leaves using the interpolated data.

3.3.3. Estimation of the Linear Extent of the Leaf Basal Meristem

1. Estimate the size of the meristematic zone of the leaves by locating the most distal mitosis in the cell files of interest. Sample the basal part of the growth zone in a fashion similar to the approach used for cell-length measurements. Again, sampling should occur during the steady-state growth and without damaging the most basal part. Place the samples in 3:1 (v/v) absolute ethanol:acetic acid for fixation of cell walls and clearing of chlorophyll. Keep the samples in this solution at 4°C from 24 h up to several weeks.
2. Rinse the samples with a buffer solution containing 50 mM NaCl, 5 mM EDTA and 10 mM TRIS-HCl, pH7.
3. Visualise the nuclei by incubating the samples in the dark for 1–20 min in the same buffer solution containing DAPI at a concentration of 1 μg/mL (for an alternative stain, *see* **Note 22**). Avoid too intense staining, since only staining of the epidermal cells and not of underlying cell layers is desired. Therefore, check the samples for fluorescent signal emission after a short incubation.
4. According to this first microscopic assessment, stop the reaction when a satisfactory signal is achieved. To stop the reaction, remove the DAPI by rinsing in the buffer solution. Sample re-incubation is always possible if it is necessary to achieve a higher signal level.
5. Mount the samples in a drop of the buffer solution on a microscope slide and cover with a coverslip. For image analysis, the same setup as for the cell-length measurements is required; a digital camera connected to a personal computer with image-analysis software enabling measurements if possible directly on the live acquired image.
6. Starting from the leaf base, score recognisable mitotic cells (metaphase, anaphase and telophase) in cell files adjacent to stomatal rows. In these stomatal rows, asymmetrical divisions take place at the more distal end of the division zone. They represent a convenient landmark for locating the region where the last divisions in the adjacent files can be found. Since mitosis is a relative rare event, it is necessary to examine at least 10 cell files per leaf to find the most distal mitotic event.

7. Determine the distance between the most distal mitotic cell, considered as a proxy for distal boundary of the meristem, and the base of the leaf. Draw and measure a straight line on the images with image-analysis software, between the most distal mitotic cell and the base of the leaf. When the meristem does not fit into a single microscope frame, add up the measurements between landmarks visible in successive overlapping microscopic frames.

3.3.4. Calculation of Overall Kinematic Parameters

1. Based on the measured LER, cell-length profile and meristem size, calculate the kinematic parameters for leaf elongation. The calculations are similar to the ones described above for root growth. Calculate the number of cells in each zone (meristem N_{mer} and elongation zone N_{el}), the mature cell length (l_{mat}), cell production of the meristem (P), the average cell-division rate (D_{avg}) and the average cell-cycle duration (T_c) exactly with the formulae, as described (*see* **Section 3.2.5**).
2. The size of the different zones is calculated in a slightly different way. Estimate the size of the growth zone (L_{gz}) as the distance from the leaf base to the position where the cells reach 95% of their mature length on the smoothed cell-length profile.
3. Estimate the meristem size (L_{mer}) in leaves as the distance at which mitotic cells occur with respect to the base of the leaf (*see* above).
4. Calculate the size of the elongation zone (L_{el}) as the difference between L_{gz} and L_{mer}.
5. In addition, the local cell-elongation rate $R(x)$ can be estimated in monocotyledonous leaves from the cell-length profile. Calculate this parameter for all positions as the position derivative of cell length multiplied by the cell production: $R(x) = P \times \frac{\partial l}{\partial x}$. Because in the meristem the flux, which is estimated by P, is not constant, this formula cannot be used in the meristem.

3.3.5. Flow Cytometry

Similar to the approach used for the Arabidopsis leaves, a deeper insight into the role of cell-cycle progression in the leaf elongation can be obtained by complementing the kinematics with flow cytometry. This technique allows the relative quantification of DNA nuclear content in the different regions of the growth zone and thereby it enables inferences of cell-cycle progression status in the meristem and endoreduplication in the elongation zone (13). However, bear in mind that this approach uses the entire leaf rather than the specific cell type analysed using the microscopy, so that some caution needs to be taken into account with the interpretation.

4. Notes

1. To facilitate regular spacing of seeds, plates with a grid on the bottom are convenient. These provide 32 full squares on a 12-cm dish. For analysis of young leaf material up to 7 DAG, the seeds can be sown densely with four seeds per square. For the time points 8–11 DAG, two seeds are sown per square while at later time points only one seed per square can be used. This way the plants have enough space to develop without overlapping with each other.
2. A drawing tube is a mirror system mounted on the microscope in the light path between the objective and the oculars that allows to view the specimen mounted on the microscope stage and at the same time a sheet of paper on the desk next to the microscope. This setup is used for outlining the epidermal cells in cleared leaf specimen and has much higher accuracy, particularly drawing the small pavement cells around the stomata, compared to using digital images, because it allows for focussing while drawing the cell outlines.
3. As the calculations of expansion and division rates involve local derivatives of velocity and cell-length data, small systematic errors may influence the data. One common source of error is the use of lenses with barrel or pincushion distortion. This type of lenses is not suitable for the measurements required for these kinematic analyses. PLAN lenses are designed to avoid this problem and need to be used, particularly for the analyses of cell lengths and velocity along an axis.
4. Plating for the analysis of leaf growth on horizontal agar plates and root growth on vertical plates requires an opposite strategy with regards to placing the seeds on the agar: While it is desirable for the root analysis to have the root growing on the surface of the agar and not penetrate it, for optimal and homogenous leaf growth the roots need to be able to penetrate the medium. To facilitate entry into the agar on horizontal plates, slightly touch the agar when placing the seeds, making small cracks in the surface that serve as an entry point for the roots. Avoid this when preparing vertical plates so the roots do not find an entry point and instead remain growing at the agar surface.
5. Sometimes the leaf material does not clear completely after transfer to lactic acid. This is typically due to the presence of relatively high amounts of starch, which makes the imaging of the epidermal cell walls very difficult. To improve the

clearing of the leaves, they can be transferred to Hoyer medium. The duration of clearing in Hoyer medium (6) depends on the stage of the leaf material and amount of starch present. For young leaves, 15–30 min are usually sufficient, while older leaves may require up to 2 h. It takes some trial and error to determine the desired clearing time. If the leaf material is kept too long in Hoyer medium, the clearing is excessive and the cell walls are no longer visible using DIC. Finally, the material needs to be transferred back to lactic acid after Hoyer treatment and mounted on object slides using lactic acid as mounting medium. The cell analysis needs to be done shortly (hours) after the Hoyer clearing, as the effect of this clearing dissipates and clearing needs to be redone.

6. For the earliest time points, it is not possible to determine the leaf area using a binocular. In these cases the outline of the primordium is drawn using a DIC microscope and drawing tube (*see* **Note 2**).
7. For practical reasons, we have opted to work on the abaxial epidermis of the first leaf pair as it does not have trichomes that complicate the cell measurements, and cell divisions are in transverse direction only. However, conceptually it is also possible to perform the calculations for the adaxial epidermis by either ignoring or taking into account the trichomes and accompanying cell complex. The palisade parenchyma could be analysed either by ignoring the divisions perpendicular to the surface that increase the number of mesophyl layers or by including measurements of the number of palisade cell layers or by including measurements of the number of palisade cell layers. In the spongy mesophyll, individual cell layers are not easily defined, complicating the analysis probably beyond what is feasible to work on. For addressing specific research questions it may be interesting to compare calculations on multiple cell layers.
8. In absence of a drawing tube, images of cells can also be made using a digital camera mounted to the microscope and tracing the walls on a computer using the mouse or a drawing pad. The disadvantage of this method is that some areas of the image may not be focussed optimally and/or the resolution of the camera is poor at low enough magnification to view sufficient numbers of cells, resulting in loss of accuracy.
9. Smoothing and interpolation of the primary cell-length and velocity data is an important step in the calculations. By using a local polynomial approach instead of fitting

a predefined function, no assumptions as to the overall shape of the curve needs to be made. The basis of the methods is a local quadratic fitting of a polynomial to a small section of the data, which allows smoothing the data and calculating the local differential. In Excel, this requires setting up three columns: 1. time (in days after germination/sowing), 2. the quadrate of time and 3. the natural log of average leaf area. Typically a five-point quadratic fitting is used to calculate adjusted leaf area and its derivative (*RLER*) for the mid-point. Because this is not possible for the first and last two points of the series, these are calculated from the same fit as the third from the beginning and end of the series, respectively. Calculations start from the third time point: In columns 4–7, the polynomial coefficients will be calculated using the function 'Linest'; the first argument for this function is the *y*-values for 5 points (C2:C6) and the second argument are the X and X^2-values for 5 points (A2:B6). The third and fourth arguments are 'TRUE' for the use of a constant and 'FALSE' for the output of the fitting statistics (these co-ordinates for cells are assuming that the data columns contain a row of headers). This results in the following formula: =Linest(C2:C6,A2:A6,TRUE,FALSE). To invoke the array calculation after entering, instead of pushing 'Enter', the combination 'Ctrl-Shift-Enter' needs to be used. This will write in the three selected cells the values for the coefficients a, b and c of the quadratic fitting ax^2+bx+c. This array calculation can then be copied down up to the third last point. Now it is possible to calculate the smoothed *y*-values $aX^2 + bX + c$ and RLER, the derivative $2aX + b$, in two extra columns (*see* **Note 9** for the rationale behind this smoothing approach). The first and last 2 points are calculated from their own X values and the coefficients of the fitting for the third points. When the X values are given in days, the obtained *RLER* needs to be divided by 24 to obtain the rates as $mm^2/mm^2/h$.

10. Flow cytometry can be used to analyse the (nuclear) DNA content of the cells of the leaf using DAPI as a fluorochrome. This allows estimating the relative cell-cycle phase duration during cell proliferation. Because the standard preparation involves chopping the entire leaf blade, this analysis is not restricted to a single tissue as for the case of cell measurements. Nevertheless, we found that the transitions from proliferation to expansion and from expansion to maturity as defined by the kinematic analysis are closely reflected in the DNA profile (13), so that with some caution these data can effectively be combined.

11. Optimise the scanning conditions to get sufficient contrast between the root and the background. Often opening the lid or putting a dark piece of paper between the plates and the lid improves the scannings. Modern scanners using LED light typically do not give good results.
12. Although the time-lapse observations can be done manually as described here, a number of automated image-analysis programs have been developed that can be used to analyse image sequences generated at much shorter intervals (14, 15).
13. It is well known that roots quickly respond to gravistimulus associated with their altered orientation. As this may impact the growth rate of the root, the plate has to be kept in a vertical orientation as much as possible. One way to minimise the potential effect of this perturbation is to place the microscope used for the time-lapse observations in the growth chamber and place it horizontally, so that the stage is in vertical orientation. Many microscopes have a flat back that allows this; otherwise the microscope needs to be fitted with an extra set of supports. It may also be required to make some adjustments to the stage to easily fit the plates and increase the resistance of the stage control to avoid gravity move the stage down during the observations. Although this setup is ideal, it is possible to make the observations at a horizontal stage. This requires keeping the observation time short enough to preclude effects on the growth. This can be tested by comparing the elongation rates calculated from the time lapse with those obtained from the control set of roots by marking the rate of displacement of the tip of the root. As also other experimental perturbations during the kinematic observations can influence the growth rate, it is good practice to include this quality control.
14. Several commercial photo-stitching programs are available, recent versions of Photoshop (Adobe) have a routine to do this (file/automate/photomerge) and there are also public domain ImageJ plug-ins available that can do this. When any of these routines are used make sure that there is no deformation of the images to make them fit better as this will lead to errors in the measured cell sizes.
15. ImageJ has a nice feature for this as it allows to leave one of the ends of the line selection in place while moving the other allowing to use the software like a digital calliper, which avoids systematic overlap or gaps between subsequent measurements.

16. Several problems are often encountered that make it impossible to measure all cells in a single file. Firstly, the root is often not entirely straight so that the file that is being measured rotates out of focus and another one replaces it. Secondly, there can be some problem in part of the specimen that prevents cells to be unambiguously measured. In such cases, it is a useful trick to measure the distance from the last unambiguous cell wall to the first one that is again unmistakable. This distance is then measured, but a note has to be made that this measurement does not correspond to the length of a cell. This trick also allows 'jumping' from one cortical file to the next.
17. For roots that are not growing steady-state (either accelerating or decelerating growth), additional cell-length measurements on a set of parallel grown roots need to be done at an interval of 1 or 2 days before and after the velocity measurements. Add the local rate of cell-density change into the calculation of cell-production rate: $P(x) = \frac{\mathrm{d}F(x)}{\mathrm{d}x} + \frac{\mathrm{d}\rho(x)}{\mathrm{d}t}$. Calculate this rate as the difference in cell density $\rho(x) = \frac{1}{l(x)}$ obtained from the extra cell-length measurement after the time of the velocity measurements minus the cell density of the extra measurement prior to the time of velocity measurements divided by the time difference between these two observations. Note that as the cell-length measurements are destructive these adjustments can only be done at a population level rather than for each individual root.
18. A more advanced alternative to the daily manual (ruler) measurements of leaf length is using electronic systems, based on linear velocity displacement transducers (LVDTs). Using LVDTs, the measurements of leaf extension are acquired at much higher time resolution (minutes or even seconds) (14, 15).
19. To ensure that the plants grow at steady-state, they should preferentially be grown under controlled conditions to avoid variations of leaf-elongation rate. Especially fluctuations in temperature, light intensity or humidity should be minimised, because they strongly influence growth rates (12, 13).
20. For maize, the growth zone of leaf 4 of maize inbred line B73 grown at 25°C spans about 100 mm (13), which can be used as a reference for the analysis of maize leaf growth.
21. The easiest and fastest way to handle these measurements is to directly measure the cells on the images captured 'live' by a digital camera connected to a PC running image-

analysis software. For ImageJ, plug-ins for a range of cameras are available from the program Web site. Alternatively, the same method as for the root can be used; capturing a series of overlapping picture, which are then merged together before measuring the cell lengths off-line.

22. An alternative staining protocol for nuclei visualisation is using Feulgen staining, which allows doing the analysis without the need of an epifluorescence microscopy (15).

References

1. Poorter, H. and Garnier, E. (1996) Plant growth analysis: An evaluation of experimental design and computational methods. J Exp Bot **47**, 1343–1351.
2. Goodwin, R. H. and Stepka, W. (1945) Growth and differentiation in the root tip of *Phleum pratense*. Amer J Bot **32**, 36–46.
3. Erickson, R. O. and Sax, K. B. (1956) Rates of cell division and cell elongation in the growth of the primary root of *Zea mays*. Proc Am Phylos Soc **100**, 499–514.
4. Silk, W. K. and Erickson, R. O. (1979) Kinematics of plant growth. J Theor Biol **76**, 481-501.
5. Fiorani, F. and Beemster, G. T. S. (2006) Quantitative analyses of cell division in plants. Plant Mol Biol **60**, 963–979.
6. Candela, H., Martínez-Laborda, A., and Micol, J. L. (1999) Venation pattern formation in *Arabidopsis thaliana* vegetative leaves. Dev Biol **205**, 205–216.
7. Donnelly, P. M., Bonetta, D., Tsukaya, H., Dengler, R. E., and Dengler, N. G. (1999) Cell cycling and cell enlargement in developing leaves of Arabidopsis. Dev Biol **215**, 407–419.
8. Beemster, G. T. S., De Vusser, K., De Tavernier, E., De Bock, K., and Inzé, D. (2002) Variation in growth rate between *Arabidopsis thaliana* ecotypes is correlated with cell division and A-type cyclin dependent kinase activity. Plant Physiol **129**, 854–864.
9. Beemster, G. T. S. and Baskin, T. I. (1998) Analysis of cell division and elongation underlying the developmental acceleration of root growth in *Arabidopsis thaliana*. Plant Physiol **116**, 515–526.
10. Dolan, L., Janmaat, K., Willemsen, V., Linstead, P., Poethig, S., Roberts, K., and Scheres, B. (1993) Cellular organisation of the *Arabidopsis thaliana* root. Development **119**, 71–84.
11. Erickson, R. O. (1976) Modeling of plant growth. Ann Rev Plant Physiol **27**, 407–434.
12. Ben-Haj-Salah, H. and Tardieu, F. (1995) Temperature affects expansion rate of maize leaves without change in spatial distribution of cell length. Analysis of the coordination between cell division and cell expansion. Plant Physiol **109**, 861–870.
13. Beemster, G. T. S., De Veylder, L., Vercruysse, S., West, G., Rombaut, D., Van Hummelen, P., Galichet, A., Gruissem, W., Inzé, D., and Vuylsteke, M. (2005) Genome-wide analysis of gene expression profiles associated with cell cycle transitions in growing organs of Arabidopsis. Plant Physiol **138**, 734–743.
14. Basu, P., Pal, A., Lynch, J. P., and Brown, K. M. (2007) A novel image-analysis technique for kinematic study of growth and curvature. Plant Physiol **145**, 305–316.
15. van der Weele, C. M., Jiang, H. S., Palaniappan, K. K., Ivanov, V. B., Palaniappan, K., and Baskin, T. I. (2003) A new algorithm for computational image analysis of deformable motion at high spatial and temporal resolution applied to root growth. Roughly uniform elongation in the meristem and also, after an abrupt acceleration, in the elongation zone. Plant Physiol **132**, 1138–1148.

Chapter 15

Flowering Time Control

Yvonne Möller-Steinbach, Cristina Alexandre, and Lars Hennig

Abstract

A dramatic change in the life cycle of plants is the transition to flowering, which is triggered by both environmental signals, such as temperature and photoperiod, and endogenous stimuli. The dicotyledonous annual plant *Arabidopsis thaliana* is widely used as a model organism to study how these different signals are integrated into a developmental response. The existence of a diverse collection of Arabidopsis flowering time mutants is particularly useful to understand the genetics of flowering time control. This chapter gives an overview of flowering time analysis, including protocols to measure flowering time in Arabidopsis and wheat. For Arabidopsis, the experimental design necessary to assign flowering time mutants to a specific pathway is described.

Key words: Flowering time, photoperiod, vernalization, *Arabidopsis thaliana*, grasses, annuals, biennials, perennials.

1. Introduction

Flowering plants can be divided into several categories, according to their flowering regime (*see* **Fig. 15.1A**). Annual plants flower only once in their life cycle, which is completed within 1 year. Biennial plants, on the other hand, germinate and grow vegetatively in the first year and flower only in the second. Perennial plants such as bushes and trees live for several years and can flower only once (monocarpic) or repeatedly (polycarpic) at intervals of one or more years. In temperate polycarpic species, flower initiation often takes place in autumn and flower development only in the following spring (1).

In plants, successful sexual reproduction depends on the right time to flower. The transition from vegetative to reproductive

L. Hennig, C. Köhler (eds.), *Plant Developmental Biology*, Methods in Molecular Biology 655,
DOI 10.1007/978-1-60761-765-5_15, © Springer Science+Business Media, LLC 2010

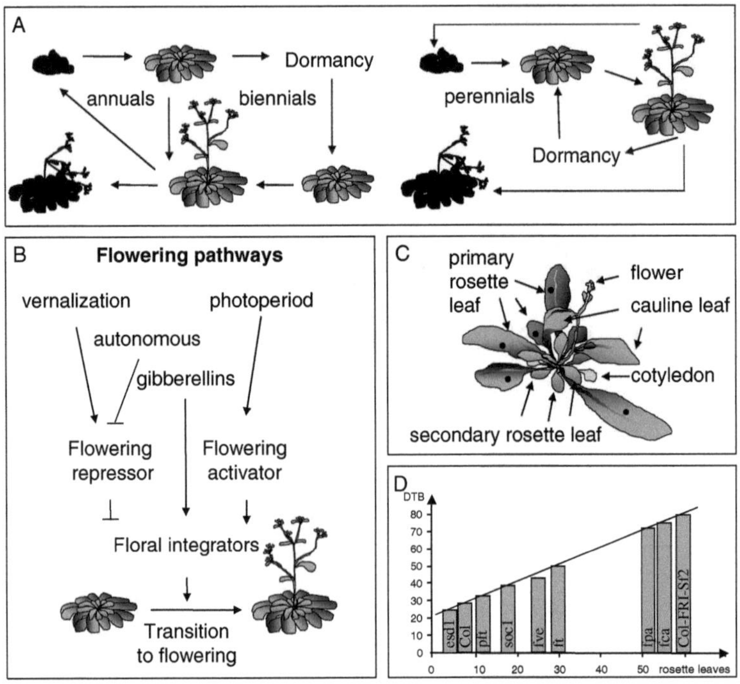

Fig. 15.1. **Control of flowering time.** (**A**) Schematic representation of annual, biennial, and perennial life cycles. Annual plants germinate, flower, and set seeds in 1 year. Biennials have an extended vegetative phase but like annuals, they senesce after flowering. Perennials on the other hand can enter dormancy after flowering, and this cycle is then repeated several times. (**B**) Major flowering time pathways in Arabidopsis (for details see text). (**C**) Guide to the various leaves that must be differentiated by the experimenter for flowering time measurements. Number of primary rosette leaves defines flowering time. Secondary leaves appear usually only after bolting. (**D**) Correlation between the number of rosette leaves and the days to bolting (DTB) as a measure of flowering time in different genotypes.

development is determined not only by environmental cues but also by endogenous signals. Plants must acquire the competence to respond to these signals, and this happens only once the plant reaches the adult phase. The mechanisms regulating transition from juvenile to adult phase are still poorly understood (2). The duration of the juvenile phase varies greatly with species, from only a couple of weeks in Arabidopsis to several years in most trees. The juvenile phase is often characterized by visible phenotypic differences in leaf morphology. In Arabidopsis, a commonly used marker for phase identity is the distribution of trichomes on the adaxial and abaxial sides of rosette leaves (3). Juvenile rosette leaves have trichomes only on the abaxial side, whereas adult rosette leaves have trichomes on both sides.

In Arabidopsis, four major genetic pathways controlling flowering time have been characterized: photoperiod, vernalization, autonomous, and gibberellin (GA) pathways (*see* **Fig. 15.1B**). Additionally, light quality, ambient temperature, and biotic/abiotic stresses can contribute to induce flowering, but they will not be discussed in detail here (for review *see* (4–7)).

Photoperiod promotes flowering in response to the day/night cycle and represents the major environmental signal regulating flowering. Although some plants are insensitive to changes in the photoperiod (day-neutral plants), in most species flowering can be induced upon exposure to a critical day-length. Plants that flower only when the day-length exceeds a critical minimum are referred to as long-day (LD) plants, and plants that flower only when the day-length falls below a critical minimum are referred to as short-day (SD) plants. Arabidopsis is a facultative LD plant that flowers earliest under inductive LD conditions, but will eventually flower under noninductive SD conditions as well. Arabidopsis mutants in the photoperiod-pathway flower late only under LD but not under SD conditions, and the late flowering phenotype cannot be suppressed by vernalization (see below) (8). Photoperiod is sensed in the leaves and promotes changes in the identity of the shoot apical meristem. In several plants, such as Arabidopsis, tomato, tobacco, and rice, the flowering time (FT) protein or its homologues moves from leaves to the shoot apex where it induces the switch to flowering. Importantly, FT can cross graft-junctions, establishing a molecular basis for the physiological florigen concept established more than 70 years ago (9, 10).

Not only is the duration of the light period important for flowering, but also the spectral composition. Different photoreceptors sensitive to far-red, red, and blue light allow the plant to perceive the light quality. Detecting variations in light quality is crucial for shade-avoidance, and has therefore an important ecological role (11). In most cases, shading by competitor plants or exposure to low red:far red ratios, as found beneath canopies, will strongly accelerate flowering (7).

Most annuals growing in temperate climates, such as Arabidopsis, barley, and wheat, developed both summer and winter varieties (12). While summer-annual Arabidopsis accessions including most laboratory strains such as Columbia and Landsberg *erecta* grow, flower, and set seeds within one spring and summer, winter-annual Arabidopsis accessions germinate in late summer, but flower only in the next spring. These latter accessions require vernalization, a prolonged cold period (corresponding to winter), in order to flower. In Arabidopsis, the major effect of vernalization is the down-regulation of the expression of the flowering repressor *FLOWERING LOCUS C* (*FLC*) (13). *FLC* is a critical flowering integrator in Arabidopsis, and its expression is maintained at intermediate levels by a group of genes that have genetically been grouped into the so-called autonomous pathway (14). It should be noted, however, that recent molecular evidence suggests that the genes of the autonomous pathway do not form a typical linear pathway, but rather act in concert to prevent excessive *FLC* expression.

Another endogenous signal affecting flowering time in Arabidopsis and many other LD rosette plants is gibberellin (10). Gibberellins are plant growth regulators, which can act directly at the level of meristem identity genes to induce flowering. In Arabidopsis, gibberellins are dispensable for flowering under LD, but essential under SD conditions.

Although wheat, barley, and rice are all grasses (Poaceae), the flowering habit of wheat and barley is more similar to *Arabidopsis* than to the closely related rice (15, 16). Both wheat and barley are LD plants, with winter varieties requiring vernalization. The switch initiating floral primordia formation determines an arrest in leaf production at the main stem, with the final number of leaves usually varying from 6 to 10 for most varieties. The extension of stem internodes and the appearance of the ear are closely connected with leaf appearance. The flag leaf appears shortly before heading (the appearance of the ear above the flag leaf). Although floral fate per se has been acquired long before, for practical reasons, heading is taken as the signal for flowering transition, and therefore, reproductive development is often quantified in terms of days to heading (occasionally, days to anthesis), normally from sowing date (17). In rice (*Oryza sativa*), the heading date at which panicles emerge from the flag leaf is also used as a measure for time of flowering. In contrast to the temperate cereals, which are LD plants, the tropical annual grass rice is a model for SD plants (18). Despite the contrasting photoperiodic requirements, rice and Arabidopsis use similar molecular strategies to respond to photoperiod (19).

In the next section, we will describe in detail how to measure flowering time in Arabidopsis and in cereals and propose a set of experiments to assess the role of Arabidopsis mutants in flowering time control.

2. Material

1. Soil.
2. Pots.
3. Greenhouse or growth chamber with the appropriate light regime (*see* **Note 1**).

3. Methods

3.1. Arabidopsis

3.1.1. Measuring Flowering Time

1. Grow a number of plants sufficient for statistical analysis (*see* **Notes 2** and **3**) under defined growth conditions (see below) until the plants bolt (*see* **Note 4**).
2. Take a note of the date and rosette diameter (*see* **Note 5**).
3. Observe your plants carefully, note any peculiar phenotype, and count the number of primary rosette leaves per plant (*see* **Fig. 15.1C, Notes 6, 7**).
4. Express the results in terms of mean and standard deviation of days to bolting and of rosette leaves formed for all plants (*see* **Table 15.1** for an example).

Table 15.1
Example of flowering time results (mean and SD) for the Columbia accession of Arabidopsis

Genotype, condition[a]	Rosette leaves	Days to bolting
Col, LD	9 ± 2	28 ± 2
Col, SD	60 ± 8	94 ± 9
Col, SD, + vernalization	30 ± 4	61 ± 5
Col *FRI*, LD	73 ± 11	79 ± 10
Col *FRI*, LD, + vernalization	12 ± 3	29 ± 4

[a]Col, Columbia; Col *FRI*, Columbia with an introgressed active *FRIGIDA* allele from the late-flowering, winter-annual Arabidopsis accession Sf-2 (21); LD, long-day photoperiods of 16 h of light; SD, short-day photoperiods of 8 h of light

3.1.2. Determining the Juvenile to Adult Phase Transition

1. Grow a number of plants sufficient for statistical analysis under defined growth conditions (see below) until bolting.
2. Define the length of the juvenile phase by counting the leaves with only adaxial trichomes and the length of the adult phase by counting the leaves with both adaxial and abaxial trichomes (*see* **Note 8**). Check carefully not to miss the first abaxial trichomes.
3. Express the result in terms of number of juvenile and adult leaves. Compare with flowering time data (*see* **Note 8**).

3.1.3. Assigning a Mutant to the Photoperiod Pathway

1. Grow mutant and wild-type plants under both LD and SD conditions (*see* **Note 1**).
2. Count the rosette leaves at bolting under both LD and SD.

A photoperiod-pathway mutant will flower late under LD, but cannot discriminate between LD and SD anymore, and therefore will flower with the same number of rosette leaves both under SD as under LD (day-neutral). Additionally, the phenotype of photoperiod-pathway mutants cannot usually be suppressed by vernalization.

3.1.4. Assigning a Mutant to the Vernalization Pathway

1. Vernalize seedlings (1–2 days after germination) for 6 weeks at 4°C (*see* **Note 9**).
2. Transfer seedlings to a growth cabinet and grow under SD (*see* **Note 10**) for 10–12 days before transfer to pots.
3. Count the number of rosette leaves at bolting. Compare mutant and wild-type before and after vernalization.

After vernalization, the wild-type should show a strong reduction in leaf number at bolting. If the mutant is impaired in the vernalization response, it will flower later than the wild-type after vernalization.

3.1.5. Assigning a Mutant to the Autonomous Pathway

1. Measure flowering time of wild-type and mutant plants under LD and SD.
2. Extract RNA and measure *FLC* expression in wild-type and mutant plants.
3. Measure flowering time of wild-type and mutant plants with and without vernalization.

Autonomous pathway mutants are extremely late-flowering under all photoperiods, but they are not day-neutral. Bona fide mutants from the autonomous pathway have increased *FLC* transcript levels. Therefore, they are extremely responsive to vernalization. Introducing an *flc*-null allele completely rescues the late flowering phenotype (20).

3.2. Cereals

3.2.1. Measuring Flowering Time in Wheat and Barley

1. Grow a number of plants sufficient for statistical analysis (*see* **Note 2**) under LD conditions.
2. Count the number of days until the appearance of the flag leaf.

3.2.2. Measuring Flowering Time in Rice

1. Grow a number of plants sufficient for statistical analysis (*see* **Note 2**) under SD conditions.
2. Count the number of days until the appearance of the flag leaf.

4. Notes

1. It is important to use the same soil for all plants in one experiment, and also for comparison between different experiments. All plants should have the same pot size to create equal humidity, water availability, and space for the plants. The light quality must be optimized by mixing cold fluorescent and incandescent light (22°C, 120–150 μmol/m^2/s) of the appropriate light regime: SD (8 h light/16 h dark) or LD (16 h light/8 h dark). Climate-controlled growth cabinets are always preferred over greenhouses.
2. Since growth space is often limiting, a reasonable compromise must be achieved. Use at least 7 (better 14) plants per genotype. For more troublesome genotypes, increasing the number of plants is generally advised.
3. Spacing and position of the pots can influence flowering time as well, so the plants should be distributed randomly. Only mutants with the same genetic background, grown at the same time, and under identical conditions are comparable, because flowering time greatly differs between ecotypes as well as with growth conditions. In reverse genetic approaches using transposon, T-DNA insertion or TILLING mutants, at least two different mutant alleles are necessary for a reliable analysis of the mutant phenotype.
4. Count rosette leaves when the inflorescence stem has 1-cm height. In this way, it is easier to distinguish between the last rosette leaf and the first cauline leaf. For very late flowering plants (flowering with $\geq$ 80 leaves), count rosette leaves before bolting. By the time plants bolt, the first rosette leaves can already be decomposed. Mark the counted leaves with color dots carefully spotted without harming the leaf (wounding can affect flowering time).
5. The number of days until bolting usually correlates well with the number of rosette leaves formed (*see* **Fig. 15.1D**), but in some mutants it may not. It is a good practice to keep track of both. Alternatively, days to anthesis, i.e., the stage when the first anther opens to release its pollen, can be used as a measure of flowering time. Note that days to anthesis and days to bolting usually correlate strongly, but that in mutants with defects in suppression of internode elongation such as in certain photoreceptor mutants, bolting is not readily recognizable.

6. At the base of the inflorescence stem, axillary meristems, which initially are inactive, can be activated to produce additional leaves, known as secondary leaves, especially under SD or in very late-flowering mutants. Secondary leaves usually appear in pairs, are smaller than the rosette leaves, and have a different orientation in relation to the inflorescence stem (*see* **Fig. 15.1C**). It is helpful to start counting the rosette leaves before bolting, as secondary leaves tend to appear mostly at/after bolting.
7. It is easier to cut away the leaves as one counts them, however, if the plants are still needed afterwards, use a marker pen and put a dot onto the leaves already counted.
8. Once the first leaf with trichomes on both sides appears (first adult leaf), all the following leaves will also be adult leaves.
9. For vernalization of Arabidopsis, a treatment of 6 weeks at 4°C is the common standard. Shorter periods can result in an incomplete effect. Note that some Nordic accessions can require considerably longer treatments (e.g., 9 weeks) for full effects. It is easiest to vernalize seeds or very young seedlings (1 or 2 days after germination) sown on MS plates. The seedlings can then be left in the dark (in a fridge or cold room) or they can be placed in any light regime. Vernalization with light is more physiological, increases the survival rate of the seedlings after vernalization treatment, and makes it easier to transfer them to soil. However, it requires a refrigerated growth room, which might not be easily available. If working with Petri dishes, then a small-scale growth cabinet can be built and simply installed in a cold room.
10. The response to vernalization can be observed under both LD and SD, but because under LD the number of rosette leaves produced is in many common laboratory accessions already very small even without vernalization, the effect of vernalization is often easier to detect under SD conditions. Avoid vernalization under LD conditions if plants will be grown under SD conditions after vernalization.

Acknowledgments

Research in the authors' laboratory is supported by grants from the Swiss National Science Foundation [3100AO-116060] and ETH [TH-16/05-2].

References

1. Ausin, I., Alonso-Blanco, C., and Martinez-Zapater, J. M. (2005) Environmental regulation of flowering. *Int J Dev Biol* **49**, 689–705.
2. Poethig, R. S. (2003) Phase change and the regulation of developmental timing in plants. *Science* **301**, 334–336.
3. Telfer, A., Bollman, K. M., and Poethig, R. S. (1997) Phase change and the regulation of trichome distribution in *Arabidopsis thaliana*. *Development* **124**, 645–654.
4. Boss, P. K., Bastow, R. M., Mylne, J. S., and Dean, C. (2004) Multiple pathways in the decision to flower: Enabling, promoting, and resetting. *Plant Cell* **16** Suppl, 18–31.
5. Komeda, Y. (2004) Genetic regulation of time to flower in *Arabidopsis thaliana*. *Annu Rev Plant Biol* **55**, 521–535.
6. Samach, A. and Wigge, P. A. (2005) Ambient temperature perception in plants. *Curr Opin Plant Biol* **8**, 483–486.
7. Thomas, B. (2006) Light signals and flowering. *J Exp Bot* **57**, 3387–3393.
8. Koornneef, M., Hanhart, C. J., and J. H. van der Veen (1991) A genetic and physiological analysis of late flowering mutants in *Arabidopsis thaliana*. *Mol Gen Genet* **229**, 57–66.
9. Zeevaart, J. A. (2006) Florigen coming of age after 70 years. *Plant Cell* **18**, 1783–1789.
10. Zeevaart, J. A. (2008) Leaf-produced floral signals. *Curr Opin Plant Biol* **11**, 541–547.
11. Franklin, K. A. and Whitelam, G. C. (2005) Phytochromes and shade-avoidance responses in plants. *Ann Bot* **96**, 169–175.
12. Laurie, D. A. (1997) Comparative genetics of flowering time. *Plant Mol Biol* **35**, 167–177.
13. Alexandre, C. M. and Hennig, L. (2008) *FLC* or not *FLC*: the other side of vernalization. *J Exp Bot* **59**, 1127–1135.
14. Simpson, G. G. (2004) The autonomous pathway: epigenetic and post-transcriptional gene regulation in the control of Arabidopsis flowering time. *Curr Opin Plant Biol* **7**, 570–574.
15. Distelfeld, A., Li, C., and Dubcovsky, J. (2009) Regulation of flowering in temperate cereals. *Curr Opin Plant Biol* **12**, 178–184.
16. Alexandre, C. and Hennig, L. (2007) *FLC*-independent vernalization responses. *Int J Plant Dev Biol* **1**, 202–211.
17. Hay, R. K. M. and Ellis, R. P. (1998) The control of flowering in wheat and barley: What recent advances in molecular genetics can reveal. *Ann Bot* **82**, 541–554.
18. Izawa, T. (2007) Daylength measurements by rice plants in photoperiodic short-day flowering. *Int Rev Cytol* **256**, 191–222.
19. Searle, I. and Coupland, G. (2004) Induction of flowering by seasonal changes in photoperiod. *EMBO J* **23**, 1217–1222.
20. Michaels, S. D. and Amasino, R. M. (2001) Loss of *FLOWERING LOCUS C* activity eliminates the late-flowering phenotype of *FRIGIDA* and autonomous pathway mutations but not responsiveness to vernalization. *Plant Cell* **13**, 935–941.
21. Lee, I. and Amasino, R. M. (1995) Effect of vernalization, photoperiod, and light quality on the flowering phenotype of Arabidopsis plants containing the *FRIGIDA* gene. *Plant Physiol* **108**, 157–162.

Chapter 16

mRNA Detection by Whole Mount In Situ Hybridization (WISH) or Sectioned Tissue In Situ Hybridization (SISH) in Arabidopsis

Yvonne Stahl and Rüdiger Simon

Abstract

Gene expression can be analyzed at high spatial resolution via RNA in situ detection methods. For many tissues and species, these will be performed on sections of embedded and fixed plant material. When very small or fragile tissues, such as embryos or roots are being investigated, whole mount methods can be employed. Protocols for both approaches are described in detail.

Key words: RNA detection, whole mount, tissue sections, nonradioactive detection.

1. Introduction

Gene expression analysis requires the detection of specific protein or RNA species in the tissue. This can be achieved using marker lines that express an easily detectable reporter from a promoter of the gene in question. However, it is first necessary to ascertain that the reporter can indeed faithfully reproduce the expression pattern seen in the organism, and in situ RNA detection remains the method of choice. In most protocols, labeled antisense RNAs are used as probes that are easy to synthesize in vitro. Depending on the tissue to be analyzed, probes can penetrate and hybridize in situ to their corresponding sense RNAs, either without prior tissue dissection (whole mount in situ hybridization, WISH) or after embedding the tissue into wax that is fine-sectioned with a microtome and fixed on slides (section in situ hybridization, SISH).

L. Hennig, C. Köhler (eds.), *Plant Developmental Biology*, Methods in Molecular Biology 655,
DOI 10.1007/978-1-60761-765-5_16, © Springer Science+Business Media, LLC 2010

The WISH protocol is mainly based on reference (1). The SISH protocol is a modification of several protocols by David Jackson, Enrico Coen, Sabine Hantke, and other former colleagues at the John Innes Center, Norwich, UK. The basic protocols can be found here (2, 3).

2. Materials and Solutions

2.1. WISH

If not otherwise stated, solutions and buffers can be stored at room temperature.

1. RNase-free water: Add 1 mL/L diethylenepyrocarbonate (DEPC), stir for 2 h, and autoclave to inactivate DEPC. DEPC treatment is not possible for solutions containing amino groups (e.g., Tris buffer, MOPS, or EDTA). These solutions have to be prepared using DEPC-treated water.
2. 10× DIG RNA labeling mix (containing 10 mM ATP, 10 mM CTP, 10 mM GTP, 6.5 mM UTP, and 3.5 mM Digoxigenin-11-UTP, Roche).
3. Ribonuclease (RNase) inhibitor (40 U/μL, Fermentas).
4. T7, T3, or SP6 RNA polymerase (10 U/μL, Roche).
5. tRNA Stock solution: 20 mg/mL tRNA from *Escherichia coli* strain W (Sigma) in DEPC-treated water. Store in aliquots at –20°C.
6. DNase I (1 U/μL, RNase-free, Roche).
7. 0.5 M ethylenediaminetetraacetic acid (EDTA), pH 8.0. Prepare in DEPC-treated water.
8. 10 M lithium chloride (LiCl). DEPC-treated.
9. 10× MOPS (3-morpholinopropanesulfonic acid) buffer: 200 mM MOPS, 50 mM sodium acetate, and 5 mM EDTA, pH 7.0. Prepare in DEPC-treated water and store protected from light (*see* **Note 1**).
10. 2× RNA-loading buffer: Mix 100 μL of formamide, 35 μL of 37% formaldehyde, 15 μL of 10× MOPS buffer, 0.02% (w/v) bromophenol blue, and 5 μL of ethidium bromide (10 mg/mL). Prepare freshly before use.
11. Fixative: 4% (w/v) Paraformaldehyde, 15% (v/v) dimethylsulfoxide (DMSO), 0.1% (v/v) Tween 20, pH 7.4 (*see* **Note 2**). Always prepare freshly before use! *Caution*: Always prepare and use paraformaldehyde in a fume hood!
12. Phosphate-buffered saline (10× PBS): 1.3 M NaCl, 70 mM Na_2HPO_4, 30 mM NaH_2PO_4, pH 7.4 and DEPC-treated. Dilute 100 mL of stock solution with 900 mL of DEPC-treated water for use. For PBST, add 0.1% (v/v) Tween 20.

13. 20× SSC: 3 M NaCl, 0.3 M sodium citrate, pH 7.0 and DEPC-treated. Dilute 100 mL of stock solution with 900 mL of DEPC-treated water for use.
14. 75% and 80% (v/v) in DEPC-treated water.
15. 50% and 25% (v/v) ethanol in 1× PBS.
16. Proteinase K stock solution: 125 mg/mL Proteinase K (Sigma) in DEPC-treated water. Store aliquots at –20°C and avoid repeated freeze/thaw cycles.
17. Glycine stock solution: Store 200 mg/mL glycine stock in DEPC-treated water in aliquots at –20°C.
18. Hybridization solution: 50% (v/v) Formamide, 5× SSC, 0.1% (v/v) Tween 20, and 0.1 mg/mL heparin in DEPC-treated water. Freshly add denatured salmon sperm DNA from stock solution (50 mg/mL in DEPC-treated water) to a final concentration of 1 mg/mL (*see* **Note 3**).
19. 50% (v/v) Formamide, 2× SSC, and 0.1% (v/v) Tween 20 in DEPC-treated water.
20. 2× SSC and 0.1% (v/v) Tween 20 in DEPC-treated water.
21. 0.2× SSC and 0.1% (v/v) Tween 20 in DEPC-treated water.
22. 1% (w/v) BSA in PBST. Prepare freshly before use!
23. 1 M $MgCl_2$ in DEPC-treated water.
24. 2 M Levamisol (Sigma) in DEPC-treated water. Store in aliquots at –20°C.
25. 10× Alkaline phosphatase (ALP) buffer: 1 M Tris pH 9.4 and 1 M NaCl. For 1× ALP buffer: add 10 mL of 10× ALP stock, 5 mL of 1 M $MgCl_2$, and 100 μL of 2 M Levamisol to 84.9 mL of DEPC-treated water freshly before use.
26. NBT stock solution: 50 mg/mL Nitroblue tetrazolim chloride (Roche) in dimethylformamide (DMF), store at 20°C.
27. BCIP stock solution: 50 mg/mL 5-Bromo-4-chloro-3-indoxylphosphate (Roche) in DMF, store at –20°C.
28. Staining solution: Add 45 μL of NBT stock solution and 35 μL of BCIP stock solution to 10 mL of 1× ALP buffer. Prepare freshly before use!
29. Clearing solution: 70% (w/v) chloral hydrate and 10% (v/v) glycerol.
30. Baskets for holding back samples (e.g., from Intavis with mesh size of 100 μm for seedlings and roots and 35 μm for embryos) in multiwell plates (e.g., 24 wells hold the medium-sized baskets), alternatively falcons and falcon strainers.
31. Optional: InsituPro VS liquid handling robot (Intavis AG).

2.2. SISH

If not otherwise stated, solutions and buffers can be stored at room temperature.

1. NBT stock: 100 mg/mL Nitroblue tetrazolium salt in 70% dimethylformamide (DMF), store at –20°C.
2. BCIP stock: 50 mg/mL 5-Bromo-4-chloro-3-indolylphosphate in DMF, store at –20°C.
3. Formamide has to be deionized using a mixed bed resin.
4. Triethanolamine: Dilute in water to 2 M and adjust the pH to 8 with HCl.
5. RNAse A stock solution: Prepare stock solution at 10 mg/mL in water. Store in aliquots at –20°C.
6. Pronase stock solution: Prepare a stock of 40 mg/mL in water. Predigest the enzyme for activation by incubating for 4 h at 37°C. Store at –20°C in small aliquots.
7. Fixative: 4% (w/v) Paraformaldehyde in PBS (*see* **Note 4**).
8. 10× PBS: 1.3 M NaCl, 70 mM Na_2HPO_4, and 30 mM NaH_2PO_4.
9. 2× Carbonate buffer: 80 mM $NaHCO_3$ and 120 mM Na_2CO_3, pH 10.2.
10. TMS-buffer: 10 mM Tris-HCl pH 7.5, 10 mM $MgCl_2$, and 50 mM NaCl.
11. 10× Pronase buffer: 0.5 M Tris-HCl pH 7.5 and 50 mM EDTA.
12. 10× Hybridization salts: 3 M NaCl, 0.1 M Tris-HCl pH 6.8, 0.1 M $NaPO_4$-buffer, and 50 mM EDTA.
13. 20× SSC: 3 M NaCl and 0.3 M Na_3citrate.
14. Wash buffer: 2× SSC and 50% formamide (no need to deionize it).
15. 10× NTE-buffer: 5 M NaCl, 100 mM Tris-HCl pH 7.5, and 10 mM EDTA.
16. 10× Buffer 1: 1 M Tris-HCl pH 7.5 and 1.5 M NaCl.
17. Buffer 2: 1× Buffer 1 with 0.5% (w/v) Boehringer blocking reagent (*see* **Note 5**).
18. Buffer 3: 1× Buffer 1 with 1% (w/v) BSA and 0.3% (v/v) Triton X100.
19. Buffer 4: Buffer 3 with anti-DIG antibody, alkaline phosphatase coupled (Roche,) in a concentration of 1:3000. Prepare freshly before use.
20. Buffer 5: 100 mM Tris-HCl pH 9.5, 100 mM NaCl, and 50 mM $MgCl_2$.
21. Buffer 6: Buffer 5 with 10% (w/v) polyvinylalcohol (MW 70.000 – 100.000, e.g., SigmaP1763) (*see* **Note 6**).

22. Glass vials for tissue incubation.
23. Incubation ovens or heating blocks.
24. Microtome.
25. Slide racks and incubation boxes.

3. Methods

3.1. WISH

Critical steps are the quality of probe, tissue fixation, and permeabilization of tissues. Concentrations of probes have to be tested empirically; also adjustments to the concentration of Proteinase K can be necessary. The manual WISH protocol (excluding probe synthesis) can be carried out in three consecutive working days; however, if you use the automated system (in situ pro liquid handling robot, Intavis AG), the protocol is usually finished within 36 h. It is advisable to use a sense probe or no probe as a negative control for the duration of staining, as background staining will occur after prolonged staining times. If possible, use the appropriate knock-out line as a control for probe specificity. In the root, only the meristematic regions of main and lateral roots can be penetrated by the probes. Embryos until the older torpedo stages can be penetrated by the probe and also cotyledons, young leaves, and shoot apical meristems of young seedlings.

3.1.1. RNA Probe Preparation

1. Prepare standard plasmid restriction digest or RT-PCR on cDNA with primers containing T7, T3, or SP6 promoter sequences at the 3′ end of your gene of interest (*see* **Note 7**).
2. Clean up PCR product or digested plasmid (containing T7, T3, or SP6 promoter) with GFX PCR DNA and Gel Band purification Kit (GE Healthcare) according to manufacturer's instructions and elute in RNase-free water.
3. Set up the following DIG-labeling reaction:
   x μL PCR product (0.5–1 μg)

   (18–*x*) μL Water (DEPC-treated)

   2.5 μL 10× Transcription buffer

   2.5 μL 10× DIG RNA-labeling mix

   1.0 μL RNase inhibitor (40 U/μL)

   1.0 μL T7, T3, or SP6 RNA polymerase (10 U/μL)
4. Incubate for 2 h at 37°C in a thermocycler (for SP6 at 40°C).
5. Add 2 μL of tRNA stock solution (20 mg/mL) and 1 μL of DNase I (1 U/μL, RNase-free) and incubate for 15 min at 37°C.

6. Transfer to ice and add 0.8 μL of EDTA (0.5 M, pH 8.0,) 1 μL of LiCl (10 M) and 75 μL of ethanol.
7. Precipitate for at least 30 min at –70°C (or > 2 h at – 20°C).
8. Centrifuge for 30 min at 16000×g, 4°C.
9. Wash pellet in 80% (v/v) ethanol.
10. Centrifuge for 10 min at 16,000×*g*, 4°C.
11. Dissolve pellet after drying in 100 μL of water (DEPC-treated).
12. Analyze 10 μL of probe plus 10 μL of 2× RNA loading buffer after incubation for 5 min at 65°C on a 1% (w/v) agarose gel containing 1× MOPS and 2.2 M formaldehyde (*caution*: add formaldehyde last to dissolved agarose in 1× MOPS buffer in a chemical fume hood!)
13. A successfully synthesized probe should be visible as a single band of the expected size. Store probe in aliquots at –70°C and avoid repeated freeze/thaw cycles.

A hydrolysis step for probe fragmentation can be carried out (see SISH,) but in our hands was not found to improve quality of WISH.

3.1.2. Tissue Fixation, Permeabilization, and Probe Hybridization (Day 1)

1. Fix plant material in a 1:1 mixture of fixative and heptane for 30–45 min (*see* **Note 8**).
2. Wash samples in methanol (two times for 5 min).
3. Wash samples in ethanol (three times for 5 min).
4. Incubate samples in 1:1 ethanol/Histoclear mixture for 30 min (*see* **Note 9**).
5. Wash samples in ethanol (two times for 5 min). (From this step onwards, samples can be processed using the liquid handling robot.)
6. Rehydrate samples in ethanol/water (DEPC-treated) (75%, v/v,) and ethanol/PBS (50%, 25%, v/v, 10 min each).
7. Incubate samples in fixative for 20 min.
8. Wash samples in PBST (two times for 10 min).
9. Incubate samples with Proteinase K (60–125 μg/mL in 1× PBS) for 15 min. For optimization, *see* **Note 10**.
10. Incubate samples with glycine (2 mg/mL) in 1× PBS for 5 min.
11. Prehybridize samples in hybridization mix for 1 h at 55°C.
12. Hybridize samples in hybridization mix with 20–100 ng/mL of denatured probe (16 h at 55°C). For preparation and optimization, *see* **Note 11**.

3.1.3. Washing and Antibody Incubation (Day 2)

Please carry out the following three wash steps at 55°C:

1. Wash samples three times in 50% (v/v) formamide, 2× SSC, and 0.1% Tween 20 for 10, 60, and 20 min each.
2. Wash samples in 2× SSC and 0.1% Tween 20 for 20 min.
3. Wash samples in 0.2× SSC and 0.1% Tween 20 for 20 min. Please carry out all following steps at room temperature:
4. Wash samples in PBST (three times for 10 min).
5. Preincubate samples in 1% (w/v) BSA in PBST for 90 min.
6. Incubate samples with antibody overnight in the dark (anti DIG-ALP conjugated antibody diluted 1:2000 in 1% (w/v) BSA in PBST).

3.1.4. Washing and Detection (Day 3)

1. Wash samples in PBST (eight times for 20 min).
2. Preincubate samples in 1× ALP buffer (two times for 10 min).
3. Stain samples in staining solution in the dark at 37°C.
4. Stop staining by two washes in DEPC-treated water.
5. Wash samples in clearing solution.
6. Mount samples in clearing solution on microscope slides. For sample preparation, *see* **Note 12**.
7. Analyze samples using Nomarski DIC optics (*see* **Note 13**).

3.2. SISH

The entire procedure, from embedding over sectioning to the hybridization steps, can be performed manually, using first small glass vessels for the tissues to be embedded, and standard heating blocks or ovens for the incubations at different temperatures. If available, an automated tissue-embedding machine should be used because this speeds up the entire procedure to less than 24 h and gives highly reproducible results. Microtome sectioning of tissue still requires many hours of work and steady hands. The hybridizations and washing steps on slides can be performed using commercially available slide racks and incubation boxes. Also robotic systems are on the market that reduce labor and overall experimental time considerably.

3.2.1. Tissue Fixation and Embedding

1. Cut tissue to 2 mm in one dimension, and less than 10 mm in all other dimensions and place into glass vial with fixative.
2. Infiltrate fixative in a desiccator connected to a vacuum pump for 10 min. Test whether the tissue sinks down after releasing the vacuum. If not, repeat vacuum treatment. When the tissue has sunk down in the vials, cool again on ice and leave it to fix overnight at 4°C.
3. Dehydrate and stain tissue in a graded alcohol series with EosinY, which facilitates sectioning. All steps are performed

in glass scintillation vials. Replace fixative with ice-cold 50% ethanol, leave for 90 min on ice; repeat step with 70% ethanol, 85% ethanol, and 95% ethanol containing 0.1% EosinY. Finally, leave overnight in 100% ethanol at 4°C.

4. Stepwise replace ethanol with tissue-clearing solution, such as Histoclear. Replace overnight ethanol with 100% ethanol and incubate for 60 min at room temperature (RT), repeat, replace with 50% ethanol/50% Histoclear, then three changes of 100% Histoclear, incubate for 1 h each time. Incubate in fresh Histoclear overnight at 45°C and add about 30% (vol/vol) paraplast embedding wax (*see* **Note 14**).
5. Melt wax pellets at 60°C and replace Histoclear/wax mixture with the freshly molten wax and incubate at 60°C. Change the wax twice daily for the next 2 days.
6. Use commercially available plastic moulds (or plastic balance trays) to create tissue blocks. Place the mould on a heating block at 60°C, pour some wax into the mould, and empty a glass scintillation vial with the tissues into the mould. The tissue can be quickly oriented in the mould using forceps that are preheated in the flame of a gas burner. When the tissue is in position, float the mould on cool water, thereby solidifying the wax. Store the embedded tissues in the fridge.

3.2.2. Sectioning

Cut small tissue blocks to size, fix to microtome holder, and section ribbons of 5–10 μm thickness. Place ribbons on coated glass slides (Superfrost Plus,) using a fine paint brush. Add sterile water so that the wax ribbons float freely. Place the slide onto a hotplate at 42°C until the sections are fully flattened. Drain off the water with a tissue paper and incubate slides on the hotplate overnight. Store at 4°C until ready for the hybridization (*see* **Note 15**).

3.2.3. RNA Probe Preparation

This method is described in detail in **Section 2.1**. For better penetration of the probes into the tissue, RNA probes are subjected to limited hydrolysis by mild alkaline treatment.

The optimum length for in situ probes is about 150 bp.

1. Add 50 μL of 200 mM carbonate buffer, pH 10.2.
2. Incubate at 60°C for the calculated length of time (*see* **Note 16**).
3. Transfer to ice. Add 10 μL of 10% acetic acid and 12 μL of 3 M sodium acetate, mix (gas bubbles should appear).
4. Add 312 μL of ethanol and incubate at –20°C for 60 min.
5. Centrifuge for 10 min, wash, air dry pellet, and dissolve in 50 μL of water and store at –20°C.

3.2.4. Tissue Pretreatments

The tissue sections require pretreatments to increase probe accessibility and reduce unspecific probe binding. Place slides into racks and pass through the following solutions:

1. 100% Histoclear, for 10 min, repeat.
2. 100% Ethanol, for 1 min, repeat.
3. 95% Ethanol, for 1 min.
4. 85% Ethanol, for 1 min.
5. 50% Ethanol, for 1 min.
6. 30% Ethanol, for 1 min.
7. Water, for 1 min.
8. 0.2 M HCl, for 10 min.
9. Water, for 5 min.
10. PBS, for 2 min.
11. Pronase (0.125 mg/mL in Pronase buffer), for 10 min.
12. Glycine (0.2% (w/v) in PBS), for 2 min.
13. PBS, for 2 min.
14. Formaldehyde (4% (w/v) in PBS), for 10 min.
15. PBS, for 2 min, repeat.
16. 1% Acetic anhydride in 0.1 M Triethanolamine pH 8.0, 10 min.
17. PBS, for 2 min.

3.2.5. Hybridization

In general, about 2 μL of the hydrolyzed probe solution should be used per slide. However, it is advisable to try hybridizations with larger or smaller amounts of probe to find an optimum probe concentration. The final hybridization mix consists of 1 part 'probe mix' and 4 parts of 'hybridization buffer.'

Probe Mix, for Each Slide:

1. 2 μL Hydrolyzed DIG-labeled RNA probe.
2. 2 μL Water.
3. 4 μL Deionized formamide.

Mix, incubate for 2 min at 80°C, and cool on ice.

Hybridization Buffer, for 25 Slides:

1. 100 μL 10× Salts.
2. 400 μL Formamide (deionized).
3. 200 μL 50% Dextransulfate.
4. 10 μL 100 mg/mL tRNA.
5. 20 μL 50× Denhardts′ solution.
6. 70 μL water.

Add 8 μL of probe mix to 32 μL of hybridization buffer, mix, distribute the 40 μL on the tissue section and cover with a

24 × 50 mm cleaned coverslip, avoiding air bubbles. Place the slides on tissue paper soaked in 2× SSC and 50% formamide in a small box. Seal the box with adhesive tape to avoid evaporation and incubate overnight at 50°C.

3.2.6. Washing

Place slides back into racks and immerse in prewarmed wash buffer at 50°C. Coverslips should slide off the slides after a few minutes.

1. Place slides into fresh wash buffer and incubate at 50°C for 60 min, repeat.
2. Wash in NTE at 37°C two times for 5 min each.
3. Incubate in NTE with 20 μg/mL RNAseA at 37°C for 30 min.
4. Wash in NTE at RT two times for 5 min each.
5. Wash in wash buffer at 50°C for 60 min.
6. Wash in PBS at RT for 5 min.

The RNAseA will digest any unspecifically bound single-stranded RNA, but will not affect the specifically bound (hybridized) and therefore double-stranded probe-RNA.

3.2.7. Detection

The hybridized probe–RNA will now be detected with an anti-DIG antibody that is coupled to alkaline phosphatase. The following steps are performed either in slide racks or in small trays to save solutions. This is recommended for the antibody incubation. Trays should be placed on a shaking platform. Trays should be changed and washed rather than just changing the solutions. All incubations are at RT.

1. Buffer 1 for 5 min.
2. Buffer 2 for 60 min.
3. Buffer 3 for 60 min.
4. Buffer 4 for 60 min.
5. Buffer 1 with 0.3% (v/v) Triton X100 four times for 20 min each.
6. Buffer 1 for 5 min.
7. Buffer 5 for 5 min.
8. Buffer 6 up to 3 days in the dark.

Buffer 6 contains the substrate for the alkaline phosphatase reaction. Incubate the slides in buffer 6 in trays with a transparent cover to avoid evaporation. This allows controlling the reaction under a microscope after 12 h. Incubations for more than 3 days will result in increased background.

1. Place slides back into slide racks.
2. Wash in water for 5 min.
3. Wash in 70% ethanol for 5 min.

4. Wash in 95% ethanol for 5 min, leave to air-dry.
5. Mount the slides by adding 2–3 drops of Entellan (or Euparal,) cover with a coverslip of suitable size, and leave to dry in the fume hood for 2 h.
6. The slides are now ready for inspection with a light microscope.

   Use a microscope that is equipped with Nomarski optics. Very faint signals can be more easily detected under dark-field illumination.

4. Notes

1. Discard if the 10× MOPS buffer turns yellow.
2. Prepare a 10% (w/v) paraformaldehyde stock solution freshly in DEPC-treated water. This solution requires careful heating and drop wise addition of 1 N NaOH to dissolve. Ensure that pH is around 7.4. Then dilute the 10% (w/v) paraformaldehyde solution to prepare fixative.
3. Prepare 50 mg/mL salmon sperm DNA stock. For denaturation, heat 50 mg/mL salmon sperm DNA stock in DEPC-treated water to 100°C for 10 min and cool immediately on ice.
4. A solution of 4% (w/v) formaldehyde, freshly prepared from parafomaldehyde, is used as fixative. Paraformaldehyde and formaldehyde solutions and vapor are toxic, so all steps involving these chemicals should be handled in a chemical fume hood. In addition, due to instability, all solutions should be freshly prepared just before use. Take 100 mL of PBS-buffer, pH 6.57, and add a small pellet of NaOH. The pH will increase to about pH 11, then heat in the microwave to 70°C, add 4 g of paraformaldehyde, and shake vigorously until dissolved. Cool on ice and bring the pH to 7 by adding H_2SO_4. Finally, add 30 μL of Tween 20. The fixative is now ready for use. Aliquot the fixative into small glass scintillation vials.
5. Dissolve at 60–70°C for 1 h on a heated stirrer. The solution remains turbid. Buffer 2 can be stored in aliquots at −20°C.
6. Dissolve the polyvinylalcohol by boiling the solution on a heated stirrer. Let it cool down and then add 1.5 μL of NBT and 1.5 μL of BCIP/mL. Make shortly before use.
7. Complete cDNA regions or smaller regions of the gene of interest can be used as a probe. We had good results with probes varying in size from 250 to 2000 bp.

8. Tissue fixation requires optimization. We found fixation of, e.g., seedling roots for 30–45 min in 1:1 mixture of fixative and heptane to be sufficient. For fixation of embryos: dissect embryos from ovules on a slide using preparation needles in fixative without heptane and fix after all embryos are collected for 2 h in a 1:1 mixture of fixative and heptane by applying vacuum for 10 min.
9. Ensure that plastic containers used are withstanding Histoclear treatment, otherwise use glass vessels for this step.
10. This step may require optimization. Use as a starting point, e.g., for seedling roots, 60 μg/mL and embryos, 125 μg/mL Proteinase K.
11. Prepare desired probe dilution in advance in hybridization buffer and heat for 10 min at 65°C, then transfer to ice immediately for denaturation. The required probe dilution has to be tested for every probe due to mRNA abundance, probe quality, and tissue accessibility.
12. Great care has to be taken when handling the samples as they are extremely fragile. Prepare slides with four dried drops of nail polish not to directly rest the coverslip on the samples as they get damaged very easily.
13. Analyze the samples as soon as possible, but sometimes one or two days after mounting of samples the tissue clearing is giving better quality images. The staining (depending on its original strength) usually withstands one or two days (sometimes even longer) in clearing solution.
14. The wax (Paraplast, or other brand names) to be used also contains plastic polymers and DMSO that shall facilitate the infiltration and sectioning. These additives are unstable at temperatures higher than 62°C. The Paraplast will solidify at 56–58°C. Be very careful to keep the temperature of the wax always at 60°C! All changes of wax and handling of the embedded materials have to be done quickly.
15. Common problems with sectioning are:
    a. Sections break up or appear brittle: Material was not properly embedded or wax was destroyed by overheating.
    b. Sections split along the ribbon: Blade is chipped or dirty-clean or replace the blade.
    c. Ribbons are not straight: Wax block is not rectangular or the long side of the block is not parallel to the blade.
    d. Sections roll up, no ribbon is formed: Change the angle of the blade.
    e. Ribbon forms, but the entire ribbon rolls up or sticks to the blade: The blade is electrostatically charged; wipe it with a wet Kleenex.

16. Hydrolysis time is calculated as follows:

$$t = \frac{\mathbf{L_i} - \mathbf{L_f}}{K \times \mathbf{L_i} \times \mathbf{L_f}}$$

t = time (min)
K = rate constant (= 0.11 kb/min)
L_i = initial length (kb)
L_f = final length (kb)
Example: If your cloned DNA fragment to be transcribed is 1.5 kb, the hydrolysis time will be:

$$t = \frac{1.5 - 0.15}{0.11 \times 1.5 \times 0.15} = 54.5 \text{ min}$$

References

1. Hejatko, J. (2006) In situ hybridization technique for mRNA detection in whole mount Arabidopsis samples. *Nat Protoc* **4**, 1939–1946.
2. Jackson, D. P. (1991) In situ hybridisation in plants. In: Molecular Plant Pathology: A Practical Approach, Bowles, D. J., Gurr, S. J., and McPherson, M., eds. Oxford University Press, England.
3. Coen, E. S., Romero, J. M., Doyle, S., Elliott, R., Murphy, G., and Carpenter, R. (1990) *floricaula*: A homeotic gene required for flower development in *Antirrhinum majus*. *Cell* **63**, 1311–1322.

Chapter 17

Immunolocalization of Proteins in Plants

Michael Sauer and Jiří Friml

Abstract

Rapid advances in the field of plant biology, especially in plant cell biology, have created the need for methods that allow the localization of proteins in situ at subcellular resolution. Although in many cases recombinant proteins with fluorescent proteins can fulfill this task, antibody-based immunological detection of proteins is a complementary technique, which avoids the risk of inducing side effects by a fusion protein, such as misexpression, mistargeting, altered stability, or toxicity. Moreover, recombinant protein techniques are applicable only to a rather limited set of model plants. The immunolocalization protocols presented here can be used to display protein localization patterns in different tissues of various plant species. This chapter describes a whole mount immunolocalization protocol, which has been extensively used in Arabidopsis roots and some above-ground tissues, and that also works in other species. Additionally, for bulky or hard tissue types, a variation of this protocol for paraffin-embedded sections is given.

Key words: Immunocytochemistry, immunolocalization, antibody-based detection, fluorescence microscopy.

1. Introduction

In the last decade, our knowledge on plant gene expression has grown tremendously for some selected model species. Much is now known about the expression of Arabidopsis genes under a wide variety of conditions and tissues, and other model species are quick to follow suit. While data on gene expression can tell us much about genetic interactions and networks, in order to assess protein function, interaction, or activity, other, complementary methods are needed. One crucial element is the reliable analysis of protein localization at cellular and subcellular resolutions. While visualization involving recombinant DNA techniques with

L. Hennig, C. Köhler (eds.), *Plant Developmental Biology*, Methods in Molecular Biology 655,
DOI 10.1007/978-1-60761-765-5_17,

fluorescent proteins such as GFP can go a long way (for a review, see reference (1)), there are certain limitations to this approach. First, recombinant DNA techniques are not available for many plant species, and second, the function and localization of these proteins may not be identical to the wild-type equivalent. This is especially true for fusions to larger reporter proteins, such as GFP or GUS. Immunolocalization of endogenous proteins can be a potent complementary approach to overcome these problems.

Here, we present a relatively simple and rapid protocol for immunolocalization of proteins in plant tissues with high resolution and specificity. A basic whole mount protocol is given, along with three variations to account for different tissue types. The method is based on previously described protocols (2–4).

2. Materials

2.1. Basic Whole Mount Protocol

1. 10 × PBS Buffer: For 1 L, 2 g KCl, 80 g NaCl, 17.8 g $Na_2HPO_4 \cdot 2H_2O$, and 2.4 g KH_2PO_4 in water (*see* **Note 1**). This solution can be autoclaved and stored for many months. Prior to use, dilute to 1 × and check if pH is 7.4. If not, adjust with KOH or HCl. (*see* **Note 2**).
2. Fixative solution: 4% Paraformaldehyde (PFA) in PBS. To prepare, weigh PFA powder (in the fume hood and wearing gloves, PFA is highly toxic and the powder very fine) and add to PBS. Adding some KOH pellets to increase the pH will facilitate dissolving. After PFA powder is completely dissolved, readjust pH to 7.4. The fixative is best prepared freshly, but can be also aliquotted and stored frozen at –20°C for some months.
3. Driselase solution: 2% Driselase powder (Sigma) in PBS. Driselase is a cocktail of cell-wall degrading enzymes. The powder will not dissolve, instead, vortex vigorously and centrifuge or let sediment. Use only the supernatant. Prepare this solution freshly prior to use. (*see* **Note 3**).
4. Permeabilization solution: 10% Dimethylsulfoxide (DMSO) and 3% IGEPAL CA-630 (old name: NP40) in 1× PBS. (*see* **Note 4**).
5. Blocking solution: 2% Bovine serum albumin fraction V (BSA) in PBS.
6. Primary antibody solution: Primary antibody is diluted in blocking solution, concentration has to be empirically determined (*see* **Note 5**).

7. Secondary antibody solution: Fluorescently labeled antibody against the host of the primary antibody is diluted in blocking solution at a concentration of typically 1:600 (*see* **Note 6**).
8. Optionally, as nuclear counter stain, 1 mg/mL 4',6-diamidino-2-phenylindole dihydrochloride (DAPI) aqueous stock solution (can be kept at –20°C), diluted 1:1000 in PBS.
9. Adhesively coated glass slides, such as Super Frost Plus or Super Frost Plus Ultra (Menzel Gläser, Germany).
10. Liquid repellent marker pen (Pap-pen).
11. Humid chamber (a box with PBS wetted papers on the bottom and a grate to place the slides with a tightly sealing lid. This is important for longer incubation steps).
12. Mounting solution: There are a number of commercial variants, some of which contain 'antifade' reagents, which claim to increase the stability of the fluorophore. We prefer mounting solutions with nonhardening formulations, such as Citifluor AF1. A more economical, while slightly less efficient alternative is a mixture of 90% glycerol, 10% PBS, and 25 mg/mL 1,4-diazabicyclo[2.2.2]octane (DABCO), at pH 9.0.
13. Optionally, a box for washing slides. A box has advantages over washing directly on the slides, as the buffer volume is much larger and the number of washing steps can be reduced. The box can be placed on a gentle shaker/rocker. However, some times the material does not adhere well to the slides, in this case, rather wash directly on the slides.

2.2. Arabidopsis Embryos

Additionally to the material in **Section 2.1**:

1. Glass Pasteur pipettes and small glass tubes or jars.
2. Double-sided adhesive tape.
3. Fine tweezers and fine syringe needles.
4. Stereomicroscope with transmitted light illumination (optional).
5. Liquid nitrogen.

2.3. Arabidopsis Above-Ground Tissues

Additionally to the material in **Section 2.1**:

1. Ethanol (EtOH) of histological quality: Pure and diluted to 75, 50, and 25% in water.
2. Methanol (MeOH): Pure and mixed 50% with EtOH.
3. Xylene: Pure, mixed 50% with EtOH, and mixed 50% with MeOH.
4. Liquid nitrogen.

2.4. Paraffin-Embedded Sections

Not needed from the material list in **Section 2.1**: Driselase and permeabilization solution.

Additionally to the material in **Section 2.1**:

1. Wax solution: 90% PEG 400 distearate and 10% 1-hexadecanol. Melt PEG 400 distearate at 65°C, add 1-hexadecanol, and stir for 3 h. This wax solution will be needed pure, and diluted with EtOH, to wax concentrations of 75, 50, and 25%. Prepare the wax just before the embedding step and keep at 37°C.
2. Microtome and water bath with 25°C (preferentially next to the microtome).
3. Ethanol (EtOH) of histological quality, pure and dissolved to 90, 75, 50, and 25% in PBS.
4. As an alternative fixative instead of paraformaldehyde, ice cold MeOH:acetic acid (3:1) can be used.

3. Methods

The immunolocalization protocol relies on the recognition of the protein of interest by a primary antibody, which itself is then recognized and bound to by a secondary antibody. The secondary antibody is usually conjugated with a fluorophore, which can be detected by fluorescence microscopy.

Cell walls and membranes of plant cells prevent the free diffusion of antibodies inside the tissue; therefore, the cells have to be made permeable. In a first, enzymatic step, the cell wall is degraded; then, membranes are made more permeable by a detergent treatment. To reduce background by unspecific binding, the specimen is blocked with an excess of nontarget protein, usually albumin, but fat-free milk powder can be used instead. After incubation with both primary and secondary antibodies, the samples have to be washed extensively to remove unspecific-bound antibodies, which would otherwise lead to increased background.

Here we present a basic protocol for whole mount immunolocalization, along with three variations for Arabidopsis embryos, (young) above-ground tissue, and a protocol for paraffin-embedded sections. The basic principle is the same in each case, only the sample preparation is different to account for the various tissue types.

Paraffin sections are undoubtedly more demanding to do, as sectioning with a microtome is definitely a matter of experience and also a time-consuming process. However, for some tissues, especially more mature or bulky types, this is the only possibility.

It must be noted that the protocol leaves a lot of room for modifications, the authors are mainly working on Arabidopsis and probably many parameters can be changed to improve the method for other species and tissues. Also, in many steps, the timing is not absolutely critical.

In order to facilitate troubleshooting the method, here are some suggestions for controls.

The best control is a knock-out mutant, which lacks the protein of interest, or an over-expressing line, which should give more and/or ectopic signal.

Also, it is advisable to check the expression of the gene by an alternative method, like *promoter::GUS* fusions or in-situ hybridization and compare the results to what you get from the immunolocalization. Keep in mind that the immunolocalization may not work equally well in all tissue types of your specimen.

To test for excessive background caused by the secondary antibody, you can include a sample without the primary antibody. This is especially recommended if you are unfamiliar with immunolocalization, as it gives you a feel for the kind of background you have to expect.

If you want to test a new primary antibody, include also a sample incubated with preimmune serum of that particular host animal instead of primary antibody.

One frequent source of high background problems is that samples fell dry once during the procedure. Make sure that during longer incubation steps, the samples remain covered in the respective solution.

If you suspect that the primary antibody is not specific for your protein of interest, for example, because there are highly similar proteins, you can try to purify the antibody by affinity-chromatography with an activated sepharose column to which the antigen has been covalently bound.

These protocols, especially the whole mount ones, can be converted to certain liquid-handling robots, such as the InsituPro provided by Intavis AG, Germany (www.intavis.com).

3.1. Basic Whole Mount Protocol

This basic protocol has been successfully used for Arabidopsis, pea, and tobacco roots (the latter two were hand-cut in half longitudinally after Step 2). We expect that this protocol will work for vibratome sections as well; however, we have not tested this. For Arabidopsis embryos, as well as for Arabidopsis above-ground tissue, there are some changes to the basic protocol, which are given step-by-step in **Sections 3.2** and **3.3**.

1. Fix tissue by immersing in fixative solution. As guidance, Arabidopsis roots are fixed for 1 h at room temperature. Other tissues may require shorter or longer times. Infiltration with vacuum can help in cases where the intercellular spaces are air-filled, like in most above-ground tissues. Also, if there are problems with floating of the material, 0.1% Triton X 100 can be added to the solution. This step can be performed in microwell plates or microcentrifuge tubes, according to the size of the specimen.

2. Wash material 3 times in PBS for 10 min.
3. Wash material 2 times in water for 10 min.
4. Mount material on adhesive slides. Put a droplet of water on the slide and arrange the material on it. Let the slides dry well, either overnight at RT (room temperature) or for 1 h at 37°C. Drying is critical for good adhesion. The dried slides can be stored at –20°C for several days and processed later.
5. Surround the area of the sample with the liquid repellent marker pen to create a hydrophobic barrier around your sample. This way, solutions stay in place and the volumes used can be much smaller.
6. Rehydrate the samples by pipetting PBS on them and incubate for at least 10 min at RT.
7. Driselase treatment: Prepare driselase solution; remember to only use the supernatant. Pipette a sufficient amount (for one slide, typically between 100 and 200 μL) of the supernatant on each slide and incubate in the humid chamber at 37°C for 30 min to 1 h (*see* **Note 7**).
8. Wash slides 4 times with PBS for 10 min.
9. Pipette 100–200 μL of permeabilization solution to each slide and incubate in the humid chamber for 1 h at RT.
10. Wash extensively with PBS, at least 6 times for 10 min.
11. Pipette 100–200 μL of blocking solution on each slide and incubate in the humid chamber for 2 h at 37°C. Alternatively, blocking can be done overnight at 4°C.
12. Carefully remove blocking solution and apply primary antibody solution. Incubate for at least 4 h at 37°C in the humid chamber; alternatively, incubation can be done longer at lower temperatures, e.g., overnight at 4°C. We generally recommend incubation at 37°C, though.
13. Wash extensively with PBS, at least 6 times for 10 min.
14. Pipette secondary antibody solution onto each slide. Incubate for 3 h at 37°C in the humid chamber.
15. A) If you do not want to counter-stain nuclei, wash 6 times with PBS, for 10 min each, and continue to Step 16.
    B) If you want to counter-stain nuclei, wash 3 times with PBS for 10 min, then pipette DAPI solution on each slide, and incubate for 15 min at 37°C. After this, wash 4 more times with PBS, 10 min each.
16. Remove PBS and pipette mounting solution on each slide. For a 24 × 60-mm coverslip, use about 75 μL; for a 22 × 22-mm coverslip, usually 30 μL is enough. Carefully place the coverslip over the specimen and take care to prevent air-bubble formation.

17. Observe the samples as soon as possible. While it might be possible to store the samples at 4°C or –20°C for several weeks or even months without dramatic loss of signal quality, the signal is always better immediately after the procedure. This is especially true for FITC-conjugated secondary antibodies (*see* **Note 6**).

3.2. Arabidopsis Embryos

For Arabidopsis embryos, the basic whole mount procedure is followed, with a couple of changes:

1. The embryos are fixed while inside the developing seeds. For this, harvest seeds (tape siliques onto double adhesive tape and open them along the replum. Thus, you can easily open up the two valves of the silique and carefully remove the seeds). Transfer them to fixative solution supplemented with 0.1% Triton X 100. We recommend the use of small glass tubes or snap-lid jars instead of plastic microcentrifuge tubes, because the seeds are prone to stick to plastic surfaces. While collecting seeds, keep the tube on ice. When you are done collecting, vacuum infiltrate and keep at RT for 1 h. Continue with Steps 2 and 3 of the basic protocol, but return to this section for the mounting.
2. For mounting, transfer some (20–40) seeds into a small drop of water on an adhesive slide, using a glass Pasteur pipette. Cover with a 22 × 22-mm coverslip and carefully apply pressure on it with a pointed instrument on each ovule to release the embryo. A transmitted light stereoscope helps to monitor the process. It also helps to cut/damage the seeds before mounting the coverslip with fine syringe needles, especially for very young embryos.
3. To remove the coverslip, immerse the slide in liquid nitrogen for 15 s. Remove and immediately crack apart the coverslip (insert a scalpel or razorblade between slide and coverslip and pry apart), while the water is still frozen. Then let sample dry as in Step 4 of the basic protocol and continue with the basic protocol starting from Step 5.

3.3. Arabidopsis Above-Ground Tissues

This protocol has been successfully used for Arabidopsis hypocotyls, young primary leaves, and cotyledons. It includes several steps to simultaneously remove cuticular waxes and chlorophyll.

1. Carry out Steps 1 and 2 of the basic protocol; use microcentrifuge tubes and vacuum infiltration for Step 1.
2. Remove PBS and add pure MeOH, incubate 10 min at 37°C, and repeat this step at least 2 times.
3. Remove MeOH, pipette EtOH/xylene (1:1), and incubate for 10 min at 37°C, repeat 2 times.

4. Remove EtOH/xylene, pipette pure xylene, and incubate for 10 min at 37°C, repeat 2 times.
5. Remove xylene, pipette EtOH/xylene (1:1), and incubate for 10 min at RT, repeat 1 time.
6. Remove EtOH/xylene, pipette 96% ethanol, and incubate for 10 min at RT, repeat 1 time.
7. Rehydrate gradually in an EtOH series of 75, 50, and 25% EtOH in water, each step for at least 5 min at RT.
8. Wash material with water 2 times for 5 min at RT.
9. Mount material on adhesive slides. Put a droplet of water on the slide and arrange the material on it. Cover the material with a coverslip and immerse the slide in liquid nitrogen. Remove and let thaw, then freeze again. Repeat freezethawing for at least 5 times (*see* **Note 8**).
10. To remove the coverslip, immerse the slide in liquid nitrogen for 10 s. Remove and immediately crack apart the coverslip (insert a scalpel or razor blade between slide and coverslip and pry apart), while the water is still frozen. Then let sample dry as in Step 4 of the basic protocol and continue with the basic protocol starting from Step 5.

3.4. Paraffin-Embedded Sections

1. Fix tissue by immersing in fixative solution and infiltrate with vacuum for 1 h at RT. If there are problems with floating of the material, 0.1% Triton X 100 can be added to the solution. Use microwell plates or microcentrifuge tubes, according to the size of the specimen. Alternatively, tissue can be fixed by incubation in MeOH:acetic acid (3:1) for several hours at –20°C.
2. Wash material 4 times in PBS for 10 min.
3. Dehydrate specimen in an EtOH series of 25%, 50%, 75%, and pure EtOH, each step for 1 h at RT. At any step, the procedure can be interrupted and continued the next day.
4. Replace EtOH with fresh EtOH and incubate 10 min at 37°C
5. "Paraffinize" specimen in a gradual series of 25%, 50%, 75%, and pure wax solution, each step at least for 1 h at 37°C. Proper infiltration with wax is crucial for successful sectioning. Depending on the bulk and density of your specimen, it may be necessary to prolong each incubation step considerably (for up to 24 h each step).
6. In an appropriate prewarmed container, such as a small Petri dish, arrange the specimen for embedding, fill up with pure wax solution, and let solidify at 4°C. Samples can be stored at –20°C for several days to weeks.

7. For sectioning, follow the instructions of the microtome supplier. Adjust section thickness to 6–8 μm and use a knife appropriate for paraffin-embedded material.
8. Place strips of sections on a 25°C water bath and let them straighten out a bit, then transfer them to adhesive slides and let them dry completely, e.g., for 30 min at 37°C or overnight at RT. In this form, the samples can be stored at –20°C for several days to weeks.
9. For dewaxing and rehydration, incubate the slides in pure EtOH, 90% EtOH, 50% EtOH, 25% EtOH (all in PBS), and 2 times in pure PBS, each step for 10 min at RT.
10. Pipette 100–200 μL of blocking solution on each slide and incubate in the humid chamber for 30 min h at RT.
11. Carefully remove blocking solution and apply primary antibody solution. Incubate at least 2 h at 37°C in the humid chamber, alternatively, incubation can be done longer at lower temperatures, e.g., overnight at 4°C.
12. Wash extensively with PBS, at least 6 times for 10 min.
13. Pipette secondary antibody solution onto each slide. Incubate for 1–2 h at RT in the humid chamber.
14. A) If you do not want to counter-stain nuclei, wash 6 times with PBS, for 10 min each, and continue to Step 16.
    B) If you want to counter-stain nuclei, wash 3 times with PBS for 10 min, then pipette DAPI solution on each slide and incubate for 10 min at 37°C. After this, wash 4 more times with PBS, 10 min each.
15. Remove PBS and pipette-mounting solution on each slide. For a 24 × 60 mm coverslip, use about 75 μL. Carefully place the coverslip over the specimen and take care to prevent air bubble formation.
16. Observe the samples as soon as possible. While it is often possible to store the samples at 4°C or –20°C for several weeks or even months without dramatic loss of signal quality, the signal is always better immediately after the procedure. This is especially true for FITC-conjugated secondary antibodies (*see* **Note 6**).

4. Notes

1. Throughout this chapter "water" refers to bi-distilled or reverse osmosis water.
2. Alternatively, another buffer can be used instead of PBS, in some cases it may yield better results. 10 × Microtubule Sta-

bilizing Buffer (10 × MTSB): for 1 L: 15 g PIPES, 1.9 g EGTA, 1.32 g $MgSO_4 \cdot 7H_2O$, and 5 g KOH. Adjust pH to 7.0 before autoclaving and dilute to 1-fold prior to use.

3. Required driselase concentrations may vary, depending on the tissue type. Also, there is significant batch-to-batch variation, so optimal concentration and incubation time have to be determined empirically. Instead of driselase, which is relatively "mild," it is also possible to use other cell-wall degrading enzymes, such as macerozyme, cellulase or pectolyase, or mixes of these. Protocols for generating protoplasts may be helpful to find enzymes that work (5). The aim is to digest cell walls just enough to facilitate antibody penetration, but still preserve tissue integrity.
4. The optimal IGEPAL concentration is, similar to driselase concentration, a matter of experimentation, but 3% is a good starting point for most applications.
5. Primary antibody concentration determines to a great extent the success of your experiment. To determine the optimal concentration of a new antibody, it is advisable to check a spectrum of at least 4 dilutions, like 1:100, 1: 200, 1:400, and 1:1000. Sometimes, even higher antibody concentrations have to be used.
6. Choice of the fluorophore for the secondary antibody must be based on the features of your microscope (filters, lasers, etc.) and also the autofluorescent properties of the tissue. Quite often, plant tissues exhibit autofluorescence in the yellow-greenish part of the spectrum, when excited with blue or ultraviolet light, which in some cases may hinder the use of a green fluorescent dye. In our experience, CY3 has been a good all-round red fluorophore, which can be combined with strong green dyes, such as ALEXA 488 or FITC, or far-red dyes, such as CY5 for colocalization studies. Be aware that different fluorophores have different chemical properties, like sensitivity to pH and half-life, strongly depending on the mounting solution used. For example, in our hands, FITC fluorescence is initially bright and contrasty, but after a few days, the signals become diffuse. If you are in doubt, try several different secondary antibodies and settle for the one, which gives best results.
7. Incubation times have to be determined empirically. As a guideline, Arabidopsis roots require 30–40 min.
8. This repeated freezing and thawing serves to create small cracks and fissures in the material, which helps antibody penetration. However, at the same time it degrades tissue integrity. You will have to weigh these two factors against each other.

Acknowledgments

This work was funded by Human Frontiers Research Organization Long Term Fellowship to M.S.

References

1. Brandizzi, F., Fricker, M., and Hawes, C. A. (2002) A greener world: The revolution in plant bioimaging. *Nat Rev Mol Cell Biol* 7, 520–530.
2. Lauber, M. H., Waizenegger, I., Steinmann, T., Schwarz, H., Mayer, U., Hwang, I., Lukowitz, W., and Jürgens, G. (1997) The Arabidopsis KNOLLE protein is a cytokinesis-specific syntaxin. *J Cell Biol* 139, 1485–1493.
3. Friml, J., Benková, E., Mayer, U., Palme, K., and Muster, G. (2003) Automated whole-mount localization techniques for plant seedlings. *Plant J* 34, 115–124.
4. Paciorek, T., Sauer, M., Balla, J., Wisniewska, J., and Friml, J. (2006) Immunocytochemical technique for protein localization in sections of plant tissues. *Nat Protoc* 1, 104–107.
5. Yoo, S. D., Cho, Y. H., and Sheen, J. (2007) Arabidopsis mesophyll protoplasts: A versatile cell system for transient gene expression analysis. Nat Protoc 2, 1565–1572.

Chapter 18

Detection of Small Non-coding RNAs

Tamas Dalmay

Abstract

Gene expression is regulated at several levels in plants, and one of the most recently discovered regulatory layers involve short RNAs. Short RNAs are produced through several pathways and target either mRNAs or genomic DNA. Different classes of short RNAs have slightly different sizes and detection of their accumulation is an important step in validating and studying non-coding short RNAs. Northern blotting is routinely used to detect short RNAs because it gives information about both the amount and size of the analysed short RNAs. Choice of the right RNA extraction protocol is crucial when short RNAs are being studied, because several routinely used commercial RNA extraction kits do not yield any short RNAs. This chapter describes optimised RNA extraction methods, which give good yields of short RNAs, and separation, transfer and hybridisation protocols to study the accumulation of short RNAs.

Key words: Short non-coding RNAs, microRNAs, ta-siRNAs, heterochromatin siRNA, gene-expression regulation, gene silencing, RNAi, RNA silencing.

1. Introduction

It became apparent in the last 10 years that endogenously expressed short non-coding RNAs play an important role in regulating gene expression in plants (1). Short RNAs regulate developmental processes (2) as well as responses to environmental changes (3). There are at least four classes of plant short RNAs: 1, microRNAs (miRNAs), which are mainly 21-nt long, are generated by Dicer Like (DCL) 1 from hairpin-structure precursor RNA and target mRNAs (1); 2, trans-acting small interfering RNAs (ta-siRNAs), which are mainly 21-nt long, are generated by DCL4 from RDR6-synthesised double-stranded RNA (dsRNA) that is produced after an miRNA cleaves the primary non-coding

L. Hennig, C. Köhler (eds.), *Plant Developmental Biology*, Methods in Molecular Biology 655,
DOI 10.1007/978-1-60761-765-5_18, © Springer Science+Business Media, LLC 2010

TAS RNAs (4, 5); 3, natural-siRNAs (nat-siRNAs), which exist in two forms: primary nat-siRNAs (24-nt long, generated by DCL2 from dsRNA formed by complementary overlapping 3' untranslated regions of mRNAs) and secondary nat-siRNAs (21-nt long, produced by DCL1 from RDR6-generated dsRNA following the cleavage by the primary nat-siRNA) (6); and 4, heterochromatin siRNAs, which are mainly 24-nt long and are generated by DCL3 from dsRNA synthesised by RDR2 on RNAs derived from transposons and other repeat elements (7). Efficient, quantitative and reliable detection of short RNA accumulation in various plant tissues is important to characterise these molecules and understand their biological role.

2. Materials

1. Lysis buffer (10×): 1 M glycine, 100 mM EDTA and 1 M NaCl. Wrap up the bottle to avoid direct sunlight.
2. Extraction buffer A: Mix 7 volume of sterile water, 1 volume of 10× lysis buffer and 2 volume of 10% SDS (sodium dodecyl sulphate).
3. Extraction buffer B: 2% CTAB (cetyltrimethylammonium bromide), 20 mM Na_2EDTA, 0.2 M boric acid, 0.8 M NaCl, adjusted to pH 7.6 with Tris and 1% 2-mercaptoethanol added just before use.
4. SSTE buffer: 1.0 M NaCl, 0.5% SDS, 10 mM Tris-HCl (pH 8.0) and 1 mM Na_2EDTA.
5. FDE buffer: Mix 10 mL of formamide (deionised), 200 μL of 0.5 M EDTA (pH 8.0), 10 mg of xylene cyanol FF and 10 mg of bromophenol blue.
6. TBE (10×): 108 g/L Tris base, 55 g/L boric acid and 5.84 g/L EDTA.
7. 50 mL of 15% Denaturing polyacrylamide gel in TBE buffer: Mix 25 g of urea, 5 mL of 10× TBE, 18.75 mL of 40% acrylamide (38:2=acrylamide:bis-acrylamide) and 10 mL of sterile water. Heat up in microwave oven, but do not let it boil. When the polyacrylamide is completely dissolved, add more water until the total volume is 50 mL. Just before pouring the gel, add 300 μL of 10% APS (ammonium persulphate) and 20 μL of TEMED and pour gel immediately. Polymerization time is about 30 min.
8. 50 mL of 15% Denaturing polyacrylamide gel in MOPS buffer: First prepare 10× MOPS (200 mM) buffer. To make 250 mL of 200 mM MOPS/NaOH pH 7.0, dissolve

11.56 g of MOPS in 200 mL of sterile water. Adjust the pH to 7.0 with NaOH. Make up to 250 mL with sterile water. Filter sterilise (DO NOT autoclave) and cover with foil. Discard if it turns intensively yellow; light yellow (straw-like) coloured is still usable. Preparation of the gel is similar to the TBE gel. The difference is that initially you mix: 25 g of urea, 5 mL of 10× MOPS, 18.75 mL of 40% acrylamide (38:2=acrylamide:bis-acrylamide) and 10 mL of sterile water, then the protocol is the same.

9. 12-mL crosslinking solution (enough for 2 Bio-Rad Mini-Protean II gels): Mix 10 mL of sterile water and 122.5 μL of 12.5 M 1-methylimidazole. Adjust the pH to 8.0 by the addition of 1 M HCl (normally only a few drops). Dissolve 0.373 g of EDC [l-ethyl-3-(3dimethylaminopropyl) carbodiimide] in the methylimidazole, pH 8.0, solution. Make the volume up to 12 mL with sterile water.
10. Washing buffer: 0.2× SSC (0.03 M NaCl, 3 mM sodium citrate, 0.01 mM EDTA) and 0.1% SDS.
11. γ-^{32}P-ATP.

3. Methods

Detection of short non-coding RNAs starts with RNA extraction. Traditional column/membrane-based RNA extraction kits such as RNeasy (Qiagen) or SV Total RNA Isolation System (Promega) yield good-quality high-molecular weight RNA. However, short RNAs do not bind strongly enough to the column/membrane due their small size, and they are easily lost during the wash step. There are commercially available kits for purifying short RNAs such as miRVana (Ambion), PureLink miRNA isolation kit (Invitrogen) and miRNeasy kit (Qiagen) and these usually work well. The protocol described in this chapter provides a cheap alternative to these kits. Another advantage of this protocol is that it is easy to scale it up if larger amount of RNA is required. An alternative protocol is also provided for extracting RNA from tissues rich in polysaccharides/polyphenols, such as tomato fruit (8).

Once the RNA is extracted, it has to be separated on a denaturing polyacrylamide gel because agarose gel does not separate short non-coding RNAs from tRNAs and small nucleolar RNAs. The percentage of polyacrylamide gel depends on the size of the running apparatus ranging between 8 and 15% (the shorter the run, the higher the percentage). We usually use the Bio-Rad Mini-Protean II system and run 15% gels. After gel electrophoresis, the RNA is transferred to a membrane. The traditional capillary transfer works well, but we also provide a protocol for a quicker method using semi-dry blotting. Following transfer, the short RNAs need to be crosslinked to the membrane and there are two ways to do it. The traditional UV crosslinking is very quick and easy, but the hybridisation of probes to short RNAs is less efficient. This is because nucleotides crosslinked to the membrane cannot be involved in annealing to the probe molecules. Due to the small size of short RNAs, this can significantly affect the hybridisation signal, although strongly expressed short RNAs can still be easily detected by this method. An alternative protocol is described based on Pall et al. (9), where short RNAs are linked to the membrane only through their 5′ end. This method gave about ten times stronger signal in our laboratory. Please note that the crosslinking method determines the electrophoresis buffer and membrane. Finally, hybridisation takes place where usually a 5′ labelled oligonucleotide complementary to the short RNA is used as a probe. After hybridisation, the membrane is washed and exposed to either X-ray film or a phosphoimager plate and the signal is visualised. The membranes can be stripped and re-used several times (we have successfully used membranes up to ten times).

3.1. RNA Extraction (Generic Protocol)

1. Place sterile mortals and microcentrifuge tubes on ice.
2. Place 650 μL of phenol into each microcentrifuge tube.
3. Prepare extraction buffer A (keep it at room temperature).
4. Grind your sample (100–150 mg) in the ice-chilled mortar to a fine powder.
5. Add 650 μL of extraction buffer A and continue the homogenization for 10 s.
6. Transfer the homogenised sample into the microcentrifuge tube containing phenol and vortex. Continue with the remaining samples.
7. Centrifuge for 10 min at 15,000×*g* and 4°C or room temperature.
8. Place 300 μL of phenol and 300 μL of chloroform into new microcentrifuge tubes.
9. After centrifugation, remove the upper phase and transfer to the microcentrifuge tube containing phenol/chloroform. Important: do not remove or disturb the white interphase. Pipette out about 500 μL).
10. Vortex and centrifuge for 5 min, 15,000×*g* (4°C or room temperature).
11. Place 500 μL of chloroform into new microcentrifuge tubes and keep on ice.

12. After centrifugation, transfer the upper phase to the tubes containing 500 μL chloroform.
13. Vortex and centrifuge for 3 min, 15,000×*g* (4°C or room temperature).
14. Add 20 μL of 4 M Na acetate (pH 5.2) to the upper phase (this should be not more than 400 μL) and precipitate the nucleic acids with 1 mL of 96% EtOH. Do not vortex; just invert the tubes several times.
15. Place the tubes on ice for 10 min.
16. Centrifuge for 10 min, 15,000×*g* (4°C or room temperature).
17. Carefully remove supernatant.
18. Wash the pellet with 1 mL of 70% EtOH by centrifuging for 3 min, 15,000×*g* (4°C or room temperature).
19. Dry the pellet at room temperature. Important: if the RNA is completely dried, it is impossible to dissolve.
20. Place the microcentrifuge tubes on ice and dissolve the pellet in 50 μL of sterile water.
21. You can check the integrity of the extracted RNA as follows: Denature 5 μL of RNA sample by adding 5 μL of FDE and incubating at 65°C for 10 min. Load and separate 10 μL of denatured samples in 1.2% agarose gel (1×TBE).

3.2. RNA Extraction from Tomato Fruit

The method is adapted from Chang et al. (10) omitting the LiCl precipitation step (due to inefficient recovery of short RNAs). Most of the polysaccharides/polyphenols are removed and therefore this protocol is applicable to other tissues rich in polysaccharides/polyphenols.

1. Grind the tissue in liquid N_2 thoroughly with a pestle and mortar with extraction buffer B. Use 10 mL of buffer for each gram of tissue (fresh weight).
2. Heat the mixture to 65°C for 10 min to lyse tissue; then cool to room temperature. Extract once with an equal volume of chloroform.
3. Collect the upper phase and add an equal volume of isopropanol, incubate on ice for 30 min and centrifuge for 10 min, 13,000×*g* at 4°C. Wash pellet with 76%EtOH containing 10 mM NH_4Ac and air dry.
4. Dissolve the dried nucleic acid pellet in SSTE buffer. Incubation at 37°C can speed up this step.
5. Extract the dissolved nucleic acids once with an equal volume of chloroform avoiding carry-over of interface material and precipitate with an equal volume of isopropanol as above.

6. Finally wash the pellet with 70% ethanol followed by 100% ethanol and dry.

3.3. Denaturing Polyacrylamide Gel Electrophoresis

Depending on the method of crosslinking, the samples are separated on a TBE or MOPS denaturing gels.

1. Prepare the gel as described under **Subheading 2** (*see* **Note 1**).
2. Wash the electrophoresis equipment with detergent, rinse it with water and finally with sterile water.
3. Denature the RNA samples so that you load the same amount of RNA into each well. Mix the samples (5–10 μL) with the same volume of FDE buffer and denature at 65°C for 10 min. Place them on ice and keep them on ice until loading.
4. Pre-run the gel (either in 1×TBE or in 1×MOPS) at 450 V (54 mA) for 30 min (*see* **Note 2**).
5. Wash the wells with the running buffer and load the samples (*see* **Notes 3** and **4**).
6. Run the gel at 100 V until the bromophenol blue front is about 1–2 cm from the bottom of the gel. (Please note that actual voltage depends on the size of the electrophoresis apparatus; 100 V is used for the Bio-Rad Mini-Protean II).

3.4. Transfer of RNA

There are several ways to transfer the RNA from the gel to the membrane. Two alternative protocols are described here. The advantage of the traditional capillary blotting is that it does not require any equipment. The advantage of the semi-dry blotting is that it is much quicker. For choice of membranes please *see* **Note 5**.

3.4.1. Capillary Blotting

1. Soak the gel in 10 mM Na-phosphate buffer (pH 7.0) for 10 min and subsequently in 20× SSC for additional 10 min before blotting.
2. Cut the membrane (the type of membrane depends on the crosslinking method, *see* **Note 6**) to match the size of the gel. Soak it in 20× SSC for 5 min.
3. Cut two large Whatman papers to be used as the bridge and two more large pieces to be placed immediately on the bridge.
4. Place the gel on the Whatman papers and wrap the areas not covered by the gel with Saran Wrap. Carefully lay the membrane onto the gel. Carefully remove bubbles between gel and membrane with your fingers (always wear gloves when working with membranes, also *see* **Note 7**).
5. Place three pieces of Whatman paper (soaked with 20× SSC) that are the same size as the gel on the membrane. Add a

stack of towel-paper onto the Whatman papers and finally put a flat glass plate with a weight on the top.

6. Let it blot for at least 16 h, usually overnight.

3.4.2. Semi-dry Blotting

1. Cut six pieces of Whatman paper (the same size as the gel) and one piece of membrane (same size as the gel, *see* **Note 6** about type of membrane).
2. Soak Whatman papers and the membrane in 1× gel running buffer (TBE or MOPS).
3. Place three pieces of Whatman paper on top of each other in the semi-dry blotter. Roll a 5-mL pipette over the paper to remove air bubbles (*see* **Note 7**).
4. Place the membrane on the Whatman papers.
5. Place the gel on top of the membrane.
6. Add the three remaining Whatman papers on the top of the gel. Again remove air bubbles with a 5-mL pipette.
7. Transfer RNA at 12 V for 30 min for one gel (1 h for two gels) in the cold room.

3.5. Crosslinking

3.5.1. Crosslinking by UV

1. Disassemble the capillary or semi-dry blotting system and dry the membrane on a Whatman paper at room temperature for 5 min.
2. Place the membrane into a UV crosslinker and follow the manufacturer's guide. Alternatively, illuminate for up to 2 min with a benchtop UV lamp. In both cases, make sure that the RNA-carrying side of the membrane faces towards the UV source.

3.5.2. Chemical Crosslinking

1. Cut out a piece of Whatman paper slightly larger than your membrane.
2. Soak the Whatman paper on a large piece of Saran Wrap in the crosslinking solution (5 mL is sufficient).
3. Place the membrane on top of the Whatman paper with the RNA-containing side up.
4. Wrap the Saran Wrap around the membrane to make a sealed parcel.
5. Incubate the parcel for 1 h at 60°C.
6. Wash the membrane in sterile water for 10 min on a shaker.
7. Repeat the washing step.
8. Wrap the membrane in Saran Wrap and store in fridge. This membrane is now stable and can be stored for several months.

3.6. Hybridisation

3.6.1. Pre-hybridisation

1. Pre-heat hybridization oven to 37°C.
2. Place crosslinked membrane into a hybridisation bottle and cover with 5 mL Ultrahyb-oligo buffer (Ambion, cat no. 8663).
3. Rotate the bottle in hybridisation oven for at least 2 h (alternatively overnight).

3.6.2. Probe Preparation

1. Add the following into a screw-capped microcentrifuge tube: 2 μL of oligonucleotide (reverse complementary to the short RNA to be detected) from a 10 μM stock, 2 μL of 10× polynucleotide kinase (PNK) buffer (Promega), 12 μL of sterile water, 3 μL of γ-^{32}P-ATP and 1 μL of PNK (Promega) (*see* **Note 5**).
2. Incubate for 1 h at 37°C.
3. Snap cool on ice and add 20 μL of sterile water.
4. Purify through a G-25 Sephadex column to remove unincorporated γ-^{32}P-ATP. This step can be omitted; however, it is recommended because it reduces background.

3.6.3. Hybridisation

1. Add probe to hybridisation bottles. (There is no need to change hybridization buffer after pre-hybridisation, but *see* **Note 8**).
2. Hybridise overnight at 37°C by rotating the hybridization bottle (*see* **Note 9**).

3.6.4. Washing and Exposure

1. Wash the membrane 2 times for 30 min with about 20 mL washing buffer at 37°C (keep rotating the hybridization bottle, also *see* **Notes 9** and **10**).
2. Remove membrane from hybridization bottle and wrap in Saran Wrap (*see* **Note 11**).
3. Expose membrane to X-ray film or phosphoimager plate.
4. Develop film or scan phosphoimager plate.

4. Notes

1. We normally load 10–15 μg total RNA per lane. If it is difficult to detect a certain short RNA, up to 40–50 μg can be loaded, although the amount of RNA affects the running (less RNA gives better resolution and clearer bands). The sensitivity can be increased by about ten times by using the chemical crosslinking protocol. The signal is increased by

another ten times if locked nucleic acid (LNA) primers are used as probes (11).

2. We normally run 0.75-mm 15% gels and use semi-dry blotting. These conditions are usually optimal to separate 10-15 μg total RNA. If more RNA has to be loaded, thicker gels are recommended (1, 1.5 or 2 mm). However, capillary blotting is less efficient for thicker gels. Also, if capillary blotting is used, it is recommended to decrease the gel concentration to 8 or 10%. In this case, one may run a longer gel than the Bio-Rad Mini-Protean II.
3. When trying to detect for the first time a novel short RNA, it is helpful to load a DNA primer with the same sequence as a positive control for labelling, hybridisation and detection into a separate lane.
4. Equal loading is tested by re-probing the membrane with U6-specific primer. The Arabidopsis U6 probe gives good signals for RNA from tomato and other plant species as well. The primer we use for U6 detection is: GCTAATCTTCTCTGTATCGTTCC. The U6 signal is about one third from the top of the gel while the short RNA (21–24 nt) signal is usually at about two thirds from the top; therefore, the two probes can be hybridised in the same hybridisation bottle if the probe was used before and produced a single band at the 21–24 nt region.
5. The type of membrane used for transfer is determined by the crosslinking method applied. If the UV crosslinking method is followed, either positively charged (such as Hybond N+; Amersham) or neutral (such as Hybond NX; Amersham) membranes can be used. Positively charged membranes have higher capacity; therefore, more RNA can be bound to the membrane. This can be an advantage to achieve stronger signal. However, sometimes it is a disadvantage because it can also lead to higher background due to unspecific binding of probe. When the chemical crosslinking method is applied, positively charged membranes cannot be used because RNA binds only to neutral membranes under these conditions.
6. Gels can be pre-run for 20–30 min to improve the quality of the image. It is especially recommended for larger gels because the gel will be warmed up during pre-run. This helps to keep the RNA molecules denatured during the run resulting in sharper bands.
7. It is important to remove air bubbles when blotting is set up. It is usually helpful to remove the bottom 2 mm of the gel because it often crinkles up leading to air bubbled between gel and membrane. Note that this region does not contain any short RNAs.

8. In the case of high background, the pre-hybridisation buffer can be discarded and the probe should be added in fresh hybridisation buffer. In this case, it is important to heat up the new batch of hybridisation buffer to the hybridisation temperature. It is also recommended to add the probe to the buffer before it is poured onto the membrane. Pipetting the probe directly to the hybridisation bottle containing buffer and membrane may lead to spots on the membrane.
9. If an LNA probe is used, the time of hybridisation can be reduced to as little as 4 h. The hybridisation temperature can be increased up to 50°C and more stringent washing conditions can be applied (0.1× SSC and 0.05%SDS).
10. It is recommended that membranes hybridised with different probes are washed in separate containers because crosshybridisation can occur even during washing, especially when LNA probes are used.
11. It is important that membranes are not dried after final wash, but kept slightly moist. Dried membranes cannot be stripped.

References

1. Jones-Rhoades, M. W., Bartel, D. P., and Bartel, B. (2006) MicroRNAs and their regulatory roles in plants. *Annu Rev Plant Biol* **57**, 19–53.
2. Kidner, C. A. and Martienssen, R. A. (2005) The developmental role of microRNA in plants. *Curr Opin Plant Biol* **8**, 38–44.
3. Phillips, J., Dalmay, T., and Bartels, D. (2007) The role of small RNAs in abiotic stress. *FEBS Lett* **581**, 3592–3597.
4. Peragine, A., Yoshikawa, M., Wu, G., Albrecht, H. L., and Poethig, R. S. (2004) SGS3 and SGS2/SDE1/RDR6 are required for juvenile development and the production of *trans*-acting siRNAs in *Arabidopsis. Genes Dev* **18**, 2368–2379.
5. Vazquez, F., Vaucheret, H., Rajagopalan, R., Lepers, C., Gasciolli, V., Mallory, A. C., Hilbert, J. L., Bartel, D. P., and Crete, P. (2004) Endogenous *trans*-acting siRNAs regulate the accumulation of *Arabidopsis* mRNAs. *Mol Cell* **16**, 69–79.
6. Borsani, O., Zhu, J., Verslues, P. E., Sunkar, R., and Zhu, J. K. (2005) Endogenous siRNAs derived from a pair of natural *cis*-antisense transcripts regulate salt tolerance in *Arabidopsis. Cell* **123**, 1279–1291.
7. Xie, Z., Johansen, L. K., Gustafson, A. M., Kasschau, K. D., Lellis, A. D., Zilberman, D., Jacobsen, S. E., and Carrington, J. C. (2004) Genetic and functional diversification of small RNA pathways in plants. *PLoS Biol* **2**, E104.
8. Moxon, S., Jing, R., Szittya, G., Schwach, F., Rusholme Pilcher, R. L., Moulton, V., and Dalmay, T. (2008) Deep sequencing of tomato short RNAs identifies microRNAs targeting genes involved in fruit ripening. *Genome Res* **18**, 1602–1609.
9. Pall, G. S., Codony-Servat, C., Byrne, J., Ritchie, L., and Hamilton, A. (2007) Carbodiimide-mediated cross-linking of RNA to nylon membranes improves the detection of siRNA, miRNA and piRNA by Northern blot. *Nucleic Acids Res* **35**, e60.
10. Chang, S., Puryear, J., and Cairney, J. (1993) A simple and efficient method for isolating RNA from pine trees. *Plant Mol Biol Rep* **11**, 113–116.
11. Valoczi, A., Hornyik, C., Varga, N., Burgyan, J., Kauppinen, S., and Havelda, Z. (2004) Sensitive and specific detection of microRNAs by northern blot analysis using LNA-modified oligonucleotide probes. *Nucleic Acids Res* **32**, e175.

Chapter 19

Quantitative Real Time PCR in Plant Developmental Biology

Vivien Exner

Abstract

Gene expression patterns are important determinants of a cell's state, and changes in the expression profile indicate adaptation processes as a response to developmental transitions or environmental changes. Assaying gene expression can, therefore, help to elucidate mechanisms of determination and differentiation, as well as signaling networks. Several methods have been employed to determine transcript levels. The most quantitative and widely used technique is reverse transcription coupled to quantitative real time polymerase chain reaction (RT-qPCR). Live observation of fluorescence and, therefore, product increase during RT-qPCR allows the accurate determination of differences between initial template amounts. This is in contrast to the end-point analysis of conventional PCR, where initial differences in template amounts are usually masked because the analysis is done at the plateau phase. In the plateau phase, differences can no longer be distinguished due to inherent characteristics of PCR (e.g., loss of activity of the polymerase or because reaction components become limiting) that cause a drop in amplification efficiency, so that product accumulation levels out. Real time PCR circumvents this problem by shifting the analysis to an earlier stage of the amplification reaction.

Key words: Real time PCR, RT-qPCR, quantitative PCR, SYBR, TaqMan, hydrolysis probe.

1. Introduction

Determination of the expression level of a given gene or a group of genes is a key question to understand regulatory networks and identify new components that control cell differentiation. Several methods have been developed to investigate the amount of transcripts produced from a specific locus, some of them being more precise than others.

The classical method to probe transcript levels is Northern blot hybridization (1). Northern blot hybridization relies on the specific hybridization of a labeled probe to a target RNA that has

L. Hennig, C. Köhler (eds.), *Plant Developmental Biology*, Methods in Molecular Biology 655,
DOI 10.1007/978-1-60761-765-5_19, © Springer Science+Business Media, LLC 2010

been transferred and fixed to a nucleic acid-binding membrane. The disadvantage of this method is its time-intensive procedure and relatively low sensitivity. A modern version of the hybridization technique is represented by the DNA microarray technology, where the probe is fixed on a matrix while the RNA sample of interest is applied as liquid phase. Microarrays allow the simultaneous analysis of many genes instead of only one as during conventional Northern blotting; however, the sensitivity is comparably low (2).

Sensitivity has been improved with the implementation of amplification methods, i.e., polymerase chain reaction (PCR)-based assays. Here, cDNA is derived from an RNA sample by a reverse transcriptase (RT) reaction (3). The conversion of RNA into cDNA also guarantees a higher stability of the sample, as cDNA is less prone to degradation than RNA. In a second step, generated cDNA is subjected to a PCR using target-specific primers. While the end-point analysis of a conventional PCR (4) gives only qualitative information, improved assays, like semi-quantitative or quantitative-competitive RT-PCR, enable comparison of expression levels in a more quantitative way (for a review *see*, e.g., (5) and references therein). A break-through, however, was achieved by the invention of the real time PCR technique (6, 7): the RT reaction is followed by a quantitative PCR during which the formation of the PCR product is observed live by monitoring the production of a product-dependent fluorescent signal (RT-qPCR). RT-qPCR is currently considered the most accurate method to determine transcript amounts.

Several methods have been developed to monitor product formation. All of them are based on the detection of an increase in fluorescence emission of a marker molecule during product accumulation. Two main groups of markers can be distinguished: double-stranded DNA (dsDNA)-binding dyes and fluorescently labeled probes.

The first group includes molecules that have a negligible fluorescence emission in their free state, but produce a fluorescent signal when intercalated with dsDNA. Most commonly, SYBR Green I is used for this purpose (8), and a protocol for its usage is provided below. SYBR Green I is a dye that exhibits increased fluorescence after binding to dsDNA (*see* **Fig. 19.1**). The advantage of dsDNA-binding dyes is that they will bind to any double-stranded PCR product, which means that one dye can be used for any assay. However, only one product can be monitored per reaction, and amplification of the reference gene and the gene of interest must be separated.

The second group comprises a wide variety of molecules that commonly rely on fluorescence resonance energy transfer (FRET) (for an overview *see* (9)). Because of its wide usage and due to space limitations, the protocol here only focuses on the

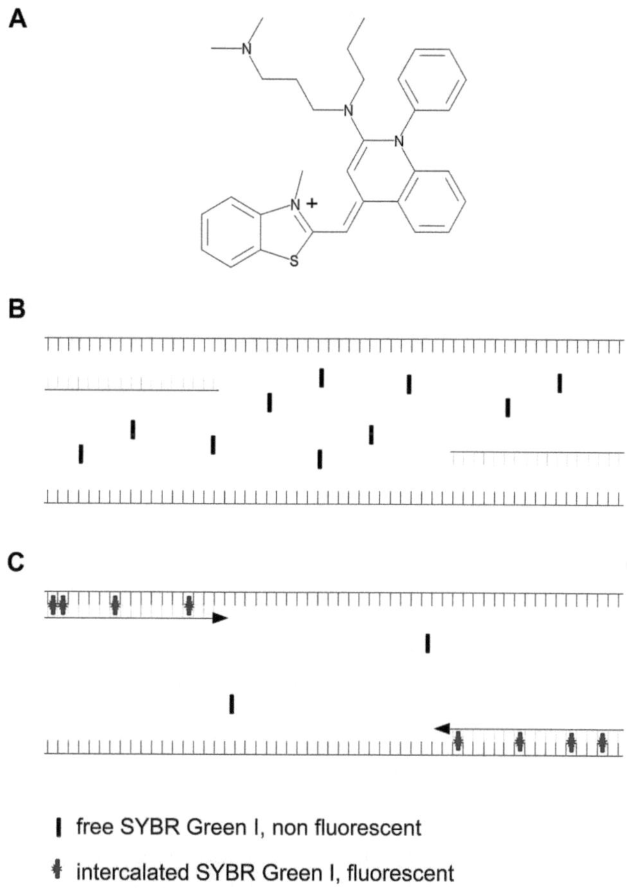

Fig. 19.1. **Real time qPCR with SYBR Green I.** SYBR Green I (2-{2-[(3-dimethylamino-propyl)-propylamino]-1-phenyl-1*H*-chinolin-4-ylidenmethyl}-3-methyl-benzothiazol-3-ium cation) is a dsDNA-binding molecule (**A**). In its free state, SYBR Green I has a negligible fluorescence (**B**), but emits a fluorescence signal upon binding to dsDNA during the amplification step (**C**).

hydrolysis (TaqMan) probe system (7, 10). Hydrolysis probes are relatively short and have a reporter molecule (often fluorescein, FAM) covalently bound to one end and a quencher molecule (often tetramethyl-6-carboxyrhodamine, TAMRA) on the other end. Due to FRET, the fluorescence of the reporter is efficiently quenched as long as the probe remains intact (*see* **Fig. 19.2**). The probe is designed to bind in between the two template-specific primers. The probe remains bound while the polymerase extends the primers. Once the polymerase reaches the hybridized probe, it degrades the probe by its 5′-exonuclease activity (11). Upon degradation, the quencher becomes separated from the reporter, which leads to an increase of the fluorescence signal.

The advantages of probe-based systems are the increase in specificity, as side products of the primer pairs, to which the probe is not complementary, do not contribute to the accumu-

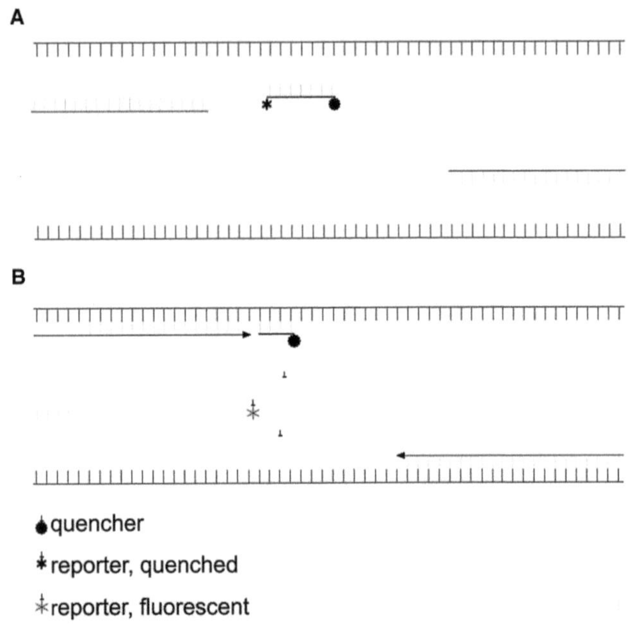

Fig. 19.2. **Real time qPCR with hydrolysis probes.** A target-specific, dual-labeled probe is included in the assay. In the intact probe, fluorescence of the reporter is quenched by the quencher (**A**). During target amplification, the probe is degraded by the 5′-3′ exonuclease activity of the polymerase (**B**), the reporter is freed from the quencher, and a fluorescent signal is produced.

lation of the fluorescence signal, and the possibility to simultaneously detect several products in a single reaction if different reporters are used for different target-specific probes. The disadvantages are the need for a specific probe for each assay and a slightly lower signal since for each DNA molecule produced only one reporter molecule is freed (in contrast to the dsDNA-binding dyes, where several dye molecules can bind to a single dsDNA fragment depending on its length).

Most of the time, expression levels detected with RT-qPCR are expressed as relative expression levels. For that purpose, the expression level of an internal control gene is taken as a reference. This control or reference gene has to be chosen carefully to make sure that the gene is really constantly expressed under all the studied conditions, either different treatments or different genotypes (12–14). While in plants glyceraldehyde-3-phosphate-dehydrogenase (*GAPDH*) or ACTIN (*ACT*) genes often have been judged as a standard reference, it is strongly advised to choose references from a wider selection of genes (14). The disadvantage of *GAPDH* and *ACT* is their high expression, which makes comparison to lowly expressed genes inaccurate. Generally, the reference gene should have a similar expression level as the target gene. When using genes encoding components of the cytoskeleton as reference genes, care should be taken that no significant alterations in cell shape or size, which could be paralleled by changing expression of the reference gene, occur under the investigated conditions. A single nontemplate control is normally included in each assay. This reaction allows distinguishing between the actual signal and background fluorescence. The other reactions are performed in replicates to balance for technical variations.

In contrast to the most frequently used single internal control gene, Vandesompele and colleagues investigated the utility of multiple internal control genes for accurate normalization of RT-qPCR assays (15). The authors argue that ideal control genes do not exist and that careful studies reveal fluctuating expression levels for all commonly used reference genes. To circumvent this problem, they developed geNorm, a Visual Basic Application for Microsoft Excel (15). The program automatically calculates a stability value for each gene in a set of several reference genes and enables the exclusion of the least suitable control gene. According to the studies of Vandesompele and coworkers, three reference genes are sufficient for most applications to determine an adequate correction factor. This more sophisticated normalization method could be especially relevant for the determination of small differences between expression levels.

Recent efforts have created data collections for high-throughput expression analyses by RT-qPCR (16–19). The studies present results on primer design and optimization and provide a helpful insight into RT-qPCR assay design for a variety of species (*see* **Table 19.1**).

Table 19.1
Community resources for high-throughput RT-qPCR analyses in different plant species

Species	Gene subset	References
Arabidopsis thaliana	Transcription factors	Czechowski et al. (16)
Oryza sativa	Transcription factors	Caldana et al. (17)
Medicago truncatula	Transcription factors	Kakar et al. (18)
Zea mays (and other *Poaceae*)	Plastome	Sharpe et al. (19)

2. Materials

2.1. RNA Extraction

1. RNA extraction buffer (e.g., TRIzol Reagent [Invitrogen Life Sciences, Basel, Switzerland]).
2. Chloroform.

3. Isopropanol, ice cold.
4. Ethanol (80%), ice cold.
5. Water, diethylpyrocarbonate (DEPC)-treated (*see* **Note 1**).
6. RNase-free DNase (e.g., Promega RQ1DNase-RNase free [Promega, Madison, WI]).
7. Phenol:chloroform, 1:1.
8. LiCl solution (4 M).
9. Optional: GlycoBlue (Ambion, Huntingdon, United Kingdom); assists in precipitating small amount of RNA and increases visibility of the pellet.
10. Optional: RNase inhibitor, e.g., RNaseOUT (Invitrogen Life Sciences, Basel, Switzerland); counteracts the activity of potential RNase contaminations.

2.2. Reverse Transcriptase Reaction

1. RNA.
2. dNTP mix.
3. Oligo(dT) (*see* **Note 2**).
4. Reverse transcriptase and a suitable reaction buffer (e.g., RevertAid First Strand cDNA Synthesis Kit [Fermentas/Lab Force, Nunningen, Switzerland]).
5. RNase inhibitor.
6. Water, DEPC-treated.

2.3. Quantitative Real Time PCR

1. Real Time PCR machine (e.g., Applied Biosystems 7,500 Fast Real-Time PCR System [Applied Biosystems, Rotkreuz, Switzerland]).
2. Reaction plates/vessels and lids suitable for your instrument of choice (*see* **Note 3**).
3. cDNA.
4. Real time PCR chemicals appropriate to the preferred system (intercalation dye or probe based; *see* **Note 4**).
5. Primers complementary to reference gene(s) and genes of interest.

3. Methods

3.1. RNA Preparation

Different RNA extraction methods exist. For efficient reverse transcription, pure and intact RNA is crucial. In our laboratory, RNA is successfully extracted from different plant species, e.g., Arabidopsis, wheat, rice, and cassava, with TRIzol (Invitrogen Life Sciences, Basel, Switzerland) using the following protocol:

1. Grind up to 0.1 g of plant tissue in liquid nitrogen (*see* **Note 5**).
2. Add 1 mL TRIzol, vortex thoroughly.
3. Incubate for 5 min at room temperature (RT) with gentle agitation.
4. Add 200 μL of chloroform, vortex for 15 s.
5. Incubate for 2–3 min at RT, then centrifuge (15 min, 16,100×*g*, 4°C).
6. Transfer the upper aqueous phase into a new tube.
7. Add 0.5 mL of isopropanol (ice cold) and vortex.
8. (Optional: add 1 μL of GlycoBlue solution and vortex; *see* **Note 6**)
9. Incubate for 10 min at RT, then centrifuge (10 min, 16,100×*g*, 4°C).
10. Discard supernatant.
11. Wash pellet with 1 mL of 80% ethanol (ice cold) and vortex briefly.
12. Centrifuge (5 min, 16,100×*g*, 4°C).
13. Discard supernatant and dry pellet carefully (*see* **Note 7**).
14. Add an appropriate volume of DEPC-treated water (e.g., 84 μL when using Promega RQ1 DNase,) dissolve for 10 min at 60°C, and transfer to ice.
15. Subject the RNA to a DNase treatment according to manufacturer's instructions (e.g., add 5 μL RQ1DNase, 10 μL 10x buffer, 1 μL RNaseOUT; incubate for 30 min at 37°C).
16. Add 1 volume of phenol/chloroform (1:1) (*see* **Note 8**), vortex and centrifuge (5 min, 16,100×*g*, 4°C).
17. Transfer upper phase into new tube.
18. Add 1 volume of chloroform, vortex, and centrifuge (5 min, 16,100×*g*, 4°C).
19. Transfer upper phase into new tube.
20. Add 1 volume of 4 M LiCl (DEPC treated) and store at 4°C overnight (*see* **Note 9**).
21. Centrifuge (10 min, 16,100×*g*, 4°C).
22. Discard supernatant, wash pellet with 200 μL of 80% ethanol (ice cold,) vortex briefly, and centrifuge (5 min, 16,100×*g*, 4°C).
23. Discard supernatant and dry pellet carefully (*see* **Note 8**).
24. Add 20–50 μL of DEPC-treated water, dissolve for 10 min at 60°C, and transfer to ice (*see* **Note 10**).

To determine concentration and purity of the RNA, the absorbance is measured at 260 and 280 nm. The ratio ($abs_{260\,nm}/abs_{280\,nm}$) should be close to 2.1 (*see* **Note 11**). Integrity of the RNA might be assayed by running 1–2 μg on a 1.5% agarose gel and inspecting the sample for the characteristic banding pattern of the rRNAs.

3.2. cDNA Synthesis

Efficiency of the RT reaction depends on the purity of the RNA and can vary with template concentration. Therefore, usually the same amount of RNA for all the samples of one experiment is subjected to RT. Any reverse transcriptase can be used according to manufacturer's instructions together with an RNase-free dNTP mix and oligo(dT) primers (*see* **Note 2**). The RNA may be protected from RNase activity by including an RNase inhibitor in the reaction.

In our laboratory, we usually use 0.5–1 μg of RNA per reaction; the applicable range, however, also depends on the reverse transcriptase used.

3.3. Primer Design

Primers for real time PCR applications have to fulfill certain criteria. Due to the standardization of the PCR program, all primers should have a melting temperature (Tm) of 58–60°C (calculated according to nearest neighbor method). The length of the amplicon usually ranges from 60 to 120 bp. Furthermore, the primers should be highly specific to their target sequence and have to be designed such that no side products (e.g., primer dimers) are formed. This is especially important if intercalating dyes are used as reporters. These dyes bind to any double-stranded DNA that is present, and side products will so feign higher levels of the target sequence than actually present. Such side products can be detected with a melting curve analysis (20) that should be performed when an assay is used for the first time.

For primer design, we made good experience with the free software PerlPrimer that has an option to design oligo nucleotides specifically for real time PCR purposes (21). If you prefer the TaqMan system and decide to use the Universal ProbeLibrary (Roche Diagnostics, Rotkreuz, Switzerland,) you can automatically design your assay online (www.universalprobelibrary.com). This online tool suggests one or more assays consisting of a primer pair and a Universal ProbeLibrary probe suitable for your gene of interest.

Primers for real time PCR applications are often designed to span an exonexon border. This ensures that the primers anneal only to cDNA and not to traces of genomic DNA that might be present in the sample.

3.4. Real Time PCR

3.4.1. qPCR with SYBR Green I (see Note 12)

Here, a protocol for RT-qPCR with SYBR Green PCR Master Mix (Applied Biosystems, Rotkreuz, Switzerland) is provided. We use 25 μL for a standard reaction, which is usually sufficient to provide a strong signal even though 50 μL per reaction is recommended by the manufacturer (the reaction volume can also be restricted by the real time PCR machine used).

1. Distribute x μL template into wells (*see* **Note 13**).
2. Set up the master mix for the reactions required. For a single reaction, use:

SYBR Green PCR Master Mix (2×)	12.5 μL
Forward primer (10 μM)	1.25 μL
Reverse primer (10 μM)	1.25 μL
Water	up to (25–x) μL

3. Add master mix to the templates.
4. Set up the real time PCR machine (for details on programming, the settings for dyes used etc., as well as for further recommendations, you should consult the guidelines of the manufacturer of the machine). The standard real time PCR program includes the following steps:

   95°C, 10 min; 1 cycle to activate the hot-start polymerase

   95°C, 15 s (denaturation)

   60°C, 1 min (primer annealing and elongation)

   40 cycles to amplify the sequence of interest (however, C_t values for reliably detectable genes are usually much lower than 40)

   Fluorescence detection is performed at the elongation step.

5. Subsequent to the amplification reaction, it is recommended to run a melting curve analysis to exclude the formation of side products. For details about the settings, please consult the guidelines of the real time PCR machine manufacturer.

3.4.2. qPCR with TaqMan Probes (see Note 12)

3.4.2.1. Probe Design

For several model organisms (including *Arabidopsis thaliana*, *Oryza sativa*, and *Zea mays*) a very convenient approach to probe and assay design is the Universal ProbeLibrary (Roche,) where probes and primers for nearly all genes can be easily found and ordered. A drawback of this approach is that the amplified region of the gene of interest has to be chosen according to the available probes, and testing several regions of a single gene such as for alternative splice variants might not be possible.

The Universal ProbeLibrary probes are only 8–9 nucleotides length. The incorporation of Locked Nucleic Acid (LNA), a duplex-stabilizing DNA analogue (22–24) increases T_m and binding specificity.

Since extension of the primers by the polymerase competes with the hybridization of the probe to the target sequence, the T_m of the probe has to be approximately 10°C higher (T_m = 68–70°C) than that of the primers to ensure that the probe binds to its target earlier than the primers, which are extended as soon as they anneal. Alternatively, molecules with a high affinity to the minor groove of the DNA can be covalently linked to the oligodeoxynucleotides, which gives rise to so-called minor groove binding (MGB) probes, which also exhibit an increased T_m (25, 26).

Conventional dual-labeled probes are 18–30, ideally 20 nucleotides long and should have a GC content of approximately 50%. The reporter is located at the 5′ end of the oligo, while the quencher is bound to the 3′ end. However, if longer probes are required, the quencher should not be added to the 3′ end but rather internally, as FRET is dependent on proximity between reporter and quencher. Nevertheless, the quencher should not be added too close to the reporter dye as efficiency of nucleolytic cleavage in between the two dyes decreases with proximity (27). In addition, attention has to be paid that the 5′ end is not a G, as Gs can efficiently quench reporter fluorescence. It is also important that the probe has no sequence complementarity to either one of the primers, nor overlaps with the primer binding sites. Furthermore, the probe sequence should be selected to avoid formation of secondary structures.

3.4.2.2. Amplification Reaction

Below, a protocol for RT-qPCR with FastStart Universal Probe Master (ROX) is provided. We use 25 μL for a standard reaction, which is sufficient to provide a good signal even though 50 μL per reaction is recommended by the manufacturer (the reaction volume can also be restricted by the real time PCR machine used).

1. Distribute *x* μL template into wells (*see* **Note 13**).
2. Set up the master mix for each gene to be tested. For a single reaction, pipette in the following order: (*see* **Note 14**).

Probe Master (2×)	12.5 μL
Probe (10 μM)	0.25 μL
Forward primer (5 μM)	1 μL
Reverse primer (5 μM)	1 μL
Water	up to (25–*x*) μL

3. Add master mix to the templates.
4. Set up the real time PCR machine (for details on programming, the settings for dyes used etc., as well as for further recommendations, you should consult the guidelines of the manufacturer of the machine). The standard real time PCR program includes the following steps:
   95°C, 10 min; 1 cycle to activate the hot start polymerase

   95°C, 15 s (denaturation)

   60°C, 1 min (primer annealing and elongation)

   45 cycles to amplify the sequence of interest (however, C_t values for reliably detectable genes are usually much lower than 45)

   Fluorescence detection is conventionally performed at the elongation step.

3.5. Data Analysis

Real time PCR machines monitor the increase of amplification products by measuring the emitted fluorescence at every cycle. For each reaction, therefore, a data set of cycle numbers and corresponding fluorescence emissions is produced. These data sets are then used to give evidence for the amounts of transcripts that were initially present in the respective RNA sample, either as an absolute value or relative to an internal reference gene or a standard.

In a real time PCR assay, the PCR cycle at which the emitted fluorescence of the reporter dye rises above predefined threshold fluorescence is used to deduce the amount of target sequences that were initially present. This parameter is often referred to as threshold cycle (C_t) and is needed for all evaluation methods. In the simplest case, it is assumed that PCR efficiency is constant over all samples and equal for all assays used, and that amplification of the target sequence follows an exponential increase, whereby the amount of product doubles in each cycle. Such an ideal case of a reaction can be described accurately by simple mathematical methods. The comparative C_t method, or $2^{-\Delta\Delta C_t}$ method, is based on these assumptions and represents the simplest method of relative quantification (28).

In reality, however, this model is applicable only within a certain range and for a limited part of the reaction. Fluorescence accumulation can be divided into three phases: the lag phase, in which no detectable increase in fluorescence occurs; the exponential phase, in which the signal rises over the background fluorescence and increases in a mathematically describable manner; and the plateau phase, in which the increase in fluorescence levels out because reaction components are being used up and become limiting. Only the exponential phase is open to mathematical modeling, and even within this phase of the reaction, PCR efficiency rarely meets the value of the ideal situation. Derivations of the $2^{-\Delta\Delta C_t}$ method correct for the authentic efficiencies that are derived from standard curves based on dilution series (28, 29).

However, the PCR efficiency might not only vary from assay to assay but also from reaction to reaction. Therefore, the efficiency calculated from a standard curve remains an approximation. If required, this problem can, however, be circumvented by retrieving the actual efficiency directly from the amplification plot. Ramakers and colleagues described a method with which the efficiency of every single reaction can be calculated separately based on the amplification plot. A computer program to perform these calculations, LinRegPCR, can be obtained for free (30).

In addition to the mathematical models, some technical aspects have to be considered. To correct for pipetting errors and well-to-well differences of the thermal cycler, technical replicates have to be performed. Triplicates in each assay can be considered as a common standard. However, since variability can not only arise from within the actual amplification reaction, but also from variations in the RT reaction, we often generate two cDNA samples from the same RNA and run duplicates of these in each assay.

Most researchers will usually perform relative quantification, where the expression levels of the genes of interest are compared to an internal reference gene. The formulas to calculate the relative expression levels are provided below. They correspond to a derivation of the $2^{-\Delta\Delta C_t}$ method (29, 31).

For each sample, the mean normalized expression (MNE) is calculated from the means and standard errors of the C_t values for reference and target gene according to equation [**1**].

$$\mathrm{MNE} = \frac{(E_{\mathrm{ref}})^{C_t^{\mathrm{ref}}}}{(E_{\mathrm{tar}})^{C_t^{\mathrm{tar}}}} \qquad [1]$$

C_t^{ref} and C_t^{tar} refer to the mean threshold cycle for the reference and the target gene assays, respectively. E_{ref} and E_{tar} are the efficiencies of the reference and target gene assays, respectively. The efficiency is derived from the slope of a linear regression in a $\log_{10}$(concentrations) vs. C_t plot of a dilution series (*see* **Note 15**).

$$E = 10^{\left(\frac{-1}{\mathrm{slope}}\right)} \qquad [2]$$

The standard error (SE) of MNE is given by equation [**3**].

$$\mathrm{SE}_{\mathrm{MNE}} = \mathrm{MNE}\sqrt{\left(\left(\ln\left(E_{tar} \times se_{tar}\right)^2\right) + \left(\ln\left(E_{ref} \times se_{ref}\right)^2\right)\right)} \qquad [3]$$

where se_{tar} and se_{ref} are the standard errors of the mean C_t value of the replicates of the target and reference assays, respectively.

If two cDNAs were included, MNE and SE_{MNE} are calculated separately for each cDNA; afterwards, the MNE values are averaged.

$$\mathrm{mMNE} = \frac{\mathrm{MNE}_1 + \mathrm{MNE}_2}{2} \quad [4]$$

The standard error of mMNE is given by equation [**5**]:

$$\mathrm{SE}_{\mathrm{MNE}} = \frac{\sqrt{\mathrm{SE}^2_{\mathrm{MNE}_1} + \mathrm{SE}^2_{\mathrm{MNE}_2}}}{\sqrt{4}} \quad [5]$$

4. Notes

1. To prevent RNA degradation, all solutions and vessels have to be free of RNase activity. Solutions can be made RNase free by overnight incubation with 0.1% DEPC (diethylpyrocarbonate) and subsequent autoclaving. DEPC is volatile and highly toxic and has to be handled with care; however, it is degraded upon autoclaving. Reagents containing primary amine groups (e.g., TRIS) cannot be treated, because the amine groups react with DEPC! Prepare diluted solvents with DEPC-treated water. Flasks are best made RNase free by filling them with water supplied with 0.1% DEPC, stirring overnight, and then autoclaving the bottle with the water. Containers that cannot be autoclaved with water can be baked at 240°C for 8 h.

   Most reaction tubes and pipette tips are provided RNase free by the manufacturer. Always wear protective gloves to prevent contamination with RNases found on human skin, and change gloves frequently. Keep the working area clean (wiping the bench with detergent and ethanol is usually sufficient; alternatively commercial RNA decontamination solutions can be used).

2. Usually, oligo(dT)$_{18}$ is used. To enhance priming at the 3′ end of the poly-A tail, primers can be designed as oligo(dT)$_{18}$V, so that a mixture of oligo(dT)$_{18}$A, oligo(dT)$_{18}$C, and oligo(dT)$_{18}$G can initiate the reverse transcription. Instead of oligo(dT) primers, random primers can be used. Oligo(dT) primers will only lead to reverse transcription of mRNA, which carries a poly-A tail. Random primers will also yield cDNAs of other RNA

species, such as many plastidic mRNAs that carry poly-A tails only transiently during break down.

3. Especially if you use 96-well plates, it is recommended to set up a pipetting scheme before. We use tables in a 96-well plate format for that purpose and found that it greatly facilitates orientation on the real plate during reaction set up.

4. SYBR Green I is available from various manufacturers. The dye is usually provided in a ready-to-use master mix containing also a hot-start DNA polymerase, nucleotides, and a passive reference dye, usually ROX. Master mixes for the probe-based systems are also commercially available but only consist of the hot-start DNA polymerase, nucleotides, and reference dye, while the probe that produces the detection signal is specific to the tested gene and is, therefore, not included in the mix. Fluorescence of the passive reference dye remains constant throughout the reaction, and the dye is added to normalize well-to-well differences that may occur by either pipetting errors or by equipment limitations.

   In our hands, SYBR Green PCR Master Mix (Applied Biosystems, Rotkreuz, Switzerland) for the SYBR Green system and FastStart Universal Probe Master (ROX) (Roche Diagnostics, Rotkreuz, Switzerland) for the probe-based system produce good results.

5. It is crucial for RNA integrity to keep the tissue frozen all the time and add TRIzol directly to the frozen powder. The powder can be stored at –80°C for several months.

6. GlycoBlue is a blue dye covalently linked to glycogen. GlycoBlue precipitates together with the RNA in ethanol precipitations. It serves as carrier to increase precipitation efficiency at low RNA concentrations and aids sample handling by marking the pellet in blue.

7. The RNA should not be over-dried, as this will interfere with redissolving. Air-drying the pellet is sufficient; vacuum drying methods have to be used with care.

8. Chloroform:phenol is highly aggressive; only use resistant vials. Store the solution at 4°C.

9. In this precipitation, the pellet cannot be stained with GlycoBlue, because GlycoBlue is not precipitated with LiCl.

10. RNA can be stored at –20°C (or –80°C for long-term storage); for complete redissolving, heat to 60°C for 10 min upon thawing.

11. Nucleic acids have a peak in the absorption spectrum at a wavelength of 260 nm. At 280 nm, the slope of the absorption spectrum of nucleic acids is very steep and protein

contaminations strongly absorb light of this wavelength. The ratio of the absorptions at 260 and 280 nm can, therefore, be used to test the sample for protein contaminations. Due to the steep slope, small variations in the wavelength around 280 nm will have a greater effect on the 260/280 nm ratio than wavelength variations around 260 nm. Consequently, the same sample may yield slightly different ratios when measured with different spectrophotometers, but each photometer will give consistent results within itself. An additional indicator for impurities is the absorption at 230 nm. Elevated absorption at 230 nm can be due to contaminations with aromatic compounds, carbohydrates, and peptides. The 260/230 nm for pure samples should be >2.0.

12. Upon reacting with light of certain wavelengths, fluorescent dyes may degrade. The chemicals should therefore not be exposed to excess of light. In addition, setting up the PCR on ice also protects the reagents from deterioration. In addition, repeated cycles of freezing and defrosting should be avoided.
13. We usually use 1/50 of a standard cDNA preparation per reaction; if very lowly expressed genes are studied, the amount of template might need to be increased. For easier handling, the cDNA for the technical replicates can be mixed with all or some of the water as a master mix and then be distributed into the wells in a larger volume that is less prone to pipetting errors. Include a nontemplate control for each assay!
14. The indicated concentrations for primers and probes are reference values. The probe concentration should be significantly lower than the primer concentration. The exact concentration, however, might vary from assay to assay.
15. The dilution series consists of at least 3, usually 4 dilutions: undiluted, 1/10, 1/100, and 1/1000. Of each concentration, use the same volume as template for a standard reaction. The reactions for the dilution series are performed in duplicates. The slope is derived from a $\log_{10}$(concentrations) vs. C_t plot: The $\log_{10}$(concentrations) are plotted against the C_t values of the dilution series. The slope is then calculated by linear regression.

Acknowledgments

I thank Bartosz Urbaniak and Ernst Aichinger for helpful discussions and critical reading of the manuscript.

References

1. Alwine, J. C., Kemp, D. J., and Stark, G. R. (1997) Method for detection of specific RNAs in agarose gels by transfer to diazobenzyloxymethyl-paper and hybridization with DNA probes. *Proc Natl Acad Sci USA* **74**, 5350–5354.
2. Schena, M., Shalon, D., Davis, R. W., and Brown, P. O. (1995) Quantitative monitoring of gene expression patterns with a complementary DNA microarray. *Science* **270**, 467–470.
3. Spiegelman, S., Watson, K. F., and Kacian, D. L. (1971) Synthesis of DNA complements of natural RNAs: A general approach. *Proc Natl Acad Sci USA* **68**, 2843–2845.
4. Mullis, K., Faloona, F., Scharf, S., Saiki, R., Horn, G., and Erlich, H. (1992) Specific enzymatic amplification of DNA in vitro: The polymerase chain reaction. *Biotechnology* **24**, 17–27.
5. Bustin, S. A. (2000) Absolute quantification of mRNA using real-time reverse transcription polymerase chain reaction assays. *J Mol Endocrinol* **25**, 169–193.
6. Higuchi, R., Dollinger, G., Walsh, P. S., and Griffith, R. (1992) Simultaneous amplification and detection of specific DNA sequences. *Biotechnology* **10**, 413–417.
7. Heid, C. A., Stevens, J., Livak, K. J., and Williams, P. M. (1996) Real time quantitative PCR. *Genome Res* **6**, 986–994.
8. Simpson, D. A., Feeney, S., Boyle, C., and Stitt, A. W. (2000) Retinal VEGF mRNA measured by SYBR green I fluorescence: A versatile approach to quantitative PCR. *Mol Vis* **6**, 178–183.
9. Giulietti, A., Overbergh, L., Valckx, D., Decallonne, B., Bouillon, R., and Mathieu, C. (2001) An overview of real-time quantitative PCR: Applications to quantify cytokine gene expression. *Methods* **25**, 386–401.
10. Gibson, U. E., Heid, C. A., and Williams, P. M. (1996) A novel method for real time quantitative RT-PCR. *Genome Res* **6**, 995–1001.
11. Holland, P. M., Abramson, R. D., Watson, R., and Gelfand, D. H. (1991) Detection of specific polymerase chain reaction product by utilizing the 5′-3′ exonuclease activity of *Thermus aquaticus* DNA polymerase. *Proc Natl Acad Sci USA* **88**, 7276–7280.
12. Vandesompele, J., De Preter, K., Pattyn, F., Poppe, B., Van Roy, N., De Paepe, A., and Speleman, F. (2002) Accurate normalization of real-time quantitative RT-PCR data by geometric averaging of multiple internal control genes. *Genome Biol* **3**, 7.
13. Pfaffl, M. W., Tichopad, A., Prgomet, C., and Neuvians, T. P. (2004) Determination of stable housekeeping genes, differentially regulated target genes and sample integrity: BestKeeper – Excel-based tool using pair-wise correlations. *Biotechnol Lett* **26**, 509–515.
14. Czechowski, T., Stitt, M., Altmann, T., Udvardi, M. K., and Scheible, W. R. (2005) Genome-wide identification and testing of superior reference genes for transcript normalization in Arabidopsis. *Plant Physiol* **139**, 5–17.
15. Vandesompele, J., De Preter, K., Pattyn, F., Poppe, B., Van Roy, N., De Paepe, A., and Speleman, F. (2002) Accurate normalization of real-time quantitative RT-PCR data by geometric averaging of multiple internal control genes. *Genome Biol* **3**, 7.
16. Czechowski, T., Bari, R. P., Stitt, M., Scheible, W.-R., and Udvardi, M. K. (2004) Real Time RT-PCR profiling of over 1400 *Arabidopsis* transcription factors: Unprecedented sensitivity reveals novel root- and shoot-specific genes. *Plant J* **38**, 366–379.
17. Caldana, C., Scheible, W.-R., Mueller-Roeber, B., and Ruzicic, S. (2007) A quantitative RT-PCR platform for high-throughput expression profiling of 2500 rice transcription factors. *Plant Methods* **3**, 7.
18. Kakar, K., Wandrey, M., Czechowski, T., Gaertner, T., Scheible, W.-R., Stitt, M., Torres-Jerez, I., Xiao, Y., Redman, J. C., Wu, H. C., Cheung, F., Town, C. D., and Udvardi, M. K. (2008) A community resource for high-throughput quantitative RT-PCR analysis of transcription factor gene expression in *Medicago truncatula*. *Plant Methods* **4**, 18.
19. Sharpe, R. M., Dunn, S. N., and Cahoon, A. B. (2008) A plastome primer set

for comprehensive quantitative real time RT-PCR analysis of *Zea mays*: A starter primer set for other *Poaceae* species. *Plant Methods* **4**, 14.

20. Ririe, K. M., Rasmussen, R. P., and Wittwer, C. T. (1997) Product differentiation by analysis of DNA melting curves during the polymerase chain reaction. *Anal Biochem* **245**, 154–160.
21. Marshall, O. J. (2004) PerlPrimer: Cross-platform, graphical primer design for standard, bisulphite and real-time PCR. *Bioinformatics* **20**, 2471–2472.
22. Letertre, C., Perelle, S., Dilasser, F., Arar, K., and Fach, P. (2003) Evaluation of the performance of LNA and MGB probes in 5′-nuclease PCR assays. *Mol Cell Probes* **17**, 307–311.
23. Mouritzen, P., Noerholm, M., Nielsen, P. S., Jacobsen, N., Lomholt, C., Pfundheller, H. M., and Tolstrup, N. (2005) ProbeLibrary: A new method for faster design and execution of quantitative real-time PCR. *Nat Methods* **2**, 313–316.
24. Braasch, D. A., and Corey, D. R. (2001) Locked nucleic acid (LNA): Fine-tuning the recognition of DNA and RNA. *Chem Biol* **8**, 1–7.
25. Lukhtanov, E. A., Kutyavin, I. V., Gamper, H. B., and Meyer, R.B., Jr. (1995) Oligodeoxyribonucleotides with conjugated dihydropyrroloindole oligopeptides: Preparation and hybridization properties. *Bioconjug Chem* **6**, 418–426.
26. Afonina, I., Zivarts, M., Kutyavin, I., Lukhtanov, E., Gamper, H., and Meyer, R. B. (1997) Efficient priming of PCR with short oligonucleotides conjugated to a minor groove binder. *Nucleic Acids Res* **25**, 2657–2660.
27. Livak, K. J., Flood, S. J., Marmaro, J., Giusti, W., and Deetz, K. (1995) Oligonucleotides with fluorescent dyes at opposite ends provide a quenched probe system useful for detecting PCR product and nucleic acid hybridization. *PCR Methods Appl* **4**, 357–362.
28. Livak, K. J. and Schmittgen, T. D. (2001) Analysis of relative gene expression data using real-time quantitative PCR and the $2^{(\text{-Delta Delta C(T)})}$ method. *Methods* **25**, 402–408.
29. Pfaffl, M. W. (2001) A new mathematical model for relative quantification in real-time RT-PCR. *Nucleic Acids Res* **29**, 2002–2007.
30. Ramakers, C., Ruijter, J. M., Deprez, R. H., and Moorman, A. F. (2003) Assumption-free analysis of quantitative real-time polymerase chain reaction (PCR) data. *Neurosci Lett* **339**, 62–66.
31. Simon, P. (2003) Q-Gene: Processing quantitative real-time RT-PCR data. *Bioinformatics* **19**, 1439–1440.

Chapter 20

Luciferase and Green Fluorescent Protein Reporter Genes as Tools to Determine Protein Abundance and Intracellular Dynamics

András Viczián and Stefan Kircher

Abstract

To get insight into molecular mechanisms governing plant development, the dynamics of abundance and cellular localisation of signalling components need to be understood. Luciferase and green fluorescent protein (GFP)-derived reporters are suitable markers to determine dynamic signalling processes in vivo. Here, analysis of phytochrome A (phyA) photoreceptor dynamics during early seedling development is used as an example of how in vitro and in vivo luciferase assays as well as GFP-imaging can be used to probe signalling dynamics.

Key words: Photomorphogenesis, phytochrome photoreceptor, green fluorescent protein (GFP), luciferase (LUC), epifluorescence microscopy, protein degradation.

1. Introduction

In this chapter, we describe how luciferase (LUC) and green fluorescent protein (GFP)-derived reporter genes can be utilised to analyse protein abundance and intracellular dynamics in response to external stimuli. As example, the light-driven dynamics of the phytochrome A photoreceptor is presented. Phytochromes (phy) are the only photoreceptors sensing the red to far-red part of the spectrum, which supplies not only quantitative but also spectral information about the local light environment (1–3). PhyA photoreceptor signalling is involved in plant developmental processes such as seed germination, early seedling development or flowering induction (4). In contrast to

L. Hennig, C. Köhler (eds.), *Plant Developmental Biology*, Methods in Molecular Biology 655,
DOI 10.1007/978-1-60761-765-5_20, © Springer Science+Business Media, LLC 2010

the other members of the phytochrome photoreceptor family (phyB to E in Arabidopsis), phyA is an extremely light-labile protein. After light perception, the active Pfr form of phyA undergoes a rapid destruction with a half time of about 30 min in etiolated Arabidopsis seedlings (5). Additionally, changes of intracellular localisation patterns of phyA are the fastest detectable molecular responses of the photoreceptor after photoconversion and binding to its specific nuclear import proteins FHY1 (far-red-elongated hypocotyl 1) and FHL (FHY1-like) (6–8). PhyA translational fusions tagged with LUC and GFP are utilised to determine changes in abundance and localisation of the receptor, respectively, in response to light stimuli. Therefore, throughout this chapter notable attention has been taken on proper light environments during experiments. These precautions may not apply in studies planned on other types of signalling.

The luciferase (LUC) protein from the North-American firefly (*Photinus pyralis*) catalyses the oxidative decarboxylation of luciferin substrate to oxyluciferin in the presence of ATP and O_2, releasing yellow-green light (λ_{max} ~560 nm) (9). These emitted photons can be perceived by sensitive equipment and therefore be used to measure ATP, luciferin or luciferase levels (10). The first attempts creating transgenic plants expressing luciferase proved, that neither the enzyme, nor its substrate are toxic for the organism (11). The internally present pools of ATP and O_2 together with the property of luciferin to easily penetrate plant tissues promoted luciferase as candidate of an ideal non-invasive plant reporter (12). After removal of a peroxisomal localisation signal and optimizing codon usage (13), a modified luciferase (LUC+) became widely used and is also used in this protocol. Luciferase is a relatively stable monomeric protein of 61 kDa. After reacting with its substrate, luciferase regenerates so slowly that it can be considered as inactive (12). In consequence, after luciferin application, the luciferase molecules available in the system undergo a single catalytic cycle with photon emission and will not generate any detectable signals immediately afterwards.

Figure 20.1 shows the kinetics of luminescence in an etiolated seedling, harbouring a *PHYA:PHYA–LUC* transgene. Immediately after spraying of luciferin, increasing luminescent signals are detectable for some minutes. After the signal reaches the maximum (*see* **Fig. 20.1**, point 1), a slower decrease starts and the steady-state level (about 6% of the maximum signal intensity) is reached in about 12–13 h after the luciferin application (*see* **Fig. 20.1**, point 2). Very similar data were plotted in a different experimental setup by (14) indicating the general appearance of this phenomenon, which allows two ways of measuring luciferase amounts.

(1) In case the luminescence is measured at the maximum value, about 5–8 min after the substrate application, the

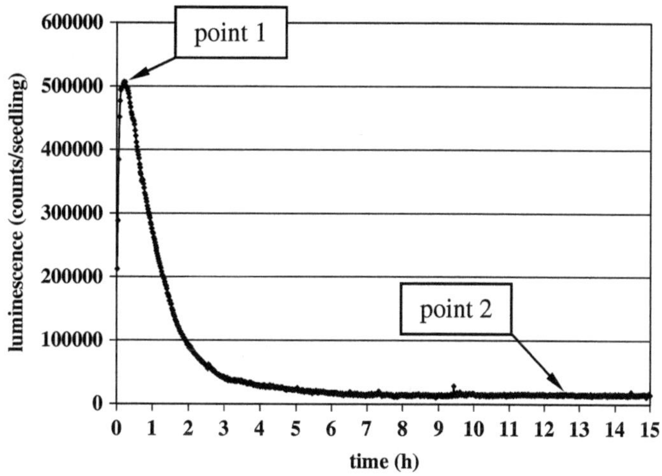

Fig. 20.1. **Luminescence, emitted by a transgenic Arabidopsis *PHYA:PHYA–LUC* seedling.** The luminescent signal from a 4-day-old etiolated seedling expressing a phyA–LUC chimera protein under the control of the phyA promoter was recorded by a VersArray XP CCD camera. The time-course measurement was started immediately after luciferin application. Exposure times were 1 min and images were taken automatically immediately after the data saved from the previous exposure.

obtained photon number is proportional to the LUC protein amount, present in the system. This method is called "flash" measurement and can be used to measure relative protein amounts by expressing a translational fusion of protein of interest and LUC.

(2) The decrease in the luminescent signal is due to the deactivation of the luciferase by its substrate. **Figure 20.1**, point 2, represents the state, where equilibrium between the newly synthesised LUC and its immediate deactivation is reached. The signal that can be measured here is proportional to the synthesis rate of luciferase (promoter activity, transcription, translation, possible modifications, etc.). Measuring of luminescence at this point is called "pre-spray" method and is often applied to assay promoter activity using promoter:LUC reporter constructs.

"Pre-spray" measurements are widely used and described in numerous scientific papers (e.g. 14–16). Many protocols discussing the details of the application of this method are available (e.g. 17, 18). The purpose of the protocol, presented here is to show, how the "flash" method can be used for protein amount measurement. While results, using the invasive in vitro protein amount measurement, have already been described elsewhere (19, 20), a novel non-destructive in vivo protocol is also presented here, using the phytochrome A photoreceptor as an example.

GFP (green fluorescent protein) from the jellyfish *Aquorea victoria* and its derivatives, as well as fluorescent proteins isolated from other sources are widely used as tags to localise proteins under in vivo situations applying diverse fluorescence microscopic techniques (21). In 2008, the Nobel Prize for chemistry was awarded to O. Shimomura, M. Chalfie and R. Tsien for the discovery and development of GFP. The following properties of modern GFP types characterise them as powerful in vivo markers: (i) as relatively small proteins (~27 kDa) they can be encoded as genetic information fused to a sequence of interest, (ii) fluorophore formation is an autocatalytic process, but with a certain maturation time, (iii) oxygen, but no co-factors are needed during this process, (iv) no toxicity has been reported besides generation of radicals by excessive photobleaching and (v) modern GFP derivates do not exhibit an inherent cellular localisation preference but distribute relatively even in cytosol and nucleoplasm. Additionally, spectrally distinct versions of fluorescent proteins allow performing in vivo co-localisation studies of two or more proteins of interest (22).

The experiments presented here are based on transgenic plant material. Arabidopsis lines, harbouring a single copy transgene of *PHYA:PHYA–GFP* or *PHYA:PHYA–LUC*, were created. The expression levels of the chimeric proteins were tested with protein blot analysis, and their functionality was assayed with mutant complementation. The phyA–GFP and phyA–LUC proteins are expressed at wild-type levels, they are functional photoreceptors and their degradation mimics the degradation of the endogenous phyA (data not shown). This indicates that the behaviour of these chimeric proteins reflects the abundance and intracellular dynamics of the phyA photoreceptor.

2. Materials

2.1. In Vivo Protein Amount Measurement Using the Luciferase Reporter Gene

1. MS (Murashige-Skoog Medium) for Arabidopsis (1,000 mL): Dissolve 4.32 g of MS (Murashige-Skoog) powder (Sigma) (23) and 30 g of sucrose in 1 L of deionised water (*see* **Note 1**). Adjust the pH to 5.6–5.8 with 1 M KOH. Add 10 g of agar and autoclave for 20 min at 15 psi, 121°C. Pour the medium into plastic Petri dishes in a laminar hood.
2. Sterilisation solution: 5% Sodium hypochlorite and 0.05% Tween 20 in water. Alternatively, commercially available diluted bleach (e.g. 30% Domestos) can be used. This solution can be stored up to 1 month at 4°C.

3. Luciferin solution: Dissolve D-Luciferin (Biosynth) in 100 mM TRIS-H_3PO_4 (pH 8.0) to prepare a 50 mM stock solution. Aliquots can be stored in the dark at –20°C for several months. Dilute stock aliquot to 5 mM luciferin concentration in 0.01% Triton X-100. This solution can be stored for several days at 4°C. In case of sterile work, filter the solution sterile. Use a small pump spray to spread the luciferin on the seedlings. The pump spray can also be sterilised by washed and filled with 100% ethanol and rinsed subsequently with sterile water (perform this procedure in the laminar hood).
4. Low-light CCD camera equipped with lens, dark box and computer with image-processing software. These components can be purchased separately from different suppliers, but also can be obtained as a complete system. Performing the presented experiment the following system was used: VersArray XP camera (Princeton Instruments) with 1024 × 1024 pixel resolution (*see* **Note 2**); Pentax SMC, 50 mm 1:1.2 lens (*see* **Note 3**); Dark box (size: 45 × 55 × 115 cm; Polytec) (*see* **Note 4**); MetaVue, (version: 6.2r6) driver and processing software (Molecular Devices) (*see* **Note 5**).
5. Rectangular 12 × 12-cm Petri dishes.
6. 1.5-mL Microcentrifuge tubes.
7. Ethanol (70%, 100%).
8. Sterile water (autoclave for 20 min at 15 psi, 121°C).
9. Micropipettes, tips.
10. Top agar (0.1% Agarose).
11. Pasteur pipette, supplied with a latex/rubber bulb.
12. Pump spray with 25–50 mL glass flask (from local pharmacy).
13. Sterile filters and syringes.
14. Filter paper for growing seedling (*see* **Note 6**).
15. Sterile bench, autoclave.
16. Data-analysis software to process data (e.g. Microsoft Excel, Sigmaplot, Origin and OpenOffice).

2.2. In Vitro Measurement of Protein Amounts in Plant Extracts Using Luciferase

1. LUC1 (extraction) buffer: Mix 100 mM KH_2PO_4 and 100 mM K_2HPO_4 to achieve pH 7.8. Add 0.05% Tween 20. Autoclave for 20 min at 15 psi, 121°C and store at 4°C. Add 1 mM dithiothreitol (DTT) before use.
2. LUC2 buffer: 80 mM glycyl-glycine (pH 7.8), 40 mM $MgSO_4$, 60 mM adenosine-5′-triphosphate (ATP). Prepare 1 mL aliquots in microcentrifuge tubes and store at –20°C.

3. Bradford reagent (24): Dissolve 100 mg of Coomassie Brilliant Blue G-250 in 50 mL of 95% ethanol. Add 100 mL of 85% (w/v) phosphoric acid. Dilute to 1 L when the dye has completely dissolved. Filter the solution and store in a dark bottle at 4°C. There are commercially available versions of this reagent (e.g. Sigma, Biorad, Fermentas).
4. Luminometer (EG&G Berthold MicroLumat LB96P luminometer) (*see* **Note** 7).
5. Microtitre plates for the luminometer (*see* **Note 8**).
6. TissueLyser with 1.5-mL tubes adaptor (Qiagen).
7. 3-mm tungsten carbide (Qiagen) or steelballs (sold for ball bearings).
8. Microcentrifuge (refrigerated).
9. Transparent microtitre plate or plastic photometer cuvettes.
10. Plate reader or photometer (e.g. BioRad Model680 Microplate Reader).
11. 10 mM luciferin solution. Dilute the stock (*see* Section 2.1) in water.
12. 20 mg/mL Bovine serum albumin (BSA) solution (Fermentas) as protein control.
13. Micropipettes, tips.
14. Liquid N_2.
15. Data-analysis software to process the results (e.g. Microsoft Excel, Sigmaplot, Origin and OpenOffice.org).

2.3. Microscopic Analysis of GFP Fusion Protein Localisation Patterns in Hypocotyl Cells

1. Petri dishes, 94 mm diameter.
2. Filter paper, 90 mm diameter.
3. Water (*see* **Note 1**).
4. White light chamber (e.g. Sanyo).
5. Light-proof containers with black cloth.
6. Epifluorescence microscope (Axisokop 2; Carl Zeiss GmbH, Jena, Germany).
7. 40×, 63× or 100× fold magnification objectives and suitable immersion media (Carl Zeiss GmbH, Jena, Germany).
8. Filter set for GFP imaging in plant cells (FITC-set no. 10; Carl Zeiss GmbH, Jena, Germany).
9. Camera system (Axiocam HR; Carl Zeiss GmbH, Jena, Germany).
10. Green filter for microscopy (custom filter combination, λ_{max} ~530 nm).
11. Microscopic slides.

12. Cover slides.
13. Water.
14. Torch equipped with green filter or based on green LEDs (λ_{max} 525 nm; Osram).

3. Methods

3.1. In Vivo Protein Amount Measurement Using the Luciferase Reporter Gene

3.1.1. Seed Surface Sterilisation

1. Place the seeds into a 1.5-mL microcentrifuge tube and add 1 mL of sterilisation solution. Work in the laminar hood.
2. Wait for 10 min, let the seeds sink down to the bottom of the tube and remove the supernatant (*see* **Note 9**).
3. Wash the seeds 4–5 times with sterile water. It is important to remove all residual bleach.
4. The sterile seeds can be dispensed with micropipette, using sterilised 1 mL-pipette tips.

3.1.2. Growing Seedlings on MS Medium

1. Place the seeds into microcentrifuge tubes and add 1 mL of water. Mix the cells by inversion and let them settle down. Keep them in the dark at 4°C for 2 days.
2. Remove the water and sterilise the surface of the seeds.
3. Distribute the seeds on the MS plate surface in a small amount of sterile water with a sterile pipette tip or Pasteur pipette. Keep the seeds floating while pipetting and remove the excess water from the plate by tilting and carefully collecting it with the pipette. Alternatively, 0.1% top agar can also be used for seed distribution.
4. Induce germination with white light (about 50–70 $\mu mol/m^2/s$) for 6–8 h at 22°C.
5. Place the plates into a plant growth chamber providing the desired light and temperature conditions.

3.1.3. Growing Seedlings on Paper

There are several advantages of growing seedlings on paper. One of these is the reduced amount of sterile work. Unfortunately, sometimes fungal contamination on the surface of the seeds can reduce germination efficiency. To solve this problem, it is recommended to perform the "Quick seed sterilization" protocol (*see* **Section 3.1.4**).

1. Place 4 layers of filter paper into a Petri dish and wet with water (*see* **Note 10**).

2. Disperse the seeds onto the paper, gently shaking them from a small piece of glossy paper. Arrange the seeds according to the planned experiment with a blunt-ended forceps. Close the Petri dishes with Saran Wrap.
3. Keep the plates in the dark at 4°C for 2 days.
4. Illuminate the seeds with white light (about 50–70 $\mu mol/m^2/s$) for 6–8 h at 22°C to induce germination.
5. Place the Petri dishes into desired light and temperature conditions. It is recommended to use temperature- and light-controlled plant growth chambers (e.g. SANYO) or custom-designed plant growth rooms. For etiolated seedlings, place the samples into a thick black cardboard box and cover them with thick black piece of textile. Alternatively, 3 layers of aluminium foil can also be used (*see* **Note 11**).
6. Let the plantlets grow for 4–5 days (*see* **Note 12**).

3.1.4. Quick Seed Sterilisation

This protocol is designed to reduce fungal contamination when seeds are sown on filter paper and does not result in sterile seed surfaces required for growth on MS plates.

1. Place the seeds into a 1.5-mL microcentrifuge tube.
2. Add 1 mL of 70% ethanol, shake gently for 10 min.
3. Decant the 70% ethanol.
4. Add 100% ethanol and shake for 5 to 8 min.
5. Pipette the seeds onto a filter paper and let them dry in a laminar hood.

3.1.5. "Flash" Measurement of Luciferase in the CCD Camera

When an experiment is being designed, the "flash" nature of this examination should be kept in mind. Disperse the seeds (either on paper or MS medium) according to the planned number of timepoints. In this presented sample experiment, monitoring the degradation of the phyA protein in etiolated seedlings during a 5-h-long red light illumination will be described. In order to measure the luminescence during the light treatment at 5 timepoints, 5 illuminated and 1 control plates will be needed. Raise the plants on the chosen medium under the desired conditions. In the presented experiment, 4-day-old etiolated, paper-grown seedlings will be examined.

1. Prepare a 12 × 12 cm-Petri dish with MS medium or paper (same system as used for seedling growth). Divide it into 6 parts with a permanent marker: "1 h illumination", "2 h illumination" and so on.
2. At time 0 h, start the illumination of the "5 h irradiation" plate; at time 2 h, the "3 h irradiation" plate and so on.
3. When the illumination of the last plate ("1 h irradiation" plate) is started, the seedlings can be arranged for the

subsequent luciferase assay in the used light field. Place 15–20 seedlings from each illuminated plate into the corresponding marked sections of the prepared 12 × 12-cm Petri dish (*see* **Note 13**). Use as many seedlings as possible to increase the accuracy of the measurement.

4. A few minutes before the end of the 5 h light treatment, transfer the rectangular Petri dish (with the seedlings to be examined) to safe green light field and add the non-illuminated dark control seedlings to it.
5. At the end of the light treatment, spray the plate with 5 mM luciferin solution. Avoid washing away the arranged seedlings. Use sterile luciferin in case the seedlings will be further grown on sterile medium.
6. Place the samples in the dark and wait 5–8 min for the luciferin diffusion to the tissues. (In case of green tissues, this dark treatment additionally reduces the phosphorescence signal of the chlorophyll, which can disturb the measurement.)
7. Record an image with the CCD camera (*see* **Fig. 20.2A,C,E**). Choose an appropriate exposure time to avoid saturated areas on the image. Short exposure times provide information of the current amount of LUC.
8. Record a reference image (*see* **Fig. 20.2B,D**). Use a green safety torch to illuminate the seedlings.
9. At the end of the experiment, record an image without samples, but with the applied exposure settings. This represents the background for all images.
10. Save the recorded images. The next steps will give some guidelines for data processing.
11. Subtract the background image from the experimental images.
12. Create individual areas of interest (these are called "region" in the MetaVue or "selection" in the ImageJ software) around each seedling. Measure the integrated intensity of the pixels inside the marked areas. This is the LUC-produced luminescent signal that was collected by the sensor through the exposure time.
13. Save the numeric results into a datasheet. In case the background correction has not been performed so far, it can also be done at this point using the MetaVue software. After marking the regions around the seedlings, transfer them into the dark background image. Save the integrated intensity values (background) and subtract them from the corresponding signal values.

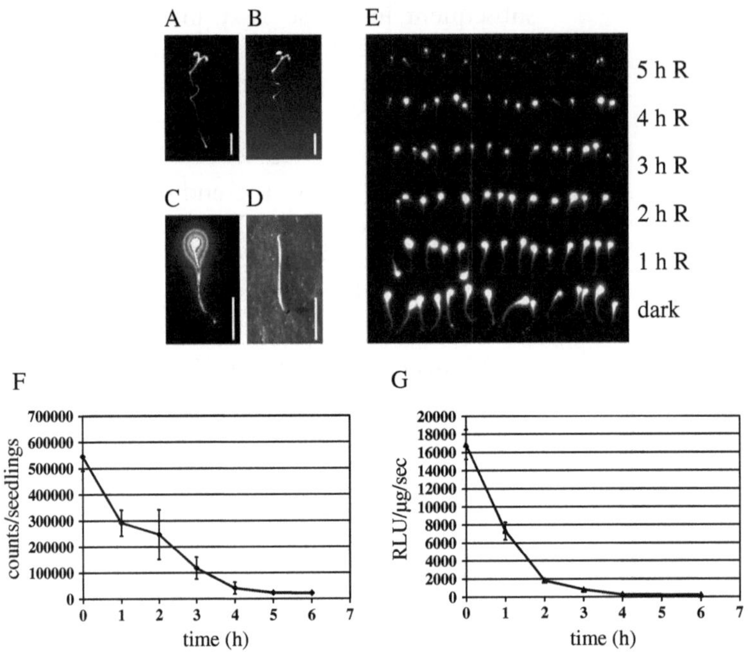

Fig. 20.2. "Flash" measurement of luminescence emitted by transgenic Arabidopsis seedlings measured by low-light CCD camera and luminometer. (**A**) Seedlings harbouring a *PHYA:PHYA–LUC* construct were grown for 7 days under 12 h white-light photoperiods on MS medium. The luminescence image was taken with 15-min exposure time. (**B**) The seedling, presented on image (**A**), was illuminated with a safety green light torch. Exposure time: 0.5 s. (**C**) Luminescence of a transgenic 4-day-old etiolated seedling was recorded with 5-min exposure time. (**D**) To obtain a bright-field reference image of the seedling shown in panel C, 0.5-s exposure time was applied. (**E**) Kinetics of light-induced degradation of a phyA–LUC fusion protein using a low-light CCD camera. Transgenic seedlings were grown for 4 days on filter paper in the dark and illuminated for 1–5 h using red light (R). The control seedlings were not treated with light (*dark*). The seedlings were arranged on wet filter paper, sprayed with luciferin, incubated 7 min in the dark and exposed for 5 min. During data processing, different lookup tables (LUT) were chosen for optimal presentation. LUT was used on images (**A**) and (**C**), where the scale extends from black (minimum signal) to white (maximum signal). On images (**B**), (**D**) and (**E**), monochrome LUT visualisation was used. Images (**A–D**) were taken with a +6 diopter close-up lens. Scale bar: 5 mm. (**F**) In vivo kinetics of light-induced degradation of a phyA–LUC fusion protein in 4-day-old etiolated seedlings using a VersArray XP CCD camera. (**G**) In vitro kinetics of light-induced degradation of a phyA–LUC fusion protein in 4-day-old etiolated seedlings using with a Berthold MicroLumat LB96P luminometer.

14. Calculate mean and standard error of the integrated pixel intensities of all seedlings from the same treatment.
15. Due to the "flash" type of measurement, each seedling is measured only once. This is one of the main differences from the promoter:LUC-based expression studies. This also means that absolute normalisation of the data cannot be done (normalise the results to the number of seedlings measured) and the standard error values may be high (*see* **Fig. 20.2F**). For better precision, increase the number of seedlings assayed. The result can be effected by different size or different position (e.g. photons directed away from the camera) of the seedlings.

The next subchapter describes a method for more precise LUC-based measurements of protein amounts; however, the plants will not survive this procedure because it is invasive and in vitro.

3.2. In Vitro Measurement of Protein Amount in Plant Extracts Using Luciferase

3.2.1. Preparation of Native Protein Extract from Seedlings

1. Grow seedlings on paper or MS plate under the desired growth conditions (*see* **Sections 3.1.2** and **3.1.3**). Prepare three parallel sample sets (triplicates) (*see* **Note 14**).
2. Sample collection: Carefully remove all water with a paper tissue from the seedlings and place them into 1.5-mL microcentrifuge tubes, containing one 3-mm stainless steelball.
3. Immediately transfer the samples to liquid N_2 and store them at –80°C until processing.
4. Mount the tubes into the previously cooled (–20°C) tube-holding adaptors of the TissueLyser and shake them two times for 30 s at maximum speed (30 s^{-1}). Cool the samples between the shakings in liquid N_2.
5. Place the tubes on ice and add 300 μL of LUC1 buffer.
6. Mix by shaking manually. Do not vortex.
7. Centrifuge for 15 min at about 20,000×*g* and 4°C.
8. Transfer the supernatant to new cold tubes and keep them on ice.
9. If possible, complete the luciferase measurements on the day of the protein extraction. Otherwise, freeze samples in liquid N_2 and store at –80°C. After freezing, thaw the samples on ice.

3.2.2. Measurement of Luminescence with an Injector Supplied Luminometer

1. Prior to the measurement, it is advisable to run the washing/cleaning program of the instrument (*see* **Note 15**), and then fill up the injector system with 10 mM luciferin solution.
2. Thaw the LUC2 buffer on ice. This buffer must be kept on ice because its ATP component is heat sensitive.
3. Add the following components into the wells of an opaque 96-well microtitre plate (*see* **Note 8**):

   5–50 μL of native protein extract (*see* **Section 3.2.1**) (*see* **Note 16**).

   50 μL of LUC2 buffer.

   LUC1 buffer up to 150 μL.

4. Let the plate warm up to room temperature.
5. Set the luminometre to inject 100 μL of luciferin solution into each well. Select an integration time between 1 and 120 s, depending on the expression level, protein concentration and instrument sensitivity. Generally, 15 s is sufficient, and a pre-measurement delay is not necessary.
6. Select the wells to be measured and run the injection/measurement assay.
7. Transfer and save your data to data-analysis software (e.g. Microsoft Excel, Sigmaplot, Origin and OpenOffice).
8. Unload the luciferin solution, clean and empty the injector system (*see* **Note 15**).

3.2.3. Measurement of Protein Amount in a Plate Reader (see Note 17)

The Bradford assay is a fast and accurate assay allowing high-throughput analysis (*see* **Note 18**). The basis of the assay is a shift in absorbance of Coomassie Brilliant Blue G-250 from 465 to 595 nm when it binds to protein (24).

1. Let Bradford reagent warm up to room temperature.
2. Add 190 μL of Bradford reagent to the wells of a transparent 96 well microtitre plate.
3. Add 10 μL of LUC1 buffer to the control wells and mix them by pipetting up and down (*see* **Note 19**).
4. Prepare a dilution series of the BSA stock in LUC1 buffer by adding 0 to 10 μg of BSA in 10 μL volume to the calibration wells. Mix by pipetting.
5. Add 10 μL of native protein extract (*see* **Section 3.2.1**) to the wells and mix up by pipetting. Assay the protein extracts in duplicate or triplicate.
6. Incubate the samples for about 5 min (not more than 1 h) at room temperature.
7. Measure the absorbance of the samples in a plate reader at 595 nm.
8. Save the data to a table of data-analysis software.
9. Plate readers can be set to calculate and subtract background values of defined wells. Otherwise, do so manually.
10. Create a calibration curve using the BSA values: plot the absorbance values as a function of the protein concentration and fit the data by linear regression. Calculate the protein concentrations of the native protein extracts based on the regression line's formula.

3.2.4. Analysis of Luminescence from Protein Extracts: Data Processing

1. Collect the protein concentrations and luminometer data into one datasheet.
2. The luminometer provides relative light unit (RLU) values. Subtract the background from obtained data.

3. Divide these corrected RLU values with the photon gaining time (measurement length) in case the luminometer has not done so yet. This value is directly proportional to the luciferase concentration in the sample, thus it is necessary to normalise it with the corresponding protein concentrations to obtain the specific luminescent activity of 1 μg protein extract per second.
4. Calculate mean values and standard errors from the corresponding triplicate values. (*see* **Section 3.1.5**)
5. Plot the calculated values as RLU/μg/sec to a chart (*see* **Fig. 20.2G**).

The results obtained by the two different methods described above are shown in **Fig. 20.2F,G** and are in good agreement with earlier analyses of light-induced degradation of PHYA in non-transgenic seedlings (5). Obviously, the absolute numbers differ between methods, and the in vivo data suffer from increased variability – a price to be paid for maintaining seedlings alive for further experiments or seed collection.

3.3. Microscopic Analysis of GFP Fusion Protein Localisation Patterns in Hypocotyl Cells

3.3.1. Light Treatments of Plants

For microscopic analysis, 4-day-old etiolated seedlings grown on filter paper in Petri dishes (*see* Section 3.1.3) were either irradiated with continuous red light (cR, λ_{max} 660 nm), far-red light (cFR, λ_{max} 720 nm) or were kept in darkness (25). Pulses of red light (λ_{max} 650 nm, KG65 filter; Balzers) were given to seedlings on microscopic slides using the transmitted light path of the microscope.

3.3.2. Microscopic Analysis of GFP Fusion Protein Localisation Patterns in Hypocotyl Cells

Microscopic in vivo analysis of light-regulated localisation patterns of GFP fusion proteins is not trivial. To minimise unwanted light perception prior and during analysis seedlings have to be handled with maximal care (*see* **Note 20**).

1. Place one drop of distilled water on a microscope slide. Under green safety light, carefully take a seedling with forceps and place it vertically in the water drop on the slide (*see* **Note 21**). Take care to cover the seedling with one cover slide only. To ensure that the seedling is held in a fixed position, the cover slide should stick to the microscopic slide by adhesion forces and should not swim on the water drop.
2. Use 400–1,000 fold magnification for analysis of subcellular localisation patterns in Arabidopsis cells. When using immersion objectives, initial focussing can be simplified by putting a small drop of immersion medium right at the place on the cover slide where it covers the seedling.

3. Place the slide on the microscope stage. Carefully focus on the upper third of the hypocotyl using the dimmed transmitted light pass filtered with green light filter. This procedure is critical and should be trained because the objective should not touch the cover slide to prevent damage of its front lens (*see* **Note 22**).
4. Choose the appropriate filter set (*see* **Note 23**), re-adjust focal plane and analyse epifluorescence signals briefly. Switch the beam path to the camera port. Exposure time should be adjusted rapidly before recording an image (*see* **Note 24**). Subsequently, take a transmitted light image for reference (*see* **Fig. 20.3**).

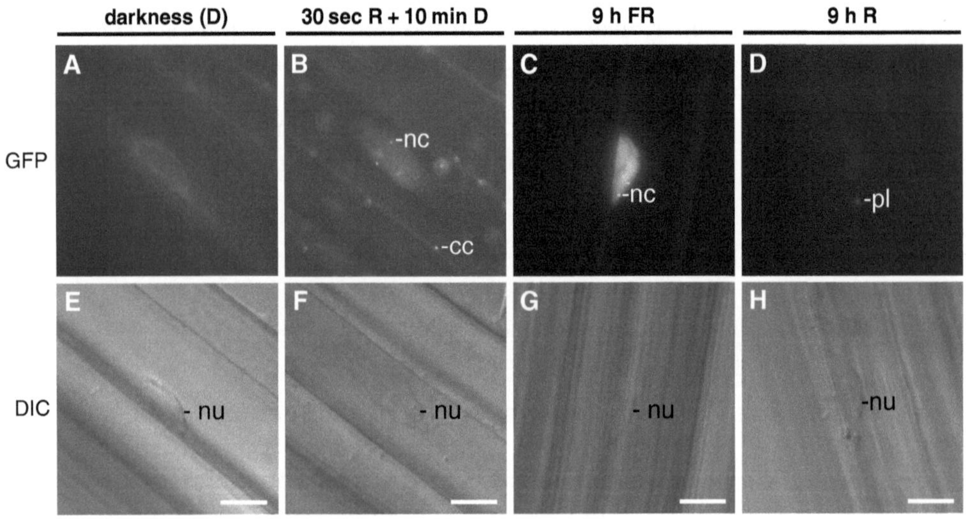

Fig. 20.3. **Light-dependent localisation patterns of phyA–GFP fusion proteins in hypocotyl cells of Arabidopsis seedlings.** Transgenic *PHYA:PHYA–GFP* Arabidopsis seedlings were grown on filter paper for 4 days in constant darkness (**A–H**) and subsequently irradiated for 9 h with red light (**D, H**), far-red light (**C, G**) or kept in darkness (**A, E**) until microscopic analysis of hypocotyl cells. Alternatively, a dark-grown seedling was mounted in dim green light on a microscopic slide. After focusing on a nuclear region of a hypocotyl cell using green light, the seedling was treated with a brief pulse (30 s) of red light obtained by filtering the transmitted light path of the microscope with a KG65 filter. After further incubation for 10 min on the microscope stage in darkness, microscopy has been performed (**B, F**). *Upper row* (**A–D**) epifluorescence images of representative cells and *lower row* (**E–H**) respective bright-field images applying DIC (differential interference contrast). Exposure times of (**A**), (**B**) and (**E**), 1.7 s in each case and exposure time of (**C**), 0.6 s. Bar represents 10 μm. nu = nucleus; pl = autofluorescence of a plastid; cc = cytosolic complex; nc = nuclear complex.

5. For data presentation, use software packages like Adobe Photoshop or the freeware GIMP. Take care not to distort the image data by manipulation. An outline about proper handling of image data is given in (26). For scaling, use the respective tool of the camera acquisition software when proper settings for applied magnification and image resolution are available. Alternatively, determine with reference slides the pixel size at the magnification and image resolution

used for analysis. Calculate from this information the pixel number equivalent to the wanted length of the scale bar and insert a respective line or box into the image using graphics software.

The results presented here (*see* **Fig. 20.3**) demonstrate the complex light-dependent intracellular dynamics of the phyA photoreceptor in etiolated seedlings at the onset of photomorphogenesis (27–29). Without illumination, phyA–GFP is detectable in the cytosol and distinct cytosolic strands surrounding the nucleus are clearly contrasting from the nuclear compartment (*see* **Fig. 20.3A**). After light activation, phyA–GFP can form complexes in the cytosol or move into the nucleus where a diffusible as well as a speckled pool becomes observable (*see* **Fig. 20.3B,C**). Due to the labile nature of activated phyA, extended irradiations with photons establishing a high proportion of Pfr lead to massive degradation of phyA–GFP resulting in hardly detectable fluorescence signals (*see* **Fig. 20.3D**).

4. Notes

1. For any part of the protocol, when water is used, it is recommended to use deionised water with resistivity of 18.2 MΩ/cm and total organic content of less than five parts per billion (Millipore).
2. Before purchasing a CCD camera, it is worth to request on-site demonstrations from various suppliers to directly test the objects of the planned research. For many applications, a liquid nitrogen or thermoelectrically cooled low-light CCD camera with high dynamic range, low signal-to-noise ratio and high resolution is suitable. Often, 16-bit image depth is essential, though this feature increases the price. Larger sensor size allows better resolution, thus enabling examinations of seedlings at organ or tissue level. Some suppliers are listed here: http://www.hamamatsu.com/, http://www.roperscientific.com/, http://www.photek.com/, http://www.ultralum.com/, http://www.princetoninstruments.com/, http://www.andor.com/.
3. Most of the CCD cameras on the market are supplied with standard C-mount to attach different optical elements. Inserting an adaptor to the C-mount allows the user to choose a suitable photographic lens best fitting for the purpose. Always use fast lenses (small f-number); prefer primes to zoom lenses. Wide-angle lenses allow imaging larger area (e.g. several Petri dishes) and macro lenses allow imaging of individual seedlings. It is also worth considering obtaining

close-up lenses (from local photographic suppliers), which can be attached to the filter thread of the primary lens. Although this setup results in loss of light and quality, it is economic and can still give satisfying results.

4. A dark box is an essential part of a luminescent camera system. It has to close properly, excluding any light contamination. The camera fitting has to be light-tight as well. The size depends on the size of plants to be examined and the lens, fitted on the CCD camera. Working with Arabidopsis seedlings, ~100 × 50 × 50 cm (height, width, depth) is sufficient.
5. It is worth to test the supplied driver software before purchase. User-friendly software makes data processing easier and faster. For data processing, ImageJ software is a free alternative (http://rsbweb.nih.gov/ij/).
6. The quality of filter paper may affect the germination efficiency of seeds. Sometimes, there are differences between batches. It is advisable to test different types from different suppliers (e.g. Macherey-Nagel, Schleicher and Schuell and Whatman).
7. It is recommended to choose a luminometer that can handle 96-well microtitre plates (some can handle 384-well plates and even stacks of plates). The best sensors generate almost no background signal and show an 8-log linear range. The presence of an injector system is required to perform luciferase measurements. In case the luminometer has two injectors, it is even possible to use the dual-luciferase reporter assay (Promega). There are numerous companies, which distribute microplate luminometers, some of them are listed below: http://www.berthold-ds.com/, http://www.promega.com, http://www.bmglabtech.com, http://www.turnerbiosystems.com/, http://www.moleculardevices.com, http://www.dynextechnologies.com.
8. Use 96-well opaque microtitre plates with low background signal. White plates usually generate higher crosstalk between the wells than black plates. Preparing the reaction mixtures, however, is easier in white than in black plates. Manufacturers of luminometers usually recommend microtitre plate types for their instrument.
9. For a proper removal of the supernatant, the seeds can be centrifuged down to the bottom of the tube. Several seconds of low-speed (<2,000×*g*) spinning does not harm the seeds.
10. The optimal amount of water depends on the size and type of paper. Too little water will reduce germination efficiency;

too much water will wash the seeds away and can reduce germination efficiency as well. In the ideal case, a drop of water should appear on the lower side of the dish when it is tilted.

11. Vertical positioning of the plates results in seedlings growing on the paper surface, which is most convenient for CCD camera imaging.
12. Arabidopsis seeds do not contain huge amounts of nutrients, thus it is not recommended to let them grow on paper for more than one week.
13. Treat the seedlings gently to avoid mechanical stress, especially when seedlings should survive the measurement. Grab them gently with forceps, preferably at the lower end of the hypocotyl. Etiolated seedlings are even more sensitive than light-grown ones. Use spring steel forceps (Carl Roth) or lift the plantlets grabbing the cotyledons with bare fingertips.
14. The amount of plant material to be collected depends on several parameters such as expression level of luciferase, developmental state of the plants and purpose of the experiment. As few as 30–50 4-day-old etiolated seedlings can give sufficient native protein sample for in vitro analysis.
15. The injector system of the luminometer must be kept clean. Always refer to the users' manual, but generally 30–50 cycles of water washes are sufficient. It is also important to remove the washing water from the system, leave it clean and empty before and after the assay.
16. The amount of native protein extract used in an assay depends on the RLU (relative light unit) produced. This value describes the output generated by the photomultiplier tube and is proportional to the photon number perceived. Each luminometer possesses different saturation values and detector characteristics. Refer to the users' manual or calibrate your luminometer with commercially available luciferase (Promega). Choose the amount of native protein extract to be used according to the expected luciferase amount in your sample. In most cases, the detectors possess conveniently broad linear detection range.
17. This assay can also be performed in a conventional spectrophotometer, using disposable 1.5-mL plastic cuvettes. In this case, increase the volume of the assay to 1 mL.
18. Due to the sensitivity of the assay, it is important to use clean plastic ware that is not contaminated with proteins. Most of the single-use tips, tubes and microtitre plates fulfil this requirement. Remember that certain buffer constituents such as Triton X-100, SDS or Chaps can

disturb the Bradford assay (see reference (30) and technical information sheets about Bradford assay kits supplied by manufacturers).

19. The best background control for the in vitro luciferase assay is the measurement of the luminescence of the protein extract, made from a non-transgenic background line. This measurement integrates the background produced by (i) the protein extract, (ii) the used assay solutions, (iii) the microtitre plate and (iv) the luminometer.

20. Controlled irradiation programs as well as sample preparation should be performed in close proximity to the microscope stand. For optimal performance, the microscopic analysis should be performed in a dark (black painted) room. As much as possible, light emitted by the microscope, camera, computer or other sources has to be shielded. Handling of plant material at the microscope has to be fast and restricted to green safety light conditions. Dimmed green light torches for sample handling and green light filters for focussing in the microscope have to be used. Even then, there remains only a very brief time span for microscopic analysis and image acquisition after onset of fluorescence excitation. A well-adjusted microscope and training are needed for optimal performance. Naturally, this is not critical for proteins not related to light.

21. Parallel orientation of the seedling to the *X*- or *Y*-axis of the microscope table makes it easier to keep track of cells in the hypocotyl.

22. Some microscopes such as the Axioskop 2 (Zeiss) are equipped with an adjustable stop position of the focus drive. Use this device for initial focussing to prevent objective crashes.

23. Make yourself familiar with the filter sets available at the local microscope systems. Analyse organs, tissues and cells of interest of wild-type seedlings side-by-side to seedlings expressing the fusion protein. Filter sets for epifluorescence analysis of a certain fluorophore such as GFP are available from different sources but may vary significantly due to different compositions of their optical components. As a consequence, autofluorescence derived by chlorophyll or other cellular components as well as "bleed-through" of potentially co-expressed fluorophores may be an issue. If necessary, further filter sets fitting your specific needs should be tested and purchased. Detectable plastid-derived autofluorescence is indicated in **Fig. 20.3D**. The filter set used in this study was pre-selected for GFP analysis in Arabidopsis hypocotyl cells.

24. An optimal exposure time intends to exploit the full dynamics of the camera system without saturation of the detector. A reasonable exposure time for a certain genotype may be determined in advance with test seedlings. For qualitative to semi-quantitative comparison of localisation patterns, exposure times should be kept constant whenever possible.

Acknowledgements

This work was supported by grants of the Deutsche Forschungsgemeinschaft (DFG) to S.K (SFB 592 and GRK 1305) and A.V. (SFB 592); and of the Alexander von Humboldt Foundation to A.V. (IV-UNG/1118446 STP).

References

1. Chen, M., Chory, J., and Fankhauser, C. (2004) Light signal transduction in higher plants. *Annu Rev Genet* **38**, 87–117.
2. Franklin, K. A. (2008) Shade avoidance. *New Phytol* **179**, 930–944.
3. Schäfer, E. and Nagy, F. (2006) Physiological basis of photomorphogenesis. In: Photomorphogenesis in Plants and Bacteria. Function and Signal Transduction Mechanisms, pp. 13–23. Schäfer, E., and Nagy, F., eds. Springer, New York, NY.
4. Fankhauser, C. and Chory, J. (1997) Light control of plant development. *Annu Rev Cell Dev Biol* **13**, 203–229.
5. Hennig, L., Büche, C., Eichenberg, K., and Schäfer, E. (1999) Dynamic properties of endogenous phytochrome A in *Arabidopsis* seedlings. *Plant Physiol* **121**, 571–577.
6. Schäfer, E., Kircher, S., and Nagy, F. (2006) Intracellular localization of phytochromes. In: Photomorphogenesis in Plants and Bacteria. Function and Signal Transduction Mechanisms, pp. 155–170. Schäfer, E., and Nagy, F., eds. Springer, New York, NY.
7. Hiltbrunner, A., Viczián, A., Bury, E., Tscheuschler, A., Kircher, S., Tóth, R., Honsberger, A., Nagy, F., Fankhauser, C., and Schäfer, E. (2005) Nuclear accumulation of the phytochrome A photoreceptor requires FHY1. *Curr Biol* **15**, 2125–2130.
8. Hiltbrunner, A., Tscheuschler, A., Viczián, A., Kunkel, T., Kircher, S., and Schäfer, E. (2006) FHY1 and FHL act together to mediate nuclear accumulation of the phytochrome A photoreceptor. *Plant Cell Physiol* **47**, 1023–1034.
9. Nakatsu, T., Ichiyama, S., Hiratake, J., Saldanha, A., Kobashi, N., Sakata, K., and Kato, H. (2006) Structural basis for the spectral difference in luciferase bioluminescence. *Nature* **440**, 372–376.
10. Wood, K. V. (2007) The bioluminescence advantage. *Promega Notes* **96**, 3–5.
11. Ow, D. W., Wet, J. R., Helinski, D. R., Howell, S. H., Wood, K. V., and Deluca, M. (1986) Transient and stable expression of the firefly luciferase gene in plant cells and transgenic plants. *Science* **234**, 856–859.
12. Van Leeuwen, W., Hagendoorn, M. J. M., Ruttink, T., Van Poecke, R., Linus, H. W., Van Der Plas, L. H. W., and Van Der Krol, A. R. (2000) The use of the luciferase reporter system for *in planta* gene expression studies. *Plant Mol Biol Rep* **18**, 143–144.
13. Sherf, B. A. and Wood, K. V. (1994) Firefly luciferase engineered for improved genetic reporting. *Promega Notes* **49**, 14–21.
14. Millar, A. J., Short, S. R., Chua, N.-H., and Kay S. A. (1992) A novel circadian phenotype based on firefly luciferase expression in transgenic plants. *Plant Cell* **4**, 1075–1084.
15. Millar, A. J., Short, S. R., Hiratsuka, K., Chua, N.-H., and Kay S. A. (1992) Firefly luciferase as a reporter of regulated gene expression in plants. *Plant Mol Biol Rep* **10**, 324–337.
16. Kay, S. A., Millar, A. J., Smith, K. W., Anderson, S. L., Brandes, C., and Hall, J. C. (1994)

Video imaging of regulated firefly luciferase activity in transgenic plants and drosophila. *Promega Notes* **49**, 22–27.
17. Southern, M. M., Brown, P. E., and Hall, A. (2006) Luciferases as reporter genes. In: *Arabidopsis* Protocols. MMB323, pp. 293–305. Salinas, J. and Sanchez-Serrano, J. J., eds. Humana Press, Totowa, NJ.
18. Hall, A. and Brown, P. (2007) Monitoring circadian rhythms in *Arabidopsis thaliana* using luciferase reporter genes. In: Methods in Molecular Biology, Circadian Rhythms: Reviews and Protocols Volume 362, pp. 143–152. Rosato, E., ed. Humana Press, Totowa, NJ.
19. Viczián, A., Kircher, S., Fejes, E., Millar, A.J., Schäfer, E., Kozma-Bognár, L., and Nagy, F. (2005) Functional characterization of phytochrome interacting factor 3 for the *Arabidopsis thaliana* circadian clockwork. *Plant Cell Physiol* **46**, 1591–1602.
20. Shen, H., Moon, J., and Huq, E. (2005) PIF1 is regulated by light-mediated degradation through the ubiquitin-26S proteasome pathway to optimize photomorphogenesis of seedlings in *Arabidopsis. Plant J* **44**, 1023–1035.
21. Brandizzi, F., Frizker, M., and Hawes C. (2002) A greener world: The revolution of plant bioimaging. *Nat Mol Cell Biol* **3**, 520–530.
22. Jach, G. (2006) Use of fluorescent proteins as reporters. In: *Arabidopsis* Protocols. MMB323. Salinas, J. and Sanchez-Serrano, J. J., eds. Humana Press, Totowa, NJ.
23. Murashige, T. and Skoog, F. (1962) A revised medium for rapid growth and bioassays with tobacco tissue cultures. *Physiol Plant* **15**, 473–497.
24. Bradford, M. (1976) A rapid and sensitive method for the quantitation of microgram quantities of protein utilizing the principle of protein-dye binding. *Anal Biochem* **72**, 248–254.
25. Schäfer, E. (1977) Kunstlicht und Pflanzenzucht. In: Optische Strahlungsquellen. Albrecht, H., ed. Lexika-Verlag, Grafenau, Germany.
26. Rossner, M. and Yamada, K. M. (2004) Whats' in a picture? The temptation of image manipulation. *J Cell Biol* **166**, 11–15.
27. Kircher, S., Kozma-Bognar, L., Kim, L., Ádám, E., Harter, K., Schäfer, E., and Nagy, F. (1999) Light quality-dependent nuclear import of the plant photoreceptors phytochrome A and B. *Plant Cell* **11**, 1445–1456.
28. Kircher, S., Gil, P., Kozma-Bognar, L., Fejes, E., Bury, E., Adam, E., Schäfer, E., and Nagy, F. (2002) Nucleocytoplasmic partitioning of the plant photoreceptors phytochrome A, B, C, D and E is regulated differentially by light and exhibits a diurnal rhythm. *Plant Cell* **14**, 1541–1555.
29. Kim, L., Kircher, S., Toth, R., Adam, E., Schäfer, E., and Nagy, F. (2000) Light induced nuclear import of phytochrome-A: GFP fusion proteins is differentially regulated in transgenic tobacco and *Arabidopsis. Plant J* **22**, 125–133.
30. Holtzhauer, M. (2006) Basic Methods for the Biochemical Lab. Springer, New York, NY.

Chapter 21

Fluorescence-Activated Cell Sorting in Plant Developmental Biology

Anjali S. Iyer-Pascuzzi and Philip N. Benfey

Abstract

Understanding the development of an organ requires knowledge of gene, protein, and metabolite expression in the specific cell types and tissues that comprise the organ. Fluorescence-activated cell sorting (FACS) is an efficient method to isolate specific cells of interest, and the information gained from this approach has been integral to plant developmental biology. The Benfey lab has developed this method to examine gene expression profiles of different cell types in the Arabidopsis root under both standard and stress conditions. In addition to gene expression, downstream applications of FACS include proteomic and metabolite analysis. This is a powerful method to examine biological functions of specific cell types and tissues with a systems biology approach.

Key words: Cell sorting, GFP, FACS, cell-type specific gene expression, root, Arabidopsis.

1. Introduction

Fluorescence-activated cell sorting (FACS) is a rapid, efficient method to isolate specific cell types of interest. Once isolated, these cells can be used for downstream applications, including transcriptomics and proteomics. The Benfey lab has used FACS of GFP-marked lines coupled with microarray analysis to create a gene expression map of the Arabidopsis root (1, 2), and 5 cell types under 2 different stress conditions (3).

FACS is a type of flow cytometry in which individual cells in a heterogenous suspension are sorted based on fluorescence and light-scattering properties. In plant developmental biology, FACS is most often used to isolate enriched populations of cells

L. Hennig, C. Köhler (eds.), *Plant Developmental Biology*, Methods in Molecular Biology 655,
DOI 10.1007/978-1-60761-765-5_21, © Springer Science+Business Media, LLC 2010

expressing a fluorescent marker (FM) under control of a cell-type or tissue-specific promoter. Transgenic plants expressing such a marker are protoplasted and the protoplasts are subsequently sorted into pure subpopulations expressing or not expressing the FM. Unlike other methods for cell isolation, FACS is a rapid, robust procedure which starts with living tissue and typically results in thousands of cells. If there is a concern about protoplasting (*see* **Note 1**), tissue can be homogenized and nuclei sorted (4). Other techniques for isolating specific cell types within plants include laser capture microdissection (LCM) and microsampling. LCM is advantageous in species where fluorescently marked cell types of interest are not available but is more labor intensive than FACS. Usually plant tissue must be sectioned and first fixed, frozen, or paraffin-embedded prior to sampling (5). Microsampling is a minimally invasive sampling method that uses glass microcapillaries to collect tissue (6), but a drawback is that it is difficult to acquire more than a small sample of tissue.

Here we describe FACS for transgenic plants expressing GFP in specific cell types within the Arabidopsis root. This is an update to the procedure first detailed in (7). This protocol describes how to prepare seeds, grow plants and protoplast for FACS. It assumes contact with a cell-sorting facility that will perform either the actual sorting or guide the researcher through that technique.

2. Materials

2.1. Seed Preparation and Sterilization

1. FM-marked lines in the cell type of interest (the researcher should check with the cell-sorting facility as to what FM can be sorted; GFP is very typical).
2. Bleach.
3. Tween 20.
4. 50-mL Falcon tubes.

2.2. Seed Plating and Growing Conditions

1. Nitex 03-100/44 mesh (Sefar, www.sefaramerica.com or 800-283-8182).
2. Square plates for growing Arabidopsis.
3. Transfer pipettes.

2.3. Protoplasting

1. Cellulysin (Calbiochem).
2. Pectolyase (Sigma).
3. Small Petri dish (35 × 10 mm, BD Falcon).
4. 70-μm cell strainer mesh (BD Falcon; *see* **Note 2**).
5. 40-μm cell strainer mesh (BD Falcon).

6. Surgical blades (such as Feather #10 or 11 or Miltex #21).
7. 50- and 15-mL Falcon tubes.
8. Tubes to sort into (*see* **Notes 3** and **4**).
9. Solution A (*see* **Note 5**): 600 mM mannitol, 2 mM $MgCl_2$, 0.1% BSA, 2 mM $CaCl_2$, 2 mM MES, 10 mM KCl, pH 5.5. Store at –20°C.
10. Solution B (*see* **Note 6**): Aliquot 50 mL of fresh solution A, add 0.75 g cellulysin and 0.05 g pectolyase. Can be stored at –20°C for up to 1 month.

3. Methods

3.1. Seed Preparation and Sterilization

The amount of seeds needed depends on where the marker is expressed, the level of expression, how many replicates are sorted, and how many cells are needed for downstream applications. If expression is in one cell type with very few cells per plant, more plants are needed than if the marker is expressed in a greater number of cells. It is best to empirically determine the amount of seed needed for individual FM lines. It is very important to do preliminary experiments with each FM line to set the window for sorting and know how many cells will result from a set number of plants (*see* **Note 7**). As a general guide, 3–5 plates per sorting replicate are typical for major cell types in Arabidopsis roots. Approximately 100–150 μL seed is necessary for five MS plates.

Seeds can be sterilized in any typical manner, but suspending seeds in sterile water in the tube prior to plating is generally used. Below is a common sterilization method:

1. Rinse seeds with water. For the amount of seed used in cell sorting, it is best to use 50 mL-Falcon tubes for this step.
2. Incubate seeds in 50% bleach/0.001% Tween solution, mixing for 5 min.
3. Centrifuge seeds briefly (10–30 s).
4. In sterile hood, rinse well with sterile water five times.
5. Stratify at 4°C now or after seed plating (below).

3.2. Seed Plating and Growing Conditions

1. Before plating, cut mesh just slightly smaller than the size of the square plates and autoclave in small batches (*see* **Note 8**). Place individual squares of mesh directly on agar in a sterile hood. Use either (very clean) hands or tweezers.
2. Use a transfer pipette to plate sterile seeds directly on mesh ('paint' them on in a line). It is convenient to plate in two rows, allowing enough space for root growth between the rows. Use about 500–1,000 seeds per row. Do not make

the line more than 5 seeds thick (2–4 seed thickness works well). Follow typical protocols for stratification and growth conditions. Sorting 5–6 days after stratification works well.

3. Grow plants vertically on MS agar. Use typical growing conditions. Stresses, such as nutrient deficiencies or toxicities, can be easily assayed by incorporating these in the MS agar.

3.3. Protoplasting

1. Before beginning, gently thaw solutions A and B in water. This takes about 30 min. Try not to shake solution B as this may disrupt enzyme activity and form excessive bubbles.
2. Place a 70-μm cell strainer in a small Petri dish. For each replicate, aliquot 6–7 mL of solution B into the dish (it should be about 2/3 full). Do not trap air bubbles under the strainer.
3. Cut roots from Arabidopsis plants using a sterile blade. Using the blade, make one clean cut through the roots across the plate, and then quickly chop the roots into smaller pieces (4–5 strokes of the blade). Place the roots in the cell strainer in solution B using the blade. You can use up to 7 plates of roots, each with two rows of plants, per 7 mL of solution B. Gently mix roots by pushing them to the side of the strainer with a plastic transfer pipette, lift the strainer a little out of the solution, and then mix roots gently again with the strainer back in solution.
4. Place Petri dish with roots on shaker (85 rpm) for 60 min. Gently mix roots every 15–20 min as above.
5. While roots are protoplasting, set up two 50-mL Falcon tubes. Place a 70-μm strainer in one and a 40-μm strainer in the other. In addition, label cell-sorting tubes. Make two sets, one labeled with ‘protoplasts’ and one with ‘collection.’ In the ‘collection’ tube, add the medium into which you want to sort protoplasts (*see* **Note 9**).
6. Remove strainer with leftover root pieces. Using a transfer pipette, aspirate some leftover roots into the bottom of a 15-mL Falcon tube (these will form a pellet during centrifugation). Then gently aspirate all the liquid on the outside of the strainer and in the Petri dish into the 15-mL tube. Discard or wash and keep strainer for the next use.
7. Centrifuge at 22°C, 200×*g* for 6 min. Take tubes out immediately when finished spinning (*see* **Note 10**).
8. Aspirate the supernatant and discard. Do not aspirate the entire supernatant, as protoplasts are at the bottom near the pellet and cannot be seen.
9. Add 350 μL of solution A to the protoplasts (*see* **Note 11**).

10. Transfer the protoplast–solution A mixture through the 70-μm strainer in the Falcon tube. Do not manually push the mixture through the strainer.
11. Add another 350 μL of solution A to the strainer (*see* **Note 11**).
12. Pass the entire protoplast–solution A mixture through the 40-μm strainer.
13. Transfer this mixture into a tube for cell sorting (such as a 5-mL polystyrene round-bottom tube). Cells can be kept at 4°C or room temperature before sorting. The type of tube used may depend on the cell-sorting machine. Know the type of tube needed prior to sorting (*see* **Notes 2** and **4**).
14. Protoplasts are now ready to be sorted. Store sorted samples on dry ice (*see* **Note 12**) while waiting for other samples to finish sorting. After sorting, store at –80°C if not using immediately. Cells can be used for RNA, protein, or metabolite extraction.

4. Notes

1. If sorted cells are used for transcriptome analysis, gene expression changes during protoplasting and sorting may be a concern. Protoplasting and sorting do affect gene expression, but we have found this effect to be fairly small. Move quickly through the protocol and limit the sort time to less than 60 min. Our typical sorts are ~20 min. Also, when testing the effect of a stress on a particular cell type, it is important to always use sorted, nonstressed cells as the control. We have examined the effect of sorting on gene expression under stress conditions by examining the effect of stress on gene expression in whole, nonsorted roots and in sorted cells from WT (non-GFP) whole roots. Expression changes caused by protoplasting and sorting do not often overlap with genes affected by the stress. When an overlap does occur, rarely is the direction of the gene expression change opposite to that occurring under the stress (3). Additionally, using stringent criteria for detecting significance in differential gene expression should alleviate concerns that gene expression changes are sorting-related.
2. Forty- and seventy-μm strainers are often on backorder. Make sure you have ordered more than enough. If you are running low and do not anticipate receiving more on time, wash and reuse the sieves. You can also use one type of sieve twice if necessary.

3. Sorting machines and facilities differ, and it is imperative that the researcher contact the cell-sorting facility prior to sorting to set up preliminary experiments and decide which buffers and tubes to use.
4. Cells are sorted into tubes, slides, or culture plates. The type of tube or plate may differ according to each facility. Common types of tubes are polypropylene or polystyrene (such as 5-mL polystyrene round-bottom tubes – BD Falcon). Before sorting, check with the cell-sorting facility for any specific guidelines.
5. Make solution A and store in aliquots in –20°C until needed. This can be freeze–thawed ~5 times.
6. 50 mL of solution B is enough for 10,000 Arabidopsis roots approximately 1 week after imbibition.
7. Do preliminary experiments to optimize sorting parameters, determine the correct sorting gate necessary to capture only FM cells and the number of plants necessary to achieve the cell number needed for downstream applications. Use a wild-type (non-FM) plant as a control to determine the baseline for autofluorescence and adjust the sorting window accordingly (7). After sorting, collect a small amount of sorted cells and check for fluorescence (note that some cells may have damaged membranes and may not fluoresce). Additionally, prior to sorting, use confocal microscopy to ensure the marker is expressed in a specific cell type under the conditions assayed.
8. A standard size post-it note is a good size to use for measuring how large to cut the mesh.
9. The medium into which cells are sorted will differ for your downstream application and may need to be empirically determined. Speak with your cell-sorting facility for recommendations. For transcriptome analysis, we sort into RLT buffer + ß-mercaptoethanol (ß-met) for later RNA extraction. RLT buffer is found in Qiagen's RNeasy RNA extraction kits, and instructions for adding ß-met are included in the kit. The amount of medium needed will depend on the number of cells. We have found that 450 μL RLT buffer is sufficient for 50,000 cells.
10. Do not spin the cells too long or leave them in the centrifuge too long.
11. The amount of solution A to add depends on the sample and sorting parameters (pressure and speed of sorting), and may need to be empirically determined.
12. After sorting, immediately place samples on dry ice, not regular ice.

References

1. Birnbaum, K., Shasha, D. E., Wang, J. Y., Jung, J. W., Lambert, G. M., Galbraith, D. W., et al. (2003) A gene expression map of the Arabidopsis root. *Science* **302**, 1956–1960.
2. Brady, S. M., Orlando D. A., Lee, J. Y., Wang J. Y., Koch, J., Dinneny J. R., et al. (2007) A high-resolution root spatiotemporal map reveals dominant expression patterns. *Science* **318**, 801–806.
3. Dinneny, J. R., Long, T. A., Wang, J. Y., Jung, J. W., Mace, D., Pointer, S., et al. (2008) Cell identity mediates the response of Arabidopsis roots to abiotic stress. *Science* **320**, 942–945.
4. Zhang, C., Barthelson, R. A., Lambert, G. M., and Galbraith, D. W. (2008) Global characterization of cell-specific gene expression through fluorescence-activated sorting of nuclei. *Plant Physiol* **147**, 30–40.
5. Galbraith, D. W. and Birnbaum, K. (2006) Global studies of cell type-specific gene expression in plants. *Annu Rev Plant Biol* **57**, 451–475.
6. Brandt, S. P. (2005) Microgenomics: Gene expression analysis at the tissue-specific and single-cell levels. *J Exp Bot* **56**, 495–505.
7. Birnbaum, K., Jung, J. W., Wang, J. Y., Lambert, G. M., Hirst, J. A., Galbraith, D. W., et al. (2005) Cell type-specific expression profiling in plants via cell sorting of protoplasts from fluorescent reporter lines. *Nat Protoc* **2**, 615–619.

Chapter 22

Laser Microdissection of Paraffin-Embedded Plant Tissues for Transcript Profiling

Robert C. Day

Abstract

High-resolution cellular analysis will help answer many important questions in plant biology including how genetic information is differentially used to enable the formation and development of the plant body. By comparing transcriptome data from distinct cell types during various stages of development, insight can be obtained into the transcriptional networks that underpin the attributes and contributions of particular cells and tissues. Laser microdissection (LM) is a technique that enables researchers to obtain specific cells or tissues from histological samples in a manner conducive to downstream molecular analysis. LM has become an established strategy in many areas of biology and it has recently been adapted for use with many types of plant tissue.

Key words: Laser-capture microdissection, laser microdissection, gene expression, paraffin sections, chemical fixatives.

1. Introduction

Plant biology has already generated a large amount of whole plant-and organ-derived transcriptome data that have provided an important starting point for identifying the role of particular structures during the life cycle of the plant (1, 2). However, these data are generally not sufficient to address the contribution of individual cells and tissues to the developmental systems operating within the organism. Furthermore, important cell type-specific transcripts may not be easily apparent in these data due to a low representation across a mixed cell population (3).

L. Hennig, C. Köhler (eds.), *Plant Developmental Biology*, Methods in Molecular Biology 655,
DOI 10.1007/978-1-60761-765-5_22, © Springer Science+Business Media, LLC 2010

LM is becoming an established method of obtaining cell- or tissue-specific expression data from plants, enabling novel insight into molecular function (reviewed in (4–7)). Whilst many studies have used the technique to gain spatial information about the expression of a small number of transcripts by PCR-based methods, recent technological advances now mean that the expression of thousands of transcripts can be measured simultaneously using platforms such as microarray and/or high-throughput sequencing (e.g. 3, 8–11). Whilst other technologies are helping to provide important strategies for sampling and reporting from specific cell-types (reviewed in (12, 13)), LM protocols are perhaps the most easily transferable across plant species and enable a wide range of cellular targets.

Several different LM apparatus exist and use strategies based on laser-capture microdissection (LCM) and/or laser-excision microdissection (LEM) (14). The strategies used by the Arcturus Pix Cell II and Leica AS LMD provide the basis for this chapter and are shown in **Fig. 22.1**. The LCM approach uses an infrared laser to capture the selected target cells onto a plastic

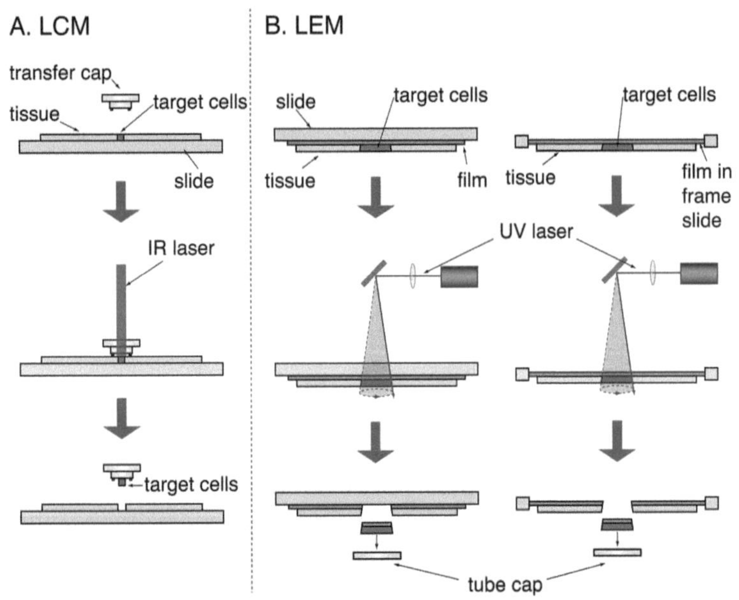

Fig. 22.1. **Cartoon of the process of LCM and LEM using the Arcturus PixCell II or the Leica AS LMD.** (**A**) LCM involves the placement of a plastic cap onto a dehydrated tissue section mounted on a glass slide. Target cells are attached to the cap by melting the thermoplastic polymer on the lower surface of the cap down onto the target cells using heat from an IR laser. When the cap is removed the target cells are pulled away from the surrounding tissues. (**B**) LEM can be performed on tissue sections mounted on a length of PEN membrane attached to a glass slide or from PEN membrane across a frame slide. The targeted area is separated from the section by ablating both the membrane and the tissues around the circumference of the target and the excised fragment detaches and falls into a tube cap.

cap by melting a thermoplastic polymer down onto the sample that is mounted on a glass slide. Withdrawal of the cap then pulls the target away from the surrounding tissues. LEM usually has the tissue mounted on a membrane, and a UV laser ablates around the perimeter of the target enabling collection into a receptacle below the slide. Both approaches have distinct technical issues with regard to tissue handling and microdissection (14, 15) and both have been used successfully on plant tissues (*see* **Table 22.1**).

Here we provide a comprehensive guide to LM, based on established protocols, which should enable plant researchers to obtain high-quality total RNA from chemically fixed paraffin-embedded plant tissues using either LCM or LEM.

2. Materials

2.1. Sample Preparation and Storage

1. RNaseZap (Sigma). CAUTION! RNase Zap is a mild irritant and so eye protection and gloves are recommended.
2. Analytical grade absolute ethanol. CAUTION! Ethanol is flammable and a mild contact hazard, so use with appropriate ventilation, wear gloves and eye protection when handling.
3. Analytical grade glacial acetic acid. CAUTION! Glacial acetic acid is harmful and should be used with appropriate ventilation and discarded in a hazardous waste container labelled “Acid Waste”. Glacial acetic acid is flammable and a strong contact hazard, so wear eye protection and acid-resistant gloves when handling.
4. Farmer’s fixative. Make fresh on the day of tissue sampling. Mix 150 mL of ethanol with 50 mL of glacial acetic acid in an RNase-free Duran bottle.
5. Tissue-processing cassettes. Heat and solvent resistant with lids (e.g. SIMPORT # M480/2 with M481).
6. Biopsy pads (Kartell # 2922040500).
7. Diethylpyrocarbonate (DEPC). CAUTION! DEPC is harmful and a contact hazard, so use with appropriate ventilation and wear eye protection and gloves when handling.
8. Sterile 0.1% DEPC-treated double distilled or Milli-Q water. Add DEPC to the water in a Duran bottle. Shake vigorously to break up the DEPC droplets. Allow it to stand overnight, then autoclave to sterilise and remove the DEPC.
9. Plastic pots. Solvent resistant with screw tops.

Table 22.1
Summary of successful laser microdissection studies carried out in plants with details of the systems used. The reports listed contain technical details of the procedures in their methods sections

Plant species	Preferred fixative	LM platform	References
Arabidopsis	NA	PALM MicroBeam	(16)
Arabidopsis	NA	Arcturus PixCell II	(17, 18)
Arabidopsis	Microwave	Leica AS LMD	(19)
Arabidopsis	Farmer's	Arcturus PixCell II	(8, 20)
Arabidopsis	Modified Farmer's	PALM MicroBeam	(21)
Arabidopsis	Farmer's	Arcturus PixCell II	(22)
Arabidopsis	Farmer's	Arcturus PixCell II	(23)
Barley	Farmer's	PALM MicroBeam	(24)
Cotton	Farmer's	PALM MicroBeam	(11, 25)
Cucumber	Farmer's	PALM MicroBeam	(26)
Maize	Methacarn	Leica AS LMD	(27)
Maize	NA	Arcturus PixCell II	(28)
Maize	Microwaved Acetone	PALM MicroBeam	(46)
Maize	Acetone	PALM MicroBeam	(10)
Maize	Farmer's	Arcturus PixCell II	(29)
Maize	Acetone	PALM MicroBeam	(30)
M.armeniacum	Farme'rs	Leica AS LMD	(31)
Olive	NA	PALM MicroBeam	(32)
Periwinkle	Farmer's	Arcturus Pix-Cell II	(33)
Potato	Farmer's	PALM MicroBeam	(34)
Rice	Acetone	SIGMA KOKI Laser Micro Cutter	(35)
Rice	Farmer's	Arcturus Veritas	(36)
Rice	Farmer's	Leica AS LMD	(37)
Rice	Methacarn	Arcturus PiCell II	(38)
Rice	Acetone	Arcturus PixCell II and Veritas	(3)
Soybean	Farmer's	Arcturus PixCell II	(39)
Soybean	Farmer's	Leica AS LMD	(40)
Soybean	Farmer's	Arcturus PixCell II	(41)
Tobacco	Farmer's	Leica AS LMD	(42)
Tomato	Methacarn	Leica AS LMD	(43)
Tomato	Farmer's	PALM Robot Combi System	(44)
Varius	Farmer's	Arcturus PixCell II	(45)

10. Ethanol (95%). Mix 190 mL of molecular grade ethanol with 10 mL of DEPC-treated water in an RNase-free Duran bottle.
11. Analytical grade isopropylalcohol. CAUTION! Isopropyl alcohol is flammable and a mild contact hazard, so wear gloves and eye protection when handling.
12. Analytical grade xylene isomers. CAUTION! Xylene is harmful by inhalation and should be used with appropriate ventilation and discarded in a hazardous waste container labelled "Xylene Waste". Xylene is flammable and a contact hazard, so wear eye protection and gloves when handling.
13. Paraplast wax (McCormick Scientific # 8889501006).
14. Frame slides with foil (Leica # 11505151) or glass slides with foil (Leica # 1150515) for UV laser cutting or glass slides (e.g. Fisher, # 12-550-14) for IR laser capture.

2.2. Laser Capture from Glass Slides and Laser Excision from Membrane Slides

1. HS LCM Caps (MDS Analytical Technologies) or Capsure LCM Macro Caps (MDS Analytical Technologies).
2. RNA extraction buffer (Arcturus Picopure RNA extraction kit, MDS Analytical Technologies).
3. PCR tubes (0.5 mL) that fit the Leica tube holders, e.g. Thermo Scientific # AB-0350.

2.3. RNA Extraction, Quantification and Quality Assessment

1. Arcturus Picopure RNA extraction kit (MDS Analytical Technologies).
2. RNase-free DNase set (Qiagen).
3. RiboGreen RNA Quantitation Reagent Kit (Molecular Probes/ Invitrogen).

3. Methods

A flow diagram of the steps required for a typical high-throughput LM transcript analysis is given in **Fig. 22.2**. Here we cover the process from plants to RNA. Subsequent use of high-throughput platforms will require amplification of the RNA, and examples of commercial RNA amplification kits already used by plant biologists are given in **Table 22.2**. The procedure given here for tissue preparation (*see* **Section 3.1**) has been used successfully for a range of plant tissues. However, variations in tissue structure and perhaps more importantly local availability of histology equipment may warrant the user to develop alternative steps. **Table 22.1** provides a list of LM studies on plants in which methods conducive to RNA analysis have been described. Variations

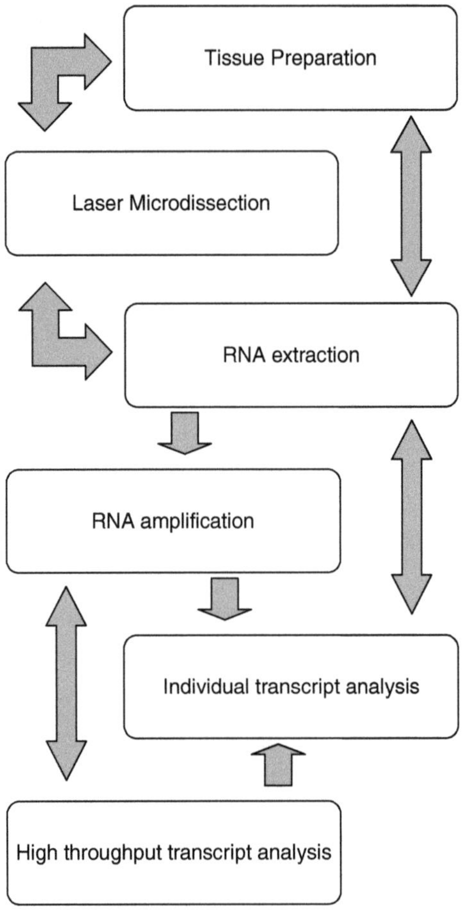

Fig. 22.2. **Flow diagram showing the steps required for a laser microdissection study involving high-throughput transcript analysis.** Bidirectional arrows represent steps that may require optimisation based on the outcome of the downstream step.

between tissue preparation protocols involve standard histological procedures. A consistent requirement throughout any LM transcript study is an RNase-free environment (*see* **Note 1**).

3.1. Sample Preparation and Storage

Paraffin embedding after chemical fixation in ethanol:acetic acid has proven to be amenable to extraction of good RNA from a wide range of tissues.

1. Prepare 200 mL of fresh Farmers's fixative and keep on ice for 15 min.
2. Ensure three glass beakers of fresh molten wax are available with sufficient depth to cover the sample cassettes. The beakers can be covered with aluminium foil until required and labelled 1–3. When required, move these to a heated vacuum chamber (*see* Step 15).
3. In a flow hood, decant a small amount of fixative into a glass Petri dish or similar vessel on a bed of ice.

Table 22.2
Examples of RNA extraction and RNA amplification kits used successfully with plant material

RNA isolation kits	Manufacturer	References
PicoPure RNA Isolation Kit	Arcturus	(10, 19, 20, 22, 23, 28, 30, 34, 37, 38, 41, 45, 46)
Dynabeads oligo d(T)25	Dynal/Invitrogen	(21)
Absolutely RNA Micro- or Nanoprep Kits	Stratagene	(26, 45, 47)
TRIzol	Invitrogen	(19, 45)
Purescript RNA Isolation Kit	Gentra systems	(16)
RNeasy Micro Kit	Qiagen	(19)
RNaqueous Micro Kit	Ambion	(28)
RNA extraction Kit	PALM	(10)
RNA Amplification Kits	**Manufacturer**	**References**
BD-SMART mRNA Amplification Kits	BD Biosciences	(16, 21, 28)
RiboAmp RNA, HS or OA Amplification Kits	Arcturus	(3, 26, 38)
Message Amp Kits	Ambion	(17, 22)
Target Amp Kit	Epicentre Biotechnologies	(46)
BioArray RNA amplification and labelling system	Enzo Life Sciences	(8, 20)

4. Use forceps and a scalpel to harvest plant tissues and immediately transfer these into the Petri dish containing fixative. Trim samples to no more than 0.4 cm in diameter. Keep a biopsy pad adjacent to the tissue in the solvent and carefully place trimmed tissues onto the pad. When enough trimmed samples have been prepared, transfer the pad to a tissue-processing cassette. Add a second pad to cover the specimens, then close the lid of the cassette.
5. Drop the cassettes into the Duran bottle containing fixative and vacuum infiltrate for 15 min (400 mm/Hg). Release the vacuum slowly.
6. Secure the lid on the Duran bottle and transfer to 4°C for 4–16 h with gentle agitation, e.g. on a gently oscillating orbital shaker in a 4°C room or refrigerator with occasional manual swirling (*see* **Note 2**).
7. Pour off the fixative and replace with 95% ethanol. Keep at room temperature for 30 min.
8. Use long forceps to remove the sample cassettes from the Duran bottle and drop them into a solvent-resistant microwavable container containing absolute ethanol.

9. Place the container in a suitable microwave that has a temperature probe, e.g. Electron Microscopy Sciences EMS-820.
10. Insert the temperature probe into the solvent through the container lid. The lid of the container will likely require a hole punching in the top to enable the probe to be inserted into the solvent. Ensure the container is stable with the probe inserted.
11. Microwave at 60°C for 10 min.
12. Replace the ethanol with isopropylalcohol and microwave at 70°C for 10 min.
13. Replace with fresh isopropylalcohol and microwave at 70°C for 10 min.
14. Replace with fresh isopropylalcohol and microwave at 74°C for 10 min.
15. Transfer sample cassettes to one of the three glass beakers of wax in the heated vacuum chamber (*see* **Note 2**). Apply a vacuum for 20 min and then slowly release.
16. Use pre-warmed long tweezers to transfer the cassettes to the second beaker of wax and re-apply a vacuum for 20 min and then slowly release (*see* **Note 3**).
17. Transfer the cassettes to the third beaker of wax and re-apply a vacuum for 20 min and then slowly release.
18. Transfer the cassettes to a hot plate or embedding centre at 65°C and release the tissue samples.
19. Place pre-warmed metal moulds on a hot plate and fill with molten wax. Transfer the tissue samples into the moulds using pre-warmed utensils. Arrange the tissues in the molten wax and transfer to a cold plate for the blocks to harden (*see* **Note 4**).
20. When fully solid, release the blocks from the metal moulds and transfer them to a refrigerator for storage (*see* **Note 5**).
21. On the day of sectioning, place microdissection slides (glass slides for LCM and glass slides with membrane or membrane Frame slides for LEM) on a slide dryer at 42°C.
22. To section, remove a block from the refrigerator and trim using a clean razor blade. Mount on a rotary microtome and cut sections of 7–20 μm depending on the target tissues required (*see* **Note 6**).
23. Pipette drops of DEPC-treated water onto the slides that cover an area slightly larger than the sections to be attached.

24. Float the sections onto the droplets and wick away excess water with the edge of a Kimwipe until the sections appear to sit directly on the slide surface.
25. Repeat this procedure until the region of the slide available for microdissection is full of sections.
26. Slides are covered with aluminium foil (ensure no contact with the specimens) and allowed to dry at 42°C for 1–2 h before storage with desiccant at 4°C (*see* **Note 7**).
27. Proceed with microdissection within a day or two of sectioning.
28. Immediately prior to microdissection, the slides need to be dewaxed and dehydrated. Pour fresh xylenes into two Coplin jars in a fume hood.
29. To remove paraffin, use forceps to submerge a slide in the first xylene solution for 2 min.
30. Transfer the slides to the second solution of xylenes for another 2 min. Remove the slide from the xylenes and place in a slide rack. Allow it to dry in a flow hood for 5–10 min.
31. Transfer the slides to a clean rack in an airtight plastic box that contains desiccant and proceed to microdissection.

3.2. Laser Capture from Glass Slides

This protocol assumes the use of an Arcturus Pix Cell II microdissection system and brightfield illumination.

1. Wipe down the work area immediately adjacent to the LM microscope with RNaseZap and confine subsequent sample and equipment processing to this zone. Also wipe down the specimen platform and the cassette holder on the LCM microscope.
2. Turn on the LCM microscope, video monitor and the control box.
3. Remove the cap cassette holder by sliding it away from the microscope platform. Press down on the flanges of the holder and push in the end locking pins to hold the plate in the load position.
4. Position a cap cartridge into the holder and slide it to the far end. A second cartridge can also be loaded if required (four caps on a cartridge).
5. Once cartridges are loaded into the holder, push down on the flanges and pull out the locking pins. The cartridges are now secure and the holder can be returned to the microscope platform by sliding it back into position.
6. Position a cap adjacent to the load position by sliding the cassette holder in its mount. When in the correct

position, the holder will click into place and the cap will be positioned for pick up by the placement arm.

7. Move the joystick to the vertical position. This will ensure optimal positioning of the cap relative to the zone of capture, thereby maximising the area of sample available for microdissection.
8. Remove a prepared specimen slide from the desiccant box. Place the slide on the stage with the left side over the vacuum hole and with specimen facing upwards.
9. Select the ×10 objective and position the slide manually so that the area targeted for sampling is central to the view, using either the oculars or the monitor. Adjust the illumination using the illumination control slider on the front left of the microscope.
10. When the slide is suitably placed, with the joystick in the vertical position, activate the vacuum by pressing the switch on the controller. This holds the slide in position during microdissection.
11. Swing the placement arm over to the cap in the load position (*see* **Note 6**). Lift the placement arm vertically to remove the cap from the cassette and rotate the arm so the cap is positioned over the sample. Upon release of the placement arm, the cap will lower onto the sample and be in the transfer position.
12. With the cap in the transfer position, activate the laser by turning the key switch on the controller and pressing the laser enable button. When the placement arm is in the transfer position, enabling the laser will activate the target beam.
13. Reduce the illumination of the sample slide until the target beam is easily apparent on the monitor.
14. Use the lever on the left side of the microscope to select the smallest spot size (7.5 μm) and use the focus adjust wheel located below the lever to focus the target beam. A properly focused beam will appear as a bright circular spot with a well-defined edge. If the target spot is fuzzy or bright concentric rings are apparent then the laser is out of focus. Once correctly focused, the laser should remain so for use with the loaded slide regardless of user alterations to the objectives and/or spot size (*see* **Note 8**).
15. Whilst still in this position, away from the tissue sample, press the red pendant button. This test fire of the laser will indicate whether the settings are sufficient to melt the cap polymer layer down onto the slide in an appropriate manner. The wetted polymer should form a clear circle in the

vicinity of the target beam. The clear circle indicates the polymer has become attached to the glass slide and is surrounded by a dark ring.

16. The power and duration of the laser burst can be altered using the up and down arrows on the front of the controller to achieve suitable wetting. Typical settings for a Capsure Macro cap are: 7.5 μm spot: 50 mW, 650 μs; 15 μm spot: 30 mW, 1.5 ms; 30 μm spot: 25 mW, 5 ms. Increasing the power and duration will usually achieve a better melt and ensure good contact with the slide/sample, but may also increase the diameter of the contact spot. By adjusting these settings, a suitable contact spot size can be obtained for most types of tissue dissection (*see* **Note 9**).

17. For microdissection, locate the cells/tissue of interest and position the target beam directly over the required region using the joystick (*see* **Fig. 22.3**). For single shots, fire the laser by pressing the red button on the pendant, reposition the target beam and repeat. For large areas of target tissue or cell layers, efficient microdissection is achieved using the "repeat" function. Rapid, multiple laser bursts can be obtained by holding the red pendant button down. The frequency of laser bursts can be altered using the "repeat" function on the controller and selecting the time between laser shots using the up and down arrows. Using the repeat function means that the microscope stage needs to be continually readjusted by the operator whilst the red fire

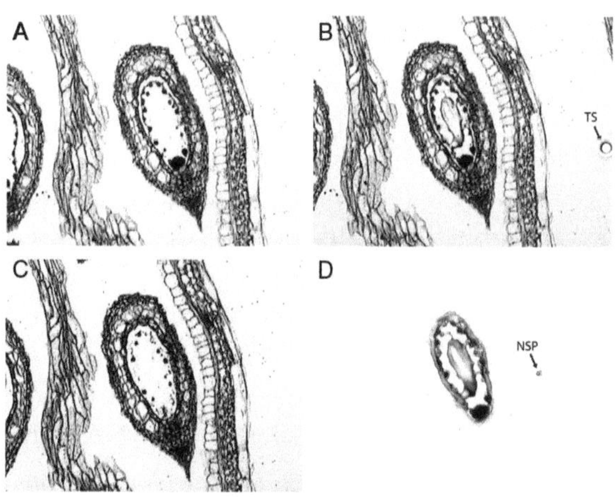

Fig. 22.3. **Laser capture microdissection of endosperm.** (**A**) The target tissue is centred in the view on the video monitor. (**B**) A test shot of the laser in an area where no tissue is apparent can confirm the laser settings are correct for capture (TS). Several laser firings can then be used to attach the targeted tissue to the Capsure. (**C**) When the Capsure device is withdrawn most of the target tissue is lifted away from the section. (**D**) The Capsure can then be imaged to confirm target capture. Some non-specific pickup may also be apparent (NSP).

button is depressed such that a continuous line of contact spots is achieved (*see* **Note 10**).

18. Once the desired amount of target tissue has been attached to the cap, it can be removed by lifting the placement arm vertically and swinging it away from the slide (*see* **Note 11**).
19. It is advisable to view the cap post-microdissection. Reposition the slide so that a tissue-free zone is apparent in the viewing area. Swing the placement arm back to deposit the cap onto this clear zone and use the joystick to position the cap for viewing. Visually assess the efficiency of capture of the selected target cells but also look for the presence of non-specific tissue adhesion (*see* **Note 12**).
20. To remove the cap from the microdissection rig, move the placement arm vertically and swing it round to the right. Position the cap over the unloading platform and slowly guide the arm down to deposit the cap in the extraction slot. Finally, swing the empty placement arm back toward the microscope stage into the rest position.
21. With the open side of the cap insertion tool facing towards the cap, slide the tool along the guide rail and over the cap. Once the cap is secured in the tool, lift the device away from the microdissection apparatus.
22. If non-specific tissue adhesion was apparent in Step 19, gently blot the cap with the tacky region of an adhesive note, e.g. "Post It" notes by 3 M (*see* **Note 13**).
23. With the device still in the insertion tool, gently slide the ends of a pair of RNase-free fine forceps along the barrel of the cap towards the polymer layer. Repeat this process around the barrel of the cap until a portion of the polymer layer peels away from the face of the cap.
24. Grip the loose polymer section with the tweezers and gently tease the remaining polymer away from the plastic cap. Transfer the polymer to a sterile RNase-free 0.5-mL microcentrifuge tube containing extraction buffer (XB) and submerge. Multiple polymer peels can be submerged into the same buffer sample, if required (*see* **Note 14**).
25. Proceed with RNA extraction or freeze the tube at −70°C for subsequent RNA extraction.
26. After all microdissections are complete, shut down the apparatus by pressing the "enable laser" button and turning the keyswitch to the *off* position. Finally, turn off the power to the microscope, controller box and the video monitor.
27. *Optional.* It is often desirable to indirectly assess the quality and yield of RNA from a tissue sample by harvesting the tissues remaining on the slide after microdissection (*see* **Note 15**).

3.3. Laser Excision from Membrane Slides

This protocol assumes the use of a Leica AS LMD microdissection system with brightfield illumination, a 4-tube collection tray and version 4.3 of the LM software.

1. Wipe down the work area immediately adjacent to the LM microscope with RNaseZap and confine subsequent sample and equipment processing to this zone.
2. Turn on the computer and the Microscope Control Box and wait for the microscope to initialise.
3. Start the Laser Microdissection software.
4. Click on the "Unload" menu button to bring up the "change specimen" screen. The AS LMD will position the collection tray for easy access.
5. Remove the tray by using the long metal handle to pull it directly away from the microscope, freeing it from its holding bracket.
6. Dismantle the collection tray by popping out the four individual tube holders and wipe the components with a Kimwipe wetted with RNaseZap (*see* **Note 16**).
7. Pull out the slide holder and wipe this and other surfaces of the collection area with RNaseZap.
8. Check if the correction settings of the objectives are appropriate for the type of slides to be used. For glass-mounted membrane or frame slides we use "0", and "1–1.2", respectively. If the correction requires changing, twist the end of the objective to the desired setting but be careful not to unscrew the lens.
9. Turn the key switch on the back of the laser box to the "on" position.
10. Load opened RNase-free tubes into the individual holders by passing the tube cap up through the central aperture. Then position the edges of the cap into the grooved holder so that the inner cap ring is facing upwards, forming a receptacle. The main body of the tube, which is still positioned under the holder, is then bent back until the tapered end fits snugly into the end slot, pointing away from the user. Load up to four tubes and place the holders back into the collection tray.
11. Pipette RNA extraction buffer into the tube cap. We use 30 μL in a 0.5-mL tube cap (*see* **Note 17**).
12. Replace the collection tray into the AS LMD apparatus being careful that it clicks back into its holding bracket.
13. Remove a membrane slide from the desiccant box and load it into the slide holder so that the membrane side, on which the sections are attached, faces downwards. Place the slide holder into position on the AS LMD.

14. Click on the "Continue" button of the "Change Specimen" screen for the AS LMD to position the collection tray, slide holder and objectives for target viewing. This will activate a live image of the slide.
15. The collection tray has five possible positions. Four cap positions A to D and a "no cap" position. Initial placement of the tray is in the "no cap" position. This facilitates imaging of the tissue sample.
16. Position a tissue section under the objective using the front spin wheels of the Smartmove mouse and focus the specimen using the rear spin wheel. Adjust illumination using the "up" and "down" buttons on the left base of the Smartmove whilst the right "up" and "down" buttons can be used to change objectives.
17. Select an objective for microdissection that enables a whole targeted region to be easily apparent on the view screen (*see* **Note 18**).
18. On the left side of the work window, click on the "Line" button. In the right side of the work window, ensure "Standard Mode" is selected in the "Cut Shape(s)" box and the "Close line(s)" of the "Draw Shapes" box is *not* selected.
19. Draw a line by holding the left mouse button down and dragging the cursor. A coloured line will appear superimposed on the image. Release the left mouse button to stop drawing.
20. Click on the "Start Cut" button in the "Cut Shape(s)" box. The laser will attempt to cut the desired line. If the laser does not cut the membrane the settings will need adjusting.
21. To adjust settings click on "Laser" on the top menu bar and then "Control". The "Laser Control" window will open. Adjust the "Aperture", "Intensity" and "Speed" settings then click on the "Apply" button. Try another cut. Keep applying more aggressive settings and test cuts until a clean cut is obtained. When a clean cut is achieved click on the "OK" button.
22. If the cut does not follow the line closely, the laser will need calibrating. Each session, we routinely re-calibrate the laser for each objective that is to be used for microdissection (*see* **Note 19**).
23. Position a targeted tissue region into view. Try to ensure the whole perimeter of the target region is apparent.
24. Select a tube position to collect the targeted tissue into by clicking on either A, B, C or D in the "Collector Device: Tube Caps" box in the work window. The selected cap will turn green on the display. Watch the collector platform to ensure the correct position is aligned.

25. Select the "Close line(s)" option in the "Draw Shape(s)" box.
26. Draw a line around the circumference of the targeted cells using the computer mouse.
27. Click on the "Start Cut" button.
28. The laser will cut around the perimeter of the targeted region. If the perimeter is not predominantly ablated, it is advisable to use more aggressive laser settings (*see* Step 21). However, even with optimised laser settings, targeted regions are likely to have a few points of connection to the surrounding membrane where particularly stubborn cell walls remain (*see* **Fig. 22.4**).

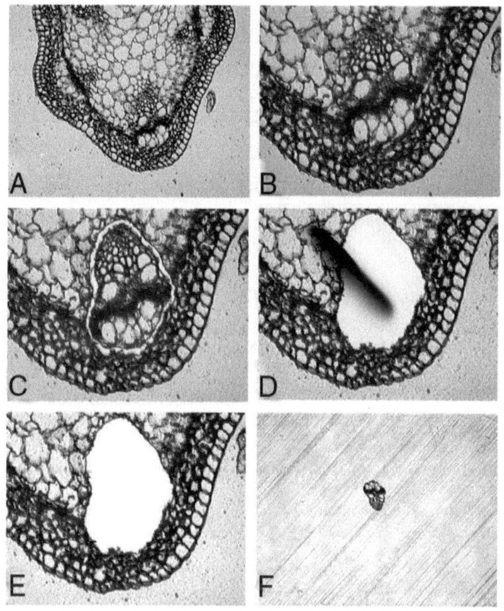

Fig. 22.4. **Laser excision microdissection of vascular bundles.** (**A**) The target tissues are located within the tissue sample at low magnification (×20 objective). (**B**) The magnification is increased (×40 objective) so that the target cells are more apparent whilst making sure the perimeter is wholly within the viewable area. (**C**) The laser ablates around the perimeter of the target cells, as defined by a line drawn by the mouse in "Standard Mode". (**D**) The target may not fall from the section after just one cut. The "Move and Cut" mode allows areas still attached to be ablated in a targeted way. Occasionally the target will hang from the section. (**E**) Ablation in the vicinity of the intersection between the perimeter and the dangling target will enable the fragment to drop. (**F**) Successful collection into a cap can be observed by using the "specimen" button in the work area of the software. Shown here with a ×10 objective.

29. Select the "Move and Cut" option in the "Cut Shapes(s)" box.
30. Position the mouse cursor to an area of attachment and left click to activate the laser. Hold the mouse button down whilst moving the cursor over the recalcitrant region until ablation is apparent. Repeat in other regions until the targeted cells fall from the slide.

31. To confirm collection of the cut fragment, select "Standard Mode" and then click on the "Collector" button. This will position the appropriate tube cap under the objective. Use the Smartmove to visualise the fragment in the tube cap and then return to a slide view by clicking on the "Specimen" button.
32. Reposition the slide and continue microdissection of target cells.
33. When you have finished collecting target cells click on the "Unload" button.
34. Remove the collection tray from the AS LMD and unload tubes from the individual tube holders being careful not to disturb the contents of the cap.
35. Hold the PCR tube cap with buffer facing upwards. Bring the body of the tube over the cap and press shut.
36. Centrifuge the PCR tubes briefly to collect the buffer in the bottom of the tubes and proceed directly to RNA extraction or store at −70°C.
37. After all microdissections are complete, turn off the laser by turning the keyswitch to the off position. Finally, turn off the microscope control box and exit the LM software.
38. Cover the microscope with a dust cover.
39. *Optional.* It is often desirable to indirectly assess the quality and yield of RNA from a tissue sample by harvesting the tissues remaining on the slide after microdissection (*see* **Note 15**).

3.4. RNA Extraction

These instructions assume the use of the Picopure RNA Isolation Kit made by Arcturus Inc. The details are similar to those provided by the manufacturer but include some minor user-based variations. RNA extraction kits used successfully by plant biologists are listed in **Table 22.2**.

1. If aliquots of extraction buffer (XB) containing harvested cells have been stored at −70°C, thaw them on ice.
2. Place the 0.5-mL sample tubes into empty RNase-free 1.7-mL microcentrifuge tubes (lids removed) for support during centrifugation (*see* **Note 20**). Centrifuge at 800×*g* for 1 min to collect reagents to the bottom of the tube.
3. Place the sample tubes in a heat block pre-heated to 42°C, cover with aluminium foil and leave for 30 min.
4. Approximately 20 min into the 42°C incubation, precondition the RNA purification columns by pipetting 250 μL of conditioning buffer (CB) onto the column filter membrane. Incubate the column for 5 min at room temperature.

5. Centrifuge the purification column in a collection tube at 16,000 × *g* for 1 min.
6. Once the 42°C, 30-min incubation of the sample at is complete (*see* 3), centrifuge the sample tubes at 800 × *g* for 1 min to collect reagents to the bottom of the tube.
7. Open the tubes and add an equivalent volume of 70% ethanol into the cell extract. Mix well by pipetting up and down.
8. Pipette the mixture into the pre-conditioned column and centrifuge for 2 min at 100 × *g*, immediately followed by 16,000 × *g* for 30 s.
9. Pipette 100 μL of wash buffer (W1) into the column and centrifuge for 1 min at 8,000 × *g*.
10. Pipette 5 μL of Qiagen DNase stock solution into 35 μL of Buffer RDD. Mix by inversion.
11. Pipette the 40 μL DNase incubation mix directly into the column membrane and incubate at room temperature for 15 min (*see* **Note 21**).
12. Pipette 40 μL of W1 buffer into the column membrane and centrifuge at 8,000 × *g* for 15 s, remove the flowthrough.
13. Replace the column into the collection tube and pipette 100 μL of wash buffer 2 (W2) into the column and centrifuge for 1 min at 8,000 × *g*.
14. Pipette another 100 μL of W2 into the column and centrifuge for 2 min at 16,000 × *g*.
15. Discard flowthrough and re-centrifuge at 16,000 × *g* for 1 min.
16. Transfer the purification column to a new 0.5-mL microcentrifuge tube.
17. Pipette 13 μL of elution buffer (EB) into the column membrane by gently touching the tip of the pipette to the surface of the membrane while dispensing.
18. Incubate the column for 1 min at room temperature.
19. Centrifuge the column for 1 min at 1,000 × *g* to distribute the EB in the column filter, then at 16,000 × *g* to elute the RNA. Pipette 2 μL of the RNA into a separate 0.5-mL microcentrifuge for quantification and quality assessment and store both tubes at −70°C for up to 6 months.
20. The pure total RNA is now ready for use with downstream analysis such as reverse transcription PCR but will likely require amplification prior to use with high-throughput platforms involving microarray analysis or high-throughput sequencing (*see* **Note 22**).

3.5. RNA Quantification

Instructions given here assume the use of the RiboGreen RNA Quantitation Reagent Kit made by Molecular Probes. This provides enough RiboGreen reagent for assay of 20,000+ samples using the low-range assay, 200 μL assay in a 96-well microplate format.

1. Defrost an aliquot of the ribosomal RNA standard and the RNA samples for quantification on ice. Also defrost the RiboGreen RNA quantitation reagent at room temperature, but shielded from light (*see* **Note 23**).
2. Pipette 200 μL of 20× TE buffer into 3.8 mL of DEPC water in a nuclease-free 15-mL Falcon tube and vortex briefly to mix. This is now the 1× TE buffer solution (TE) used to set up subsequent reagents and the final assay volume.
3. Dilute the ribosomal RNA standard (100 μg/mL) to a 100 ng/mL working stock in TE. To do this, first make a 1:50 dilution by pipetting 49 μL of TE buffer into a nuclease free 0.5-mL microcentrifuge tube and adding 1 μL of the ribosomal RNA stock solution. Mix thoroughly by vortexing. Carry out a further 1:20 dilution by taking 20 μL of the 1:50 diluted standard and pipetting it into 380 μL of TE in a new 0.5-mL microcentrifuge tube.
4. Pipette 100, 50, 10 and 2 μL of the 100 ng/mL ribosomal RNA working stock into individual wells of a black 96-well microtitre plate. Make these volumes up to 100 μL with TE and also pipette 100 μL of TE to a fifth well to act as the negative control.
5. Add a 99 μL aliquot of TE to each well that will be used to assay an RNA sample of unknown concentration. Then add 1 μL of RNA sample to its corresponding well.
6. Make a working stock of RiboGreen reagent by making a 2,000× dilution with TE. To do this, pipette 1 μL of RiboGreen reagent into 2 mL of TE in a 10-mL microcentrifuge tube, vortex thoroughly and shield from light.
7. Add 100 μL of diluted RiboGreen reagent to each well. Pipette mix the contents of each well by pipetting 100 μL of the contents up and down using the tip that dispensed the RiboGreen reagent.
8. Shield the plate from light using aluminium foil and incubate at room temperature for approximately 5 min.
9. Measure the sample fluorescence using a microplate reader or phosphoimager at standard fluorescein wavelengths (excitation ~480 nm, emission ~520 nm) (*see* **Fig. 22.5** and **Note 24**).

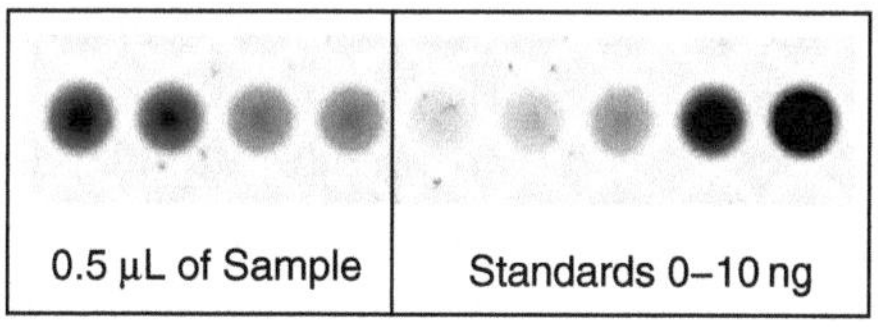

Fig. 22.5. **Quantification of total RNA from microdissected tissues.** The image shows a scan generated using a phosphoimager after incubation of the RNA with RiboGreen reagent. A small fraction of four different samples (0.5 µL from a 10 µL sample) all give a signal that falls within the range of the RNA standards thus enabling accurate downstream quantification.

10. Subtract the fluorescence value of the reagent blank from that of each of the samples and standards. Use this corrected data to generate a standard curve of fluorescence versus RNA concentration and use this to quantify the samples with unknown RNA concentration.
11. *Optional.* It is possible to assess the quality of the RNA by microcapillary electrophoresis trace generated by the Agilent Bioanalyser or the BioRad Experion platforms (*see* **Note 25**).

4. Notes

1. Throughout an LM workflow, every care should be taken to maintain an RNase-free environment. Work surfaces, utensils and other apparatus should be wiped with RNAseZap prior to use. Glassware such as beakers, coplin jars and slide racks can be washed in a 0.1% DEPC solution, rinsed in DEPC-treated water and then baked in foil for 4 h at 180–200°C. Reusable plastic ware and other non-bake materials such as biopsy pads and sample cassettes can be submersed in a DEPC solution and rinsed in DEPC-treated water. A final rinse in absolute ethanol will enable quick drying in a flow hood.
2. The amount of time required to fix the sample is both tissue and fixative dependent. Some preliminary experiments may be required.
3. Whilst working with wax, it is essential to keep metal tissue handling utensils warmed to approximately 70°C. It is advisable to have a number of each utensil warming in a heat block and/or have a Bunsen flame adjacent to the work area to reheat utensils on demand.
4. Multiple specimens of the same type can be arranged in the molten wax to maximise the amount of target cells available per section. It is essential to ensure the orientation of

the tissue will enable sectioning in the desired plane and that target cells are easily identifiable. It is often advisable to prepare samples for LM and stain and mount some of the serial sections in a conventional manner. This should help the user confirm the identity of the target cells on the dehydrated LM slides.

5. We have stored blocks at 4°C for up to a week without obvious degradation of RNA.
6. To ensure pure captures of some cell types, sections may need to be less than one cell thick (6–10 μm for most plant tissues). However, capture of whole tissues may not be limited in this way. Most UV laser-based instruments are now able to cut through sections of 50 μm or higher.
7. The floating of sections on DEPC water and subsequent slide handling are particularly prone to RNA degradation and reduced yield. Preliminary experiments may be required to try and minimise the time of slide drying. For LCM, an alternative method of transfer of thin paraffin sections from the microtome to a glass slide is provided by a Tape Transfer system. If poor RNA quality is obtained with LCM from glass slides using conventional fluid flotation transfer, this system provides an alternative that may significantly improve the quality of RNA.
8. Some tissues may occlude the target beam using normal settings. Although this is rarely the case the target beam intensity can be increased by selecting "target" on the controller and increasing the value using the up arrow button.
9. A contact spot size smaller than 750 μm may be obtained by carrying out multiple shots in the same position but using a low power setting. Multiple laser shots in exactly the same position are also a useful technique if HS caps are to be used to enable the polymer to distend the increased distance to the sample without requiring a high power setting that may widen the contact area.
10. Although large areas are probably best sampled using laser excision, it is still possible using laser capture. However, be aware that repeated melting of the polymer in a thick zone may lead to depletion if you work from the perimeter inwards. To avoid polymer depletion, it is best to start at the centre of the targeted zone and work outward.
11. The amount of target material needed to get nanogram amounts of total RNA will depend on the target type and thus varies between studies (14). This variation may also be a function of tissue preparation and handling regimes affecting the efficiency of extraction of the RNA from the collected fragments.

12. In some cases, the tissue sample may not be uniformly attached to the glass slide. In this case, loosely attached cells may become attached to non-activated regions of the cap during microdissection. To avoid this, the sample slide can be pre-treated with an LCM Prep Strip (Arcturus). This should be done immediately prior to loading the slide into the microscope for microdissection. Place the slide on a solid surface and punch out an individual Prep Strip. Peel off the blank backing and anchor the Prep Strip, non-printed side down, to one end of the sample slide with a gloved finger. Using a finger from your other hand apply the Strip across the slide using a gentle pressure away from the anchored end. Peel the Strip away from the sample using the non-anchored end with an even, upward motion. If non-specific pickup is a persistent problem there is also the option of using positively charged slides (e.g. Super-Frost Plus from Fisher Scientific # 22-034-979). We have used these successfully with Macro caps and they greatly reduce non-specific pickup on the cap. However, the trade off is that some of the targeted tissue may remain bonded to the slide after microdissection. The use of charged or even adhesive slides is only recommended if (i) your procedures is able to compensate or tolerate a reduced target yield per microdissection, and (ii) if the required contact spot size remains small enough to precisely isolate the targeted cells. Aggressive laser settings, which may increase spot size, will be required to bond the cap to the tissue in a manner conducive to pulling the upper tissue away from that bonded to the slide.
13. The cap can be reloaded into the microscope via the unloading platform if further microscopic inspection is required after blotting away unwanted tissue.
14. Whilst Macro Caps enable a relatively large capture area, they can be prone to non specific pick-up since the whole polymer surface is in contact with the tissue sample. HS caps help minimise this due to a ridge on the underside which lifts the majority of the polymer away from the sample. If used as described by the manufacturers, with an ExtracSure device, the ridge is avoided for RNA extraction further avoiding non-specific pickup but at the cost of the available area for capture. By using the polymer peel strategy described here the user gets the reduced non-specific pick-up due to the ridge arrangement whilst maintaining the maximum surface available for capture of the targeted cells. Non-specific pick-up may still occur on the ridge area but since this is raised from the collection surface, relatively aggressive blot removal of the non-specific material

is possible without dislodging the captured tissues. If HS caps are used the extra distance between the capture surface and the sample will mean more aggressive power or duration settings are required to melt the polymer down onto the targeted cells. Also worthy of note is that when sampling the external tissue layer of a section, very aggressive laser settings can be used. This will ensure strong contact between the polymer and the sample. This is achieved by accurate positioning of the sample so the targeted cell layer is clipped by the contact area but the rest of the contact spot misses the tissue.

15. When harvesting the tissues remaining on the slide after microdissection, the amount of RNA is generally much higher than that obtained from the targeted cells. This type of sample can be used to assess if the tissue preparation method and the length of a microdissection session are suitable for obtaining good quality RNA without sacrificing precious target material. It may also serve as a useful comparative sample in subsequent downstream analysis. To do this take a fine, RNase free, flat ended spatula and lightly moisten the tip with RNA extraction buffer. Use the spatula to scrape the remaining sample from the microscope slide such that it collects on the end of the spatula blade. This is then dipped into a 100 μL volume of extraction buffer in a 1.7 mL microcentrifuge tube and agitated so the tissue disperses into the fluid. A sterile tip can also be used to help enable delivery of the tissue fragments into the extraction buffer. The tissue scrape should be treated in an equivalent manner to the microdissected samples if direct comparisons or inferences are to be made. If using LEM remaining tissues can simply be excised from the membrane using a scalpel and some fine forceps and submerged in the extraction buffer.

16. Two sets of 4-tube holders are generally available with an AS LMD, for either 0.2 or 0.5 mL PCR tubes. We routinely use 0.5 mL tubes.

17. Prolonged microdissection sessions can mean the extraction buffer warms and evaporates. Regularly check the amount of buffer in the collection caps and top it up if required.

18. The ×20 or ×40 objectives are usually sufficient for sampling most cell clusters or tissue layers from plant specimens).

19. Move the slide to a position where no tissue is apparent. Click on the "Laser" menu from the top menu bar

and select "Calibrate". A message will ask if the camera is well adjusted, click "Yes". The laser will cut a cross in the top right of the viewable area. Move the computer mouse pointer so that the cross is centred on the cut cross and left click. The laser will then reposition and cut a cross in the top left position. Centre the pointer and click as before and repeat when the laser cuts a cross in the bottom right. After these three positions have been centred a message will state "calibration done", click "OK".

20. These tubes will also catch the sample if the 0.5 mL tubes fail. This tends to happen if the tubes have been stored at −70°C.
21. The addition of the enzyme mix into the filter is not always easy. If the mix does not disseminate into the filter the spin column can be transferred to a microcentrifuge and centrifugated at minimal speed for 10 s. Check for the presence of droplets and if they remain increase the centrifugation speed very slightly and repeat. Continue with this strategy until the filter appears moist, with no large drops of fluid apparent on the surface. Great care must be taken if using this instruction and *it is essential that the mix does not pass through the filter before the incubation is complete.*
22. Typical laser microdissection studies will generate nanogram amounts of total RNA. However, high-throughput studies which use microarray or library sequencing technologies will usually require microgram amounts. RNA amplification regimes are now commonplace and most commercial kits are easily available and reliable (*see* **Table 22.2**). However, some important aspects should always be kept in mind if amplification is to be used. (i) RNA amplification will introduce bias into the representative population and these biases are likely accentuated with smaller starting amounts. Do not assume that the absolute representation of transcripts in a dataset generated via global amplification techniques will be accurate. However, bias appears to be systematic so accurate expression data is achievable using relative values obtained from comparing similarly amplified samples. (ii) Different amplification regimes will identify similar but different lists of differentially expressed genes. However, uniquely identified candidates are often biologically relevant (14).
23. Some workers believe the quality of the RNA can effect the quantification. Because the RNA obtained from a slide scrape (*see* **Note 14**) is prepared in a largely equivalent manner to collected target material and is usually available in relatively large quantities it can be quantified easily

using a Nanodrop spectrophotometer. This can then be substituted for the Ribosomal RNA standard provided in the RiboGreen RNA Quantitation Reagent Kit.

24. To ensure that the sample readings remain in the detection range, alter the instrument's gain so that the sample with the highest RNA concentration yields a reading near to the fluorometer's maximum.

25. If available, microcapillary electrophoresis is ideal for use with a wide range of RNA concentrations and the high-sensitivity chips can assay as little as 100 pg/μL of total RNA. The protocols for using these apparatus do not vary from those provided by the manufacturers. Observable ribosomal peaks in the trace indicate good quality RNA.

Acknowledgements

The authors would like to thank Bronwyn Carlisle for drawing **Fig. 22.1**. Protocols were developed in the Macknight laboratory with support from the Marsden Fund.

References

1. Schmid, M., Davison, T. S., Henz, S. R., et al. (2005) A gene expression map of *Arabidopsis thaliana* development. *Nat Gen* **37**, 501–506.
2. Benedito, V. A., Torres-Jerez, I., Murray, J. D., et al. (2008) A gene expression atlas of the model legume *Medicago truncatula*. *Plant J* **55**, 504–513.
3. Jiao, Y., Lori Tausta, S., Gandotra, N., et al. (2009) A transcriptome atlas of rice cell types uncovers cellular, functional and developmental hierarchies. *Nat Gen* **41**, 258–263.
4. Balestrini, R. and Bonfante, P. (2008) Laser microdissection (LM): Applications to plant materials. *Plant Biosyst* **142**, 331–336.
5. Day, R. C., Grossniklaus, U., and Macknight, R. C. (2005) Be more specific! Laser-assisted microdissection of plant cells. *Trends Plant Sci* **10**, 397–406.
6. Nelson, T., Tausta, S. L., Gandotra, N., and Liu T. (2006) Laser microdissection of plant tissue: What you see is what you get. *Annu Rev Plant Biol* **57**, 181–201.
7. Ohtsu, K., Takahashi, H., Schnable, P., and Nakazono, M. (2006) Cell Type-specific gene expression profiling in plants by using a combination of laser microdissection and high-throughput technologies. *Plant Cell Physiol* **48**, 3–7.
8. Cai, S. and Lashbrook, C. (2008) Stamen abscission zone transcriptome profiling reveals new candidates for abscission control: enhanced retention of floral organs in transgenic plants overexpressing Arabidopsis ZINC FINGER PROTEIN2. *Plant Physiol* **146**, 1305–1321.
9. Day, R. C., Herridge, R. P., Ambrose, B. A., and Macknight, R. C. (2008) Transcriptome analysis of proliferating Arabidopsis endosperm reveals biological implications for the control of syncytial division, cytokinin signaling, and gene expression regulation. *Plant Physiol* **148**, 1964–1984.
10. Ohtsu, K., Smith, M., Emrich, S., et al. (2007) Global gene expression analysis of the shoot apical meristem of maize (*Zea mays* L.). *Plant J* **52**, 391–404.
11. Wu, Y., Machado, A. C., White, R. G., Llewellyn, D., and Dennis, E. (2006) Expression profiling identifies genes expressed early during lint fibre initiation in cotton. *Plant Cell Physiol* **47**, 107–127.

12. Galbraith, D. W. (2003) Global analysis of cell type-specific gene expression. *Comp Funct Genom* **4**, 208–215.
13. Nelson, T., Gandotra, N., and Tausta, S. (2008) Plant cell types: Reporting and sampling with new technologies. *Curr Opin Plant Biol* **11**, 567–573.
14. Day, R., Mcnoe, L., and Macknight, R. (2007) Transcript analysis of laser microdissected plant cells. *Physiol Plant* **129**, 267–282.
15. Espina, V., Wulfkuhle, J. D., Calvert, V. S., et al. (2006) Laser-capture microdissection. *Nat Protoc* **1**, 586–603.
16. Ivashikina, N., Deeken, R., Ache, P., et al. (2003) Isolation of *AtSUC2* promoter-GFP-marked companion cells for patch-clamp studies and expression profiling. *Plant J* **36**, 931–945.
17. Casson, S., Spencer, M., Walker, K., and Lindsey, K. (2005) Laser capture microdissection for the analysis of gene expression during embryogenesis of Arabidopsis. *Plant J* **42**, 111–123.
18. Spencer, M. W., Casson, S. A., and Lindsey, K. (2007) Transcriptional profiling of the Arabidopsis embryo. *Plant Physiol* **143**, 924–940.
19. Inada, N. and Wildermuth, M. C. (2005) Novel tissue preparation method and cell-specific marker for laser microdissection of Arabidopsis mature leaf. *Planta* **221**, 9–16.
20. Cai, S. and Lashbrook, C. C. (2006) Laser capture microdissection of plant cells from tape-transferred paraffin sections promotes recovery of structurally intact RNA for global gene profiling. *Plant J* **48**, 628–637.
21. Deeken, R., Ache, P., Kajahn, I., Klinkenberg, J., Bringmann, G., and Hedrich, R. (2008) Identification of *Arabidopsis thaliana* phloem RNAs provides a search criterion for phloem-based transcripts hidden in complex datasets of microarray experiments. *Plant J* **55**, 746–759.
22. Day, R. C., McNoe, L., and Macknight, R. C. (2007) Evaluation of global RNA amplification and its use for high-throughput transcript analysis of laser-microdissected endosperm. *Int J Plant Genom* 61028.
23. Galbiati, M., Simoni, L., Pavesi, G., et al. (2008) Gene trap lines identify Arabidopsis genes expressed in stomatal guard cells. *Plant J* **53**, 750–762.
24. Thiel, J., Weier, D., Sreenivasulu, N., et al. (2008) Different hormonal regulation of cellular differentiation and function in nucellar projection and endosperm transfer cells: a microdissection-based transcriptome study of young barley grains. *Plant Physiol* **148**, 1436–1452.
25. Wu, Y., Llewellyn, D., White, R., Ruggiero, K., Al-Ghazi, Y., and Dennis, E. (2007) Laser capture microdissection and cDNA microarrays used to generate gene expression profiles of the rapidly expanding fibre initial cells on the surface of cotton ovules. *Planta* **226**, 1475–1490.
26. Santi, S. and Schmidt, W. (2008) Laser microdissection-assisted analysis of the functional fate of iron deficiency-induced root hairs in cucumber. *J Exp Bot* **59**, 697–704.
27. Jiang, K., Zhang, S., Lee, S., et al. (2006) Transcription profile analyses identify genes and pathways central to root cap functions in maize. *Plant Mol Biol* **60**, 343–363.
28. Dembinsky, D., Woll, K., Saleem, M., et al. (2007) Transcriptomic and proteomic analyses of pericycle cells of the maize primary root. *Plant Physiol* **145**, 575–588.
29. Nakazono, M., Qiu, F., Borsuk, L. A., et al. (2003) Laser-capture microdissection, a tool for the global analysis of gene expression in specific plant cell types: Identification of genes expressed differentially in epidermal cells or vascular tissues of maize. *Plant Cell* **15**, 583–596.
30. Zhang, X., Madi, S., Borsuk, L., et al. (2007) Laser microdissection of narrow sheath mutant maize uncovers novel gene expression in the shoot apical meristem. *PLoS Genet* **3**, e101.
31. Nakada, M., Komatsu, M., Ochiai, T., et al. (2006) Isolation of from and analysis of its expression using laser microdissection. *Plant Sci* **170**, 143–150.
32. Corpas, F. J., Fernandez-Ocana, A., Carreras, A., et al. (2006) The expression of different superoxide dismutase forms is cell-type dependent in olive (*Olea europaea* L.) leaves. *Plant Cell Physiol* **47**, 984–994.
33. Murata, J. and Luca, V. (2005) Localization of tabersonine 16-hydroxylase and 16-OH tabersonine-16-O-methyltransferase to leaf epidermal cells defines them as a major site of precursor biosynthesis in the vindoline pathway in *Catharanthus roseus*. *Plant J* **44**, 581–594.
34. Yu, Y., Lashbrook, C. C., and Hannapel, D. J. (2007) Tissue integrity and RNA quality of laser microdissected phloem of potato. *Planta* **226**, 797–803.
35. Asano, T., Masumura, T., Kusano, H., et al. (2002) Construction of a specialized cDNA library from plant cells isolated by laser capture microdissection: Toward comprehensive

analysis of the genes expressed in the rice phloem. *Plant J* **32**, 401–408.
36. Suwabe, K., Suzuki, G., Takahashi, H., et al. (2008) Separated transcriptomes of male gametophyte and tapetum in rice: Validity of a laser microdissection (LM) microarray. *Plant Cell Physiol* **49**, 1407–1416.
37. Ishimaru, T., Nakazono, M., Masumura, T., et al. (2007) A method for obtaining high integrity RNA from developing aleurone cells and starchy endosperm in rice (*Oryza sativa* L.) by laser microdissection. *Plant Sci* **173**, 321–326.
38. Liu, H., Wang, S., Yu, X., et al. (2005) ARL1, a LOB-domain protein required for adventitious root formation in rice. *Plant J* **43**, 47–56.
39. Klink, V., Alkharouf, N., Macdonald, M., and Matthews, B. (2005) Laser capture microdissection (LCM) and expression analyses of *Glycine max* (Soybean) syncytium containing root regions formed by the plant pathogen *Heterodera glycines* (Soybean Cyst Nematode). *Plant Mol Biol* **59**, 965–979.
40. Klink, V., Overall, C., Alkharouf, N., Macdonald, M., and Matthews, B. (2007) Laser capture microdissection (LCM) and comparative microarray expression analysis of syncytial cells isolated from incompatible and compatible soybean (*Glycine max*) roots infected by the soybean cyst nematode (*Heterodera glycines*). *Planta* **226**, 1389–1409.
41. Ithal, N., Recknor, J., Nettleton, D., Maier, T., Baum, T., and Mitchum, M. (2007) Developmental transcript profiling of cyst nematode feeding cells in soybean roots. *Mol Plant-Microbe Interact* **20**, 510–525.
42. Sanders, P., Bui, A., Le, B., and Goldberg, R. (2005) Differentiation and degeneration of cells that play a major role in tobacco anther dehiscence. *Sex Plant Reprod* **17**, 219–241.
43. Balestrini, R., Gómez-Ariza, J., Lanfranco, L., and Bonfante, P. (2007) Laser microdissection reveals that transcripts for five plant and one fungal phosphate transporter genes are contemporaneously present in arbusculated cells. *Mol Plant-Microbe Interact* **20**, 1055–1062.
44. Ramsay, K., Wang, Z., and Jones, M. (2004) Using laser capture microdissection to study gene expression in early stages of giant cells induced by root-knot nematodes. *Mol Plant Pathol* **5**, 587–592.
45. Kerk, N. M., Ceserani, T., Tausta, S. L., Sussex, I. M., and Nelson, T. M. (2003) Laser capture microdissection of cells from plant tissues. *Plant Physiol* **132**, 27–35.
46. Tang, W., Coughlan, S., Crane, E., Beatty, M., and Duvick, J. (2006) The application of laser microdissection to in planta gene expression profiling of the maize anthracnose stalk rot fungus *Colletotrichum graminicola*. *Mol Plant-Microbe Interact* **19**, 1240–1250.
47. Woll, K., Borsuk, L. A., Stransky, H., Nettleton, D., et al. (2005) Isolation, characterization, and pericycle-specific transcriptome analyses of the novel maize lateral and seminal root initiation mutant rum1. *Plant Physiol* **139**, 1255–1267.

Chapter 23

Utilizing Bimolecular Fluorescence Complementation (BiFC) to Assay Protein–Protein Interaction in Plants

Nir Ohad and Shaul Yalovsky

Abstract

Protein function is often mediated by the formation of stable or transient complexes. Here we present a method for testing protein–protein interactions in plants designated bimolecular fluorescence complementation (BiFC). The advantages of BiFC are its simplicity, reliability, and the ability to observe protein–protein interactions in different cellular compartments including membranes. BiFC is based on splitting the yellow fluorescent protein (YFP) into two nonoverlapping *N*-terminal (YN) and C-terminal (YC) fragments. Each fragment is cloned in-frame with a gene of interest, enabling expression of a fusion protein. Reconstitution of the fluorescing YFP chromophore takes place upon interaction of protein pairs that are coexpressed in the same cells.

Key words: Bimolecular fluorescence complementation, BiFC, protein–protein interaction, yellow fluorescent protein (YFP).

1. Introduction

The complete genome sequence of several plants such as Arabidopsis, rice, and *Physcomitrella* and the large-scale analysis of the Arabidopsis proteome, obtained during the last few years, provide the blueprint for the putative coding sequences and their products. However, the underlying mechanisms that coordinate all cellular functions supporting the complexity of plant embryogenesis, growth, and response to the environment cannot be understood only from the knowledge of the primary sequences and identification of all proteins, but rather from knowledge about protein function.

L. Hennig, C. Köhler (eds.), *Plant Developmental Biology*, Methods in Molecular Biology 655,
DOI 10.1007/978-1-60761-765-5_23, © Springer Science+Business Media, LLC 2010

Many proteins mediate their function through the formation of stable or transient protein complexes and networks. Patterns of protein accumulation in plants are regulated spatially and temporally, and often the same protein may interact with various partners in response to different stimuli or at different developmental stages (1, 2). Moreover, protein interactions may depend on post-translational modifications (3), which could be tissue or species specific (4).

The bimolecular fluorescence complementation (BiFC) assay originally developed by (5, 6) is useful for the detection of protein–protein interactions in living cells. BiFC is based upon tethering split YFP or other GFP variants to form a functional fluorophore. The association of the split YFP/GFP/cyan fluorescent protein (CFP) molecule does not occur spontaneously and requires interaction between proteins or peptides that are fused to each of the fluorophore fragments (*see* **Fig. 23.1A**).

Upon interaction of these fused proteins/peptides, the split fluorophore fragments can interact to form a fluorescent protein that has the same spectral properties as the unsplit YFP, or other GFP variants (*see* **Fig. 23.1B, C**). If the proteins that are fused to the split fluorophore fragments do not interact, reconstitution of the YFP/GFP/CFP usually does not take place and no fluorescence is detected. In recent years, BiFC has been adopted to plants either by splitting YFP between amino acids 154 and 155 (7, 8) or between residues 174 and 175 (9).

BiFC has several major advantages as originally shown by (5, 6): (1) The assay is simple and does not require sophisticated dedicated equipment. (2) There is either no or low background signal because a fluorescing YFP would only form after interaction between proteins fused to split fragments. (3) BiFC enables determination of the subcellular localization of interacting protein complexes as well as the mutual affect of interacting partners on the subcellular localization of the complex. (4) BiFC is a sensitive assay, enabling detection of weak and transient interactions, primarily due to the stability of the reconstituted YFP complexes (5). (5) Recently, multicolor BiFC, which is based on reconstitution of split YFP and CFP (cyan fluorescent protein), enabling observation of multiple protein complexes has been successfully applied in animal cells (10) and in plants (11, 12).

BiFC suffers from several pitfalls that must be taken into account. (1) The slow maturation time of the reconstituted GFP/YFP/CFP could compromise detection of dynamic changes in protein–protein interactions in real time (5, 6, 13). This problem can be alleviated by using the Venus variant of YFP, which matures within few seconds (14, 15). (2) The stability of the reconstituted YFP complexes (5) hampers the ability to analyze the dynamics of protein–protein dissociation. This can lead to

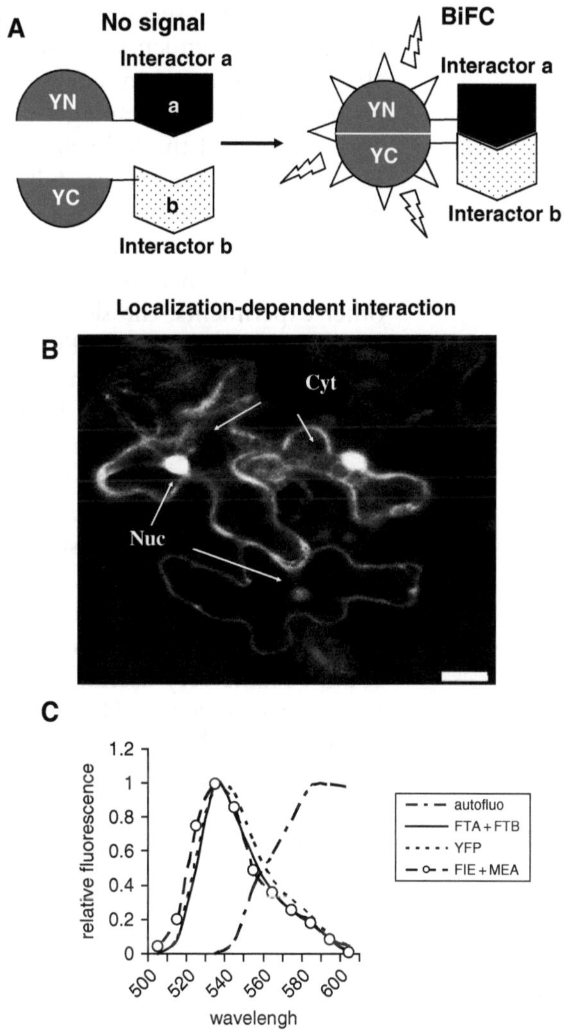

Fig. 23.1. **BiFC enables to determine the interaction between protein pairs and their subcellular localization. (A)** Schematic illustration of an interaction occurring between two proteins that are localized in the same compartment. Under physiological conditions, reconstitution of a fluorescent YFP molecule can only take place following interaction between proteins or peptides that are fused to YN and YC fragments. **(B)** Epifluorescent images of YFP fluorescence complementation. Coexpression of YN-MEA and YC-FIE. Image was taken with a polychromatic filter set 40 (Zeiss). Scale bars are 10 μm. Fluorescence emitted from interacting proteins is represented by white color. Arrows indicate the nucleus or cytoplasm (Nuc. or Cyt.), respectively. **(C)** Fluorescence emission spectra of intact and reconstituted YFPs. Fluorescence emission spectra were measured following excitation with argon laser at 488 nm together with a 500 nm beamsplitter. Identical emission spectra were formed by intact YFP (*dashed line,*) YN-FTA + YC-FTB (farnesyltransferase α and β subunits) (*continues line,*) YN-MEA + YC-FIE (*dashed line with circles,*) and autofluorescence of nontransformed tissue (*dash and doted line*). Fluorescence emission spectra were measured with the spectral detector of a Leica LCS-SL CLSM. The maximum emission of each spectrum was taken as 1, and in each curve fluorescence at a given wavelength was normalized relative to the maximum.

detection of nonspecific interactions when expression levels of the split YFP fragments are high. (3) The molecular properties of chimeric fusion proteins could be different from that of the native proteins. These disadvantages of the BiFC system require careful consideration of the following issues. Although BiFC-based systems have been used successfully for monitoring dynamic changes in Ca^{2+} concentrations (16, 17), the stability of the reconstituted YFP/GFP/CFP complexes and the maturation times of fluorescent dyes other than Venus, may compromise the results. In most other experiments, the slow maturation time should not pose a problem since typically BiFC assays are carried out using transient expression systems in which transformed tissues are typically analyzed after several hours or even 24–48 h. This allows enough time for proteins to be expressed and interact. To alleviate the problem of nonspecific interactions, expression levels should be kept low. Using negative controls such as noninteracting point mutants of tested proteins is essential. To ascertain that lack of fluorescence is not due to low expression, it is crucial to monitor expression levels of the relevant proteins with antibodies (7, 8). The quantification of the fluorescent signal could become very useful, provided it is normalized to the protein expression levels. The validity of the BiFC results should be verified by determining expression patterns, colocalization assays, and if applicable, genetic analysis and determination of protein–protein interactions in plants by an independent method (for example *see* (2)). The bottom line is that as with any other experimental system, using the right controls and proper calibration are essential.

In this chapter, we present the simplest method of transiently expressing protein pairs of BiFC constructs via leaf infiltration of *Agrobacterium tumefaciens* harboring the appropriate plasmids and monitoring their potential interaction in plants.

2. Materials

2.1. Plant Material

1. *Nicotiana benthamiana* (*N. benthamiana*) seeds are germinated on a mix of 70% soil with vermiculite, 30% sea-sand and irrigated from below with a standard gardening fertilizer until saturation. There is no need to add fertilizer later during growth. Plants are germinated and grown in environmental growth chambers under long-day conditions (16 h light/8 h dark cycles) at 25–27°C under 80 μE m^{-2}/s of light. When seedlings reach the stage of 4–6 leaves, they can

be transferred to 10-cm pots using the same soil mix. Plants should be kept under a lid while germinating and for ~1 week after transfer; the lid should then be removed to prevent accumulation of excess of humidity (*see* **Note 3**). After 2–3 weeks of additional growth, plants reach an optimal size that will best support infiltration to leaves.

2. Arabidopsis *Col-0* plants are grown in 5-cm pots. Plants are grown in soil with vermiculite and are irrigated from below. Use environmental growth chambers with long-day conditions (16 h light/8 h dark cycles) at 22°C under 80 μE m^{-2}/s of light (*see* **Note 4**).

2.2. Leaf Infiltration

1. AB mix (20×): 373.9 mM NH_4Cl, 24.34 mM $MgSO_4$, 40.23 mM KCl, 1.36 mM $CaCl_2$, 0.18 mM $FeSO_4$-$7H_2O$. Autoclave for 15 min and remove from the autoclave immediately. Store at RT.
2. Induction medium: 50 mM MES, 0.5% glucose, 1.7 mM NaH_2PO_4, 0.2 mM acetosyringone, 5% glucose, 5% 20×-AB mix. To prepare induction medium:
   a Dissolve 4.9 g of MES (morpholino ethane sulfonic acid) in 400 mL of water.
   b Adjust to pH 5.6 with NaOH.
   c Add 2.5 g of glucose and 0.14 g of NaH_2PO_4•(H_2O (powder) and mix until they dissolve.
   d Add water up to 475 mL.
   e Autoclave and let cool.
   f Add 25 mL of 20 × AB mix (ignore any orange precipitation that might appear).
   g Store at 4°C.

2.3. Vectors and Cloning Procedures

1. Different BiFC vector systems are available. The vectors generated by the presenting authors (7) are available at "The Arabidopsis Information Resource" (TAIR) (http://www.arabidopsis.org/). Additional vector sets suitable for analyzing BiFC in plants have been described by (8, 9) (*see* **Note 7**).
2. In all these vectors systems, standard cloning procedures are used to fuse coding sequences of proteins under study in-frame to either the YN or YC.
3. Vectors are transformed into *A. tumefaciens* strain GV3101/pMp90. Selected colonies can be stored as glycerol stocks in a –80°C freezer.
4. LB medium.

2.4. Protein Extraction and SDS-Polyacrylamide Gel Electrophoresis (SDS-PAGE)

1. Protein extraction buffer: 100 mM Tris pH 7.2, 10% sucrose, 5 mM $MgCl_2$, 5 mM EDTA, 40 μM β-mercaptoethanol, protease inhibitor cocktail (Roche) (2), and 2 mM PMSF.
2. Denaturing SDS-PAGE sample buffer (3×): 150 mM Tris-HCl pH 6.8, 300 mM DTT, 6% SDS, 30% glycerol, 0.3% bromophenol blue.
3. Membrane extraction buffer: Protein extraction buffer with 1% NP-40 and 0.1% SDS. Alternatively, membrane proteins can be directly solubilized in denaturing SDS-PAGE sample buffer (18). Other buffers can be used as well such as: 50 mM HEPES-KOH (pH 7.5), 10 mM KCl, 5 mM EDTA, 5 mM EGTA, 10% sucrose, 1 mg/mL phenylmethylsulfonyl fluoride, and protease inhibitor cocktail.
4. 10–12% SDS-polyacrylamide gels.
5. Nitrocellulose membrane.
6. TBS blocking buffer: 20 mM Tris-HCl, pH 7.6, and 137 mM NaCl, containing 5% (w/v) nonfat dry milk.
7. The following antibodies can be used to detect epitops present on YN or YC proteins used in the BiFC system described by (7): Monoclonal α-GFP (Covance-cat #MMS-118P; 1:1,000) can detect the YN, α-HA (Covance-cat # MMS-101R; 1:3000) or α-GLUGLU (Covance-cat #MMS-115R; 1:1000) monoclonal antibodies (mAbs) can detect the YC. Secondary antibody: HRP-horseradish peroxidase-conjugated goat α-mouse (Biorad-blotting grade; 1:10,000) (*see* **Note** 7) or goat antirabbit (Jackson. Immuono-Research cat # 111-035-003; 1:10,000).
8. Protein G Sepharose beads for precipitating antibody-antigen complexes.
9. EZ-ECL kit (Bet Haemek, Israel) for chemiluminescent protein detection. Alternatively, use homemade "development solution": Mix 9 mL of H_2O, 1 mL of 1 M Tris-HCl (pH 8.9), 50 μL of luminol (Sigma) (44 mg/mL in DMSO,) 25 μL of paracoumaric acid (Sigma) (15 mg/mL in DMSO), and 6 μL of 30% H_2O_2.
10. X-Ray film (Fugi super HR-G30).
11. TBST: TBS containing 0.05% (v/v) Tween 20.
12. Dilutions of antibodies are prepared in TBST containing 5% (w/v) nonfat dry milk.

2.5. Visualization of Fluoresces

Fluorescence or confocal microscopes can be used for visualization of fluoresces.

3. Methods

3.1. Transient Expression via A. tumefaciens Infiltraion

Transient expression of desired proteins carried out by injection of *A. tumefaciens* GV3101/pMP90 cells harboring the appropriate plasmids into *N. benthamiana* or Arabidopsis (*Col-0*) leaves (*see* **Note 2**).

3.1.1. Day 1 – Prepare Starters

Work in a clean environment (*see* **Note 1**).

1. Add to a sterile bacteria-growth tube 1.5 mL of liquid LB medium. Inoculate the medium with a colony of *A. tumefaciens* harboring a vector to be used.
2. Add appropriate antibiotics to the tube and mix.
3. Grow the bacteria overnight on a shaker at 28°C. Make sure the tube cap is loose to allow aeration for optimal growth.

3.1.2. Day 2 – Induction

Work in clean environment – near a flame!

4. Prepare a new sterile bacteria-growth tube with 3 mL of induction medium.
5. Transfer 90 μL of bacteria culture from the starter tube to the induction medium tube.
6. Grow the bacteria for 6 to 20 h on a shaker at 28°C to OD_{600} $\leq$ 0.8. Make sure the tube cap is loose to allow aeration for optimal growth.

3.1.3. Day 3 – Injection

7. Read the OD for each culture tube and dilute the culture to $OD_{600} \cong 0.2$ with induction medium.
8. Mix pairs of *Agrobacterium* strains harboring the YN and YC constructs to be assayed in a 1:1 ratio (V/V).
9. Prepare 100 μL of culture for each site of injection in a sterile plastic 1-mL syringe without a needle. Wear a lab coat, gloves, and protective goggles when injecting the *Agrobacteria*.
10. To inject, press the syringe gently to the abaxial (lower) leaf side and provide a counter-pressure with a finger from the other side of the leaf. Be careful not to apply too much pressure as this will lead to rupture of the leaf. As the injected cell culture penetrates into the leaf, it will spread in the intercellular space, which can be followed by the slight change in leaf color (*see* **Notes 3** and **4**). Leaf injection is very simple. However, one can practice beforehand by injecting water or LB into leaves.
11. Let the injected area dry for several minutes and then gently remove residual liquid with a tissue paper. Mark

the edges of the injected area with a marker so you can identify them.

12. Place the plant back in a growth chamber; make sure it is irrigated well.
13. Analyze protein expression 24–48 h after injection (*see* **Note 5**).

3.2. Fluorescence and Confocal Imaging

1. At 24–48 h after injection, cut a rectangular segment smaller than the size of a microscope slide coverslip from the marked leaf area and place it on a microscope slide in a drop of sterile water with the lower (abaxial) leaf side facing up. Place a coverslip on top.
2. Perform wide-field fluorescence imaging using a standard fluorescent microscope equipped with a cooled coupled device (CCD) camera (*see* **Note 5**).
3. For confocal imaging, use Argon laser at 488 or 514 nm to excite YFP. Set the filter sets or AOBS and spectral detector in Leica microscopes to view YFP (*see* **Note 6**).
4. Annotation of images such as adding labels, scale bar, rotation, or cropping can be done with software such as Leica LCS, Zeiss AxioVision, Adobe Photoshop 7.0, or Image J.

3.3. Protein Immunoblots

Protein immunoblot can be used to determine whether both YN and YC fusion proteins were expressed also in cases when no fluorescence is detected.

1. To extract soluble proteins, freeze leaf portions from the same sample used to monitor fluorescence in liquid N_2 and grind into powder using pestle and mortar.
2. Resuspend leaf powder (1 g) in an equal volume (w/v) of protein extraction buffer, which extracts both cytoplasmic and nuclear proteins.
3. To remove insoluble material, centrifuge the extracts at 15,000×*g* for 10 min. Collect supernatant or pellet fractions (for soluble and membrane proteins, irrespectively) and use for immunoblot analysis. Alternative protocol: Precipitate insoluble material at 100,000×*g* for 1 h. To extract membrane proteins, incubate in 1 mL of membrane extraction buffer (*see* **Note 8**) or denaturing SDS-PAGE sample buffer.
4. Resolve proteins on 10–12% acrylamide gels and electrotransfer to nitrocellulose membranes.
5. Block membranes for 1 h at room temperature in TBS blocking buffer.
6. Incubate with a primary antibody either for 1 h at room temperature or for 12–16 h at 4°C.

7. Wash blots three times in TBST for 10 min.
8. Incubate with the secondary antibody for 2 h.
9. Wash three times with TBST for 10 min.
10. Detect bound antibodies with chemiluminescent substrate and expose to X-ray film for 1 to 15 min.

3.4. Coimmunoprecipitation

In case where proteins pairs were expressed but did not result in a fluorescent signal, one can further test for their possible interaction by coimmunoprecipitation. As both YC- and YN- can be detected with the same polyclonal α-GFP antibody (Santa-Cruz, USA), they are useful as tags for protein immunoblots as well as for immunoprecipitation, but not for coimmunoprecipitation. YC or YN can be detected specifically as described under **Section 2.4**, Step 5. As some BiFC constructs are double-tagged (e.g., YC with HA), this can allow to perform coimmunoprecipitation. Otherwise, protein-specific antibodies are needed.

1. Prepare protein extracts as described above (*see* **Section 3.3** Step 1) using 1 mL of protein extraction buffer for 1 g of tissue.
2. Centrifuge extracts for 30 min at 20,000×*g* and 4°C to remove insoluble material.
3. Incubate supernatants with α-HA monoclonal antibodies at a dilution of 1:100 for 12 h at 4°C.
4. To precipitate antibody–antigen complexes, add 25 μL of protein G Sepharose beads into each tube and agitate for 1 h.
5. Precipitate immunocomplexes by centrifugation for 1 min at 10,000×*g*.
6. Wash precipitated immunocomplexes four times with 1 mL of PBS.
7. Elute proteins from the beads by boiling in 25 μL SDS-PAGE sample buffer for 5 min.
8. Equal amounts of immunoprecipitated and unbound proteins are resolved by SDS-PAGE. Load 6% of the total proteins used in the coimmunoprecipitation experiment as control onto the same gel.
9. Perform protein detection with immunoblots as described under **Section 3.3**.

4. Notes

1. When preparing *Agrobacterium* cultures, work in clean environment, preferably use a sterile hood or work on your bench next to a flame.

2. Transient expression can be performed on different plant species via different methods such as by particle bombardment of onion leaf epidermal tissue. For additional information on alternative model plants and methods for transient expression *see* reference (19).

3. Growth temperatures for *N. benthamia* plants lower than 25°C should be avoided because the plants develop thinner leaves that can not transformed efficiently. Transformation should be performed with 3–4 weeks-old plants at which time the 4th up to 6th leaves have grown large enough to support injection. Younger plants are less efficient. In our hands, no protein expression could be detected after plants have started to flower. Transformation via the leaf injection procedure should be performed on the lower (abaxial) side of the leaves. Best results are obtained when injection is made in-between major leaf veins. Injecting more then four spots on the same leaf should be avoided as it may cause the collapse of the leaf due to loss of turgor pressure.

4. Arabidopsis plants grown under short-day conditions can be used with about the same success as plants grown under long-day conditions. Short-day grown plants might be useful because they develop more and larger leaves.

5. The length of time one can view expression may vary from one sample to the other and should be determined experimentally. To observe fluorescence with a standard fluorescence microscope, we recommend using Pinkel set excitation band path filter (485/17 nm) with Zeiss triple band filter set 40 (400–440, 500–540 band path and 570-nm long-path beamsplitters and 450–470, 505–535, and 575–625-nm band path emission filters). The Zeiss filter set 40 is useful for differentiating between YFP fluorescence, which appears green, and autofluorescence, which appears as yellow-orange. Alternatively, one can use a yellow GFP filter (Chroma set #41028).

6. Measurements of fluorescence emission spectra can help determining the property of a signal observed and thus distinguish between autofluorescence and a true signal resulting from proper reconstitution of the chromophor, which should emit at a particular wavelength. Such measurements can be carried out with a confocal microscope equipped with spectral detectors as in the case of the Leica TCS-SL, Leica SP2 or SP5, and Zeiss LSM 510 Meta or LSM 700. In the Leica TCS-SL, excitation is performed at 488 or 514 nm using 500 or 514 nm beamsplitters, respectively, and emission is detected between 500 and 600 nm at 5-nm intervals. For each scan, signal intensities are normalized relative

to the wavelength (*see* **Fig. 23.1C**). A standard fluorescence spectrophotometer can also be used (7); however, due to the absence of focusing lenses, the signals are usually weak.

7. A new BiFC vector, in which the Glu–Glu– tag has been replaced with a 6 X His tag (His_6) enabling identification with the significantly cheaper α-His Abs and allowing purification with metal-chelate chromatography, has been described in reference (20).

Acknowledgment

This work was supported by the Israel Science Foundation (ISF-312/07), US–Israel Binational Research and Development fund (BARD-IS-4032-07), and the Deutschland–Israel Program (DIP-H.3.1) to SY.

NO would like to thank the Israel Science Foundation (ISF-574/04), US–Israel, the Deutschland–Israel Program (DIP-H.3.1), and Tel-Aviv University foundation for Research and Development (590168), for their support.

References

1. Hsieh, T. F., Hakim, O., Ohad, N., and Fischer, R. L. (2003) From flour to flower: How Polycomb group proteins influence multiple aspects of plant development. *Trends Plant Sci* **8**, 439–445.
2. Katz, A., Oliva, M., Mosquna, A., Hakim, O., and Ohad, N. (2004) FIE and CURLY LEAF polycomb proteins interact in the regulation of homeobox gene expression during sporophyte development. *Plant J* **37**, 707–719.
3. Hoffman, G. R., Nassar, N., and Cerione, R. A. (2000) Structure of the Rho family GTP-binding protein Cdc42 in complex with the multifunctional regulator RhoGDI. *Cell* **100**, 345–356.
4. Lavy, M., Bracha-Drori, K., Sternberg, H., and Yalovsky, S. (2002) A cell-specific, prenylation-independent mechanism regulates targeting of type II RACs. *Plant Cell* **14**, 2431–2450.
5. Hu, C. D., Chinenov, Y., and Kerppola, T. K. (2002) Visualization of interactions among bZIP and Rel family proteins in living cells using bimolecular fluorescence complementation. *Mol Cell* **9**, 789–798.
6. Hu, C. D. and Kerppola, T. K. (2003) Simultaneous visualization of multiple protein interactions in living cells using multicolor fluorescence complementation analysis. *Nat Biotechnol* **21**, 539–545.
7. Bracha-Drori, K., Shichrur, K., Katz, A., et al. (2004) Detection of protein–protein interactions in plants using bimolecular fluorescence complementation. *Plant J* **40**, 419–427.
8. Walter, M., Chaban, C., Schütze, K., et al. (2004) Visualization of protein interactions in living plant cells using bimolecular fluorescence complementation. *Plant J* **40**, 428–438.
9. Citovsky, V., Lee, L. Y., Vyas, S., et al. (2006) Subcellular localization of interacting proteins by bimolecular fluorescence complementation in planta. *J Mol Biol* **362**, 1120–1131.
10. Shyu, Y. J., Liu, H., Deng, X., and Hu, C. D. (2006) Identification of new fluorescent protein fragments for bimolecular fluorescence complementation analysis under physiological conditions. *Biotechniques* **40**, 61–66.
11. Kodama, Y. and Wada, M. (2009) Simultaneous visualization of two protein complexes in a single plant cell using multicolor fluorescence complementation analysis. *Plant Mol Biol* **70(1–2)**, 211–217.

12. Waadt, R., Schmidt, L. K., Lohse, M., Hashimoto, K., Bock, R., and Kudla, J. (2008) Multicolor bimolecular fluorescence complementation reveals simultaneous formation of alternative CBL/CIPK complexes in planta. *Plant J* **56**, 505–516.
13. Ghosh, I., Hamilton, A. D., and Regan, L. (2000) Antiparallel leucine zipper-directed protein reassembly: Application to the green fluorescent protein. *J Am Chem Soc* **122**, 5658–5659.
14. Miyawaki, A., Nagai, T., and Mizuno, H. (2003) Mechanisms of protein fluorophore formation and engineering. *Curr Opin Chem Biol* 7, 557–562.
15. Miyawaki, A., Nagai, T., and Mizuno, H. (2005) Engineering fluorescent proteins. *Adv Biochem Eng Biotechnol* **95**, 1–15.
16. Nagai, T., Sawano, A., Park, E. S., and Miyawaki, A. (2001) Circularly permuted green fluorescent proteins engineered to sense Ca^{2+}. *Proc Natl Acad Sci USA* **98**, 3197–3202.
17. Robert, V., Gurlini, P., Tosello, V., et al. (2001) Beat-to-beat oscillations of mitochondrial [Ca^{2+}] in cardiac cells. *EMBO J* **20**, 4998–5007.
18. Laemmli, U. K. (1970) Cleavage of structural proteins during the assembly of the head of bacteriophage T4. *Nature* **227**, 680–685.
19. Ohad, N., Shichrur, K., and Yalovsky, S. (2007) The analysis of protein–protein interactions in plants by bimolecular fluorescence complementation. *Plant Physiol* **145**, 1090–1099.
20. Lavy, M., Bloch, D., Hazak, O., et al. (2007) A Novel ROP/RAC effector links cell polarity, root-meristem maintenance, and vesicle trafficking. *Curr Biol* **17**, 947–952.

Chapter 24

The Split Luciferase Complementation Assay

Naohiro Kato and Jason Jones

Abstract

A split luciferase complementation assay to study protein–protein interactions within Arabidopsis protoplasts in 96-well plates is described in this protocol. Two proteins of interest, a bait and prey, which are genetically fused to amino- and carboxy-terminal fragments of *Renilla* luciferase, are transiently expressed in protoplasts. Physical interactions of these bait and prey proteins reconstitute some of the luciferase activity and result in light emission in the presence of the luciferase substrate. This luminescence is then measured by a microplate luminometer. Amounts of the bait protein accumulated in the protoplasts can be estimated by Western blotting using an antibody that recognizes the amino-terminal fragment of *Renilla* luciferase. The most advantageous aspect of this assay is its capacity of detecting both association and dissociation of a protein pair of interest in a large-scale format.

Key words: Protein–protein interactions, split luciferase complementation assay, protoplasts large-scale analysis, *Renilla reniforms*, *Arabidopsis thaliana*.

1. Introduction

The split luciferase complementation assay (SLCA) was originally invented to detect protein–protein interactions in living mammalian cells and tissues (1, 2). Unlike fluorescent proteins, which convert excitation light (lower wavelength) to emission light (higher wavelength), luciferase proteins oxidize their chemical substrate and the luciferase-bound oxidized-substrate emits light for nanoseconds known as bioluminescence (reviewed in (3)). Hence, duration and intensity of the bioluminescence depend on the availability of the substrate. In the SLCA, luciferase is genetically split into two fragments and fused to bait and prey proteins. Cells of interest are transformed with vectors that express

L. Hennig, C. Köhler (eds.), *Plant Developmental Biology*, Methods in Molecular Biology 655,
DOI 10.1007/978-1-60761-765-5_24, © Springer Science+Business Media, LLC 2010

these bait and prey proteins. Upon interactions of the bait and prey proteins, luciferase activity is partially reconstituted to emit light in the presence of the substrate. Because the SLCA does not require external light and the interaction signals are detected as simple on-or-off signals, it provides much lower background (hence higher signal-to-background ratios) than fluorescence-based interaction assays in living cells (4). On the other hand, because emission signals themselves are lower than those of fluorescent proteins, the SLCA may not be suitable for imaging at subcellular resolutions (5, 6). Moreover, unlike the FRET (free resonance energy transfer) assay, a minimum physical distance that emits an interaction signal is not known. Hence, the assay may not be suitable for studies that measure physical distances between a protein pair. Nevertheless, kinetics of association and dissociation of a protein pair that is modulated by a drug have been successfully analyzed in living mammalian cells by the SLCA (5, 7–9). Lastly, under certain conditions it could be considered a disadvantage that the SLCA requires the use of an exogenous substrate even though the substrate is nontoxic and readily permeable to cellular membranes. The interaction signals gradually reduce as a function of time, making it difficult to monitor changes of the interactions for long time. So far, four different luciferase proteins have been investigated for the SLCA: firefly (*Photinus pyralis*) (1, 2, 7), click beetle (*Pyrophorus plagiophthalamus*) (10), *Renilla* (*Renilla reniforms*) (9), and *Gaussia* (*Gaussia princeps)* luciferase (11). Firefly (61 kDa) and click beetle (61 kDa) luciferase catalyze oxidation of luciferin and emit light of 560 and 540 nm, respectively. They require a magnesium ion and ATP as cofactors. On the other hand, *Renilla* (36 KDa) and *Gaussia* (20 KDa) luciferase catalyze the oxidation of coelentazalin and emit light of 480 and 470 nm, respectively. They do not require a cofactor for catalysis.

For living plant cell analysis, firefly- and *Renilla* luciferase-based SLCAs have been developed (4, 12). In this protocol, we describe the *Renilla* luciferase-based SLCA that we have been working with for the last few years. We sought a protein–protein interaction assay in living plant cells that allows for analyzing a large number of protein sets simultaneously. To this end, we developed a method to transform protoplasts in 96-well plates so that large-scale analyses are possible (4). The preparation of Arabidopsis protoplasts as well as the transformation method are modified from the methods originally developed in the laboratory of Jen Sheen (13) to fit our needs and environment. Hence, the majority of the procedures are similar to the original protocols. In this protocol, we describe a method that is optimized for our assays. On the other hand, to make vector construction as easy as possible, we employ the Gateway® in vitro recombinant system (Invitrogen) that allows for a gene of interest to be inserted in the expression vectors in a single tube reaction. Because the detailed protocol of this system is available at the company's Web site (http://www.invitrogen.com), we omit the protocol from this article. Our expression vectors also contain Cre-lox in vitro recombinant system (Clontech) so that bait and prey proteins can be expressed from a single recombined vector that would make their expression levels similar. However, since our recent work indicated that it is not necessary to use a recombined vector in order to equally express prey and bait proteins (Kato, unpublished), we omit the protocol for the Cre-lox in vitro recombinant system as well. Details about the Cre-lox in vitro recombinant system in our vectors can be found in (4).

In our SLCA, *Renilla* luciferase is split into a 229 amino acid amino-terminal fragment and an 82 amino acid carboxy-terminal fragment. These fragments are fused to bait and prey proteins through a 2×(GGGGS) flexible linker plus the sequences required for the Gateway® in vitro recombination (4). As a consequence, the calculated sizes of the tags are 30 kDa for the amino-terminal fragment tag (NLuc) and 12 kDa for the carboxy-terminal fragment tag (CLuc), respectively. One of the concerns about these tags is to determine which end (amino- or carboxy-terminal) of the prey and bait proteins need to be fused to which tags (NLuc or CLuc). At this point, no clear explanation is available about how the luciferase structure is reconstituted in the complementation assay. Hence, unless three-dimensional structures of a protein pair (bait and prey) are known, the assay may need to be blindly conducted with all possible configurations; NLuc-bait and CLuc-prey (a head-to-head configuration), NLuc-bait and prey-CLuc (a head-to-tail configuration), bait-NLuc and CLuc-prey (a tail-to-head configuration), and bait-NLuc and prey-CLuc (a tail-to-tail configuration). It is also true that switching bait and prey proteins (i.e., NLuc-bait and CLuc-prey to NLuc-prey and CLuc-bait) may affect the reconstitution of the split luciferase activity. To overcome this problem to some degree, we generated three different expression vectors designated pDuExAn6, pDuExDc6, and pDuExD7 (4). These vectors allow the head-to-tail and head-to-head configurations. The pDuExD7 vector also allows fusing CLuc to both N- and C-terminal ends of the protein of interest so that the luciferase activity could reconstitute independent of the orientations of the interactions of the bait and prey proteins. More details about the configuration of the tag sites and the vectors can be found in (4). Vectors that allow the tail-to-tail pDuExAc6 and tail-to-head pDuExDn6 configurations have been constructed (Kato et al., unpublished) and available to the public (*see* **Note 1**).

After the SLCA, amounts of the (NLuc tagged) bait protein accumulated in the transformed protoplasts can be estimated by Western blotting using an antibody that recognizes the amino-terminal fragment of *Renilla* luciferase (4). Because protocols of Western blotting are available elsewhere and we simply follow them, in this protocol, we describe only the sample preparation and use of the antibody.

2. Materials

2.1. Vector DNA

1. CompactPrep Plasmid Midi Kit (Qiagen).
2. Bacteria stocks carrying an expression vector for bait or prey protein (*see* **Note 1**).
3. LB/ampicillin culture broth: 100 mL in 250-mL flasks.

2.2. Arabidopsis Protoplasts

1. Digestion buffer: 0.4 M Mannitol, 20 mM KCl, 10 mM $CaCl_2$, 20 mM MES, pH 5.7. Make 250 mL of 80 mM MES, pH 5.7, by mixing 3.9 g of MES hydrate in distilled water and adjusting pH to 5.7 with 1 M NaOH. Store the MES solution in a refrigerator after sterilizing it with a 0.22 μm filter. The MES solution lasts at least 3 months. Mix 18.3 g of mannitol, 0.38 g of KCl, 0.36 g of $CaCl_2$ $2H_2O$, and 62.5 mL of 80 mM MES, pH 5.7, in distilled water to make 250 mL of solution. Store the solution at room temperature after sterilizing it with a 0.22 μm filter. The buffer lasts at least 1 month.
2. Digestion enzyme mix: 0.1 g of Cellulase R10 and 0.025 g of Macerozyme R10 (Yakult Pharmaceutical Industry, Japan, http://www.yakult.co.jp/ypi/en/index.html; purchase request can be made by e-mail to yakultph@mb.infoweb.ne.jp). Put 2.0 g of Cellulase R10 and 0.5 g of Macerozyme R10 into a plastic test tube. Mix well by shaking the tube and divide into twenty 1.5-mL tubes (0.125 g of the mix per tube). Store in a refrigerator. The enzymes last at least 3 years.
3. W5 solution: 154 mM NaCl, 125 mM $CaCl_2$, 5 mM KCl, 2 mM MES, pH 5.7. Mix 9.0 g of NaCl, 18.3 g of $CaCl_2$ $2H_2O$, 0.37 g of KCl, and 25 mL of 80 mM MES, pH 5.7, in distilled water to make 1 L of solution. Store the solution at room temperature in a 1 L media bottle equipped with a 10-mL bottle top dispenser after sterilizing it with a 0.22-μm filter. The buffer lasts at least 1 month.
4. MMg solution: 0.4 M mannitol, 15 mM $MgCl_2$, 4 mM MES, pH 5.7. Mix 36.5 g of mannitol, 1.5 g of $MgCl_26H_2O$, and 25 mL of 80 mM MES, pH 5.7, in distilled water to make 500 mL of solution. Store at room

temperature in a 1 L media bottle equipped with a 10-mL bottle top dispenser after sterilizing it with a 0.22-μm filter. The solution lasts at least 1 month.

5. Four-weeks-old *Arabidopsis thaliana* (ecotype Columbia) plants grown in a chamber (16 h light and 8 h dark, humidity at 55%, and temperature at 22°C) (*see* **Note 2**).
6. Nylon cell strainers (100 μm) (BD Falcon).
7. Large-orifice 200-μL pipette tips (Fisher Scientific).
8. Hematocytometer (Hausser Scientific).
9. Single-edge carbon steel razor blades (Electron Microscopy Sciences).
10. Bottle top dispensers (10 mL) (EMD Chemicals).

2.3. Protoplast Transformation

1. Vector DNAs: Expression vector for a prey protein (5 μg), expression vector for bait protein (5 μg) (*see* **Note 1**).
2. Protoplast suspension (3×10^5 cells mL^{-1}; 20 mL).
3. PEG/Ca^{2+} solution: 40% (w/w) PEG4000, 0.2 M mannitol, 0.1 M $CaCl_2$. Mix 20 g of PEG4000, 1.8 g of mannitol, and 0.74 g $CaCl_2 2H_2O$ in distilled water to make a 50 mL solution in a 50-mL conical tube. Microwave it for 10 s and rotate for 1 h to dissolve PEG4000. Store at room temperature. The solution lasts at least 1 month.
4. WI solution: 0.5 M Mannitol, 20 mM KCl, 4 mM MES, pH 5.7. Mix 91 g of mannitol, 1.48 g of KCl, and 50 mL of thc 80 mM MES, pH 5.7, solution to make a 1 L solution. Store in a 1-L media bottle at room temperature after sterilizing with 0.22-μm filter. The solution lasts at least 1 month.
5. U-bottom white 96-well plates (white) (Nunc).
6. Plastic covers for 96-well plates (clear) (Nunc).
7. Eight-channel motorized 200-μL pipetter (Fisher Scientific).
8. Large-orifice 200-μL pipette tips (Fisher Scientific).
9. Solution basins for a multichannel piepetter (55 mL) (Labcore Products).
10. Digital vortexer equipped with a microplate holder (Fisher Scientific).
11. Thermomixer R microplate orbital shaker (Eppendorf) (*see* **Note 3**).
12. ELx404 microplate washer (Bio-Tek) (*see* **Note 4**).
13. Desk-top microplate centrifuge (nonrefrigerated) (Thermo Scientific).

2.4. Luminescence Detection

1. VVR stocks: Add 50 μL of DMSO in 5 mg of ViviRen^{TM} *Renilla* luciferase substrate (Promega) to make 120 mM stocks. Divide the solution into 1.5-mL tubes (10 μL each). These stocks last at least 6 months in a –80°C freezer. Add 90 μL of DMSO in the 120 mM stock to make 12 mM stocks. Divide the solution into 1.5-mL tubes (10 μL each). These stocks last at least 1 month in a –20°C freezer.
2. V-bottom 96-well plates (Nunc).
3. Plastic covers for 96-well plates (clear).
4. Eight-channel motorized 200-μL pipetter.
5. Thermomixer R microplate orbital shaker (Eppendorf).
6. Veritas microplate luminometer (Turner BioSystems) (*see* **Note 5**).

2.5. Sample Preparation and Antibodies for Western Blotting

1. TBS buffer: 150 mM NaCl and 20 mM Tris, pH 7.5. Mix 0.88 g of NaCl and 0.24 g of Tris to make 100 mL of solution. Adjust pH to 7.5 with 1 M HCl. Store in a 100-mL media bottle at room temperature after autoclaving. The solution lasts at least 12 months.
2. Sample buffer: 1% (w/v) Triton X-100, 150 mM NaCl, 4 M Urea, 50 mM Tris-HCl, pH 6.8, 2% (w/v) SDS, 10% (w/v) glycerol, and 2% (w/v) 2-mercaptoethanol. Make 1 M Tris-HCl, pH 6.8. Mix 0.87 g of NaCl, 24.0 g of urea, 2 g of SDS, 10 g of glycerol, and 5 mL of Tris-HCl, pH 6.8, to make 98 mL of solution. Store in a 100-mL media bottle at room temperature after sterilizing it with a 0.22-μm filter. The solution lasts at least 12 months. Take 980 μL of the solution in a 1.5-mL centrifuge tube and add 20 μL of 2-mercaptoethanol (Sigma) immediately before use.
3. Anti-*Renilla* Luciferase (clone 5B11.2, Chemicon International).
4. Immun-Star^{TM} goat antimouse-HRP conjugate (Bio-Rad).
5. Immun-Star^{TM} HRP chemiluminescent kit (Bio-Rad).

3. Methods

To analyze protein–protein interactions in Arabidopsis protoplasts in 96-well plates, four major steps are involved; (1) preparation of vector DNAs, (2) preparation of protoplasts, (3) transformation of protoplasts with vectors, and (4) detection of luminescence in a luminometer. If accumulation levels of bait proteins (tagged with NLuc) need to be assessed, the protoplasts can be collected from wells after the SLCA and subjected to Western blotting with a commercially available anti-*Renilla*-luciferase antibody that recognizes the *N*-terminal fragment of the protein.

We use 4 wells (4 independent transformations) for one protein pair to obtain statistically relevant data. A modestly trained person (i.e., undergraduate) can analyze up to four plates (96 wells × 4 plates = 384 independent transformations) using one gram of Arabidopsis leaves (about 50 leaves) in 2 days. It is important to add the same control protein pair (*see* **Note 1**) in each plate so that differences between the plates can be normalized based on RLUs (relative luminescence units) of the control protein pair. In this protocol, we will describe processes for assays with four plates, which may require some practice before the person can complete the assay comfortably. We recommend analyzing only one plate in the beginning, which is comfortably conducted for the first time.

3.1. Preparation of Vector DNA

1. Culture bacteria stocks containing an expression vector for a bait or prey protein (*see* **Note 1**) in 100 mL of LB/ampicillin broth in a 250-mL flask at 37°C for 16 h.
2. Divide the cultured solution into two 50-mL conical tubes.
3. Centrifuge the tubes at 3,800×*g* for 15 min to obtain the bacteria pellet.
4. Extract the vectors according to the protocol that the company provides (*see* **Note 6**).
5. Quantify the concentration of the vectors by a spectrometer and adjust them to 0.5 μg/μL with distilled water.
6. Divide the vector solutions into 1.5-mL tubes (10 μL each) and store them in a –20°C freezer until use.

3.2. Preparation of Protoplasts

1. Place 10 mL of the digestion buffer and 0.125 g of the digestion enzyme mix into a 50-mL conical tube. Vortex the tube for 5 s and gently shake on an orbital shaker until use (for about 10 min).
2. Bring plants from the growth chamber to the lab space (*see* **Note 7**).
3. Collect 1.0 g of leaves (about 50) from the plants using fingers and a small scissors (*see* **Note 8**).
4. Pour the digestion buffer containing the digestion enzyme mix into a Petri dish.
5. Pile 5–10 leaves in the same direction using a hand and forceps on a copy paper and gently hold the pile by a finger (*see* **Note 8**).
6. Slice the piled leaves vertically against the main leaf-vein that runs from bottom to top of the leaf using a razor

blade to make 1–2 mm strips (*see* **Note 8**) and immediately transfer the strips into the Petri dish containing the digestion buffer/enzyme mix.

7. Repeat Steps 5 and 6 until all leaves are sliced.
8. Put the Petri dish in a small box to shut out the light and then place the box in a vacuum desiccator.
9. Vacuum in the desiccator for 30 min at 7–8 psi (pound per square inch) to penetrate the enzyme solution into the leaf strips.
10. Upon completion of the vacuum, leave the box in a 25°C incubator for 6 h to digest the cell walls.
11. Upon completion of the digestion, gently shake the Petri dish orbitally by hand about 30 times to release the protoplasts from the leaf strips (*see* **Note 9**). Transfer the solution from the Petri dish to the 50-mL conical tube through a cell strainer by a transfer pipette.
12. Add 10 mL of the W5 buffer into the Petri dish and gently shake it orbitally about 20 times by hand to collect the protoplasts that remained in the leaf strips.
13. Transfer the solution from the Petri dish to the same 50-mL conical tube through the cell strainer by the disposable pipette.
14. Centrifuge the tube at $100 \times g$ for 3 min using a swing-out rotor (*see* **Note 10**).
15. Remove the supernatant using a 25-mL pipette and a motorized pipette dispenser so that about 3 mL of solution is left with dark-green protoplasts on the bottom in the tube (*see* **Note 10**).
16. Add 10 mL of the W5 buffer in the tube. Run the buffer from a dispenser tip to the inside wall of the tube to avoid forming bubbles in the protoplast suspension. Gently mix the protoplast suspension by repeating aspirations and dispenses 5 times with the disposable transfer pipette. Avoid forming air bubbles during mixing to prevent bursting of the protoplasts.
17. Repeat Steps 14–16.
18. Place the tube in a 4°C refrigerator for 30 min to help the protoplasts recover from stresses that may have occurred during the enzyme digestion.
19. Remove the supernatant using a 25-mL pipette and the motorized pipette dispenser so that about 3 mL solution is left with dark-green protoplasts that precipitate on the bottom of the tube during the incubation.

20. Add 10 mL of the MMg solution by running the buffer from a dispenser tip to the inside wall of the tube. Gently mix the protoplast suspension by repeating aspirations and dispenses 5 times with a disposable transfer pipette.
21. Centrifuge the tube at $100 \times g$ for 3 min.
22. Remove the supernatant using the 25-mL pipette and the motorized dispenser so that about 3 mL solution is left in the tube.
23. Add 20 mL of the MMg solution by running the buffer from a dispenser tip to the inside wall of the tube.
24. After gently mixing the protoplast suspension by a disposable transfer pipette to homogenize the protoplast population in the tube, take 10 μL of the suspension with a large-orifice pipette tip and count the number of protoplasts using a hemacytometer. Adjust the protoplast concentration in the tube to 3×10^5 cells mL^{-1} with the MMg solution (*see* **Note 11**).
25. Leave the tube at room temperature until use.

3.3. Protoplast Transformation

1. Mix vector DNAs in a 1.5-mL tube, 10 μL (5 μg) of bait expression vector, 10 μL (5 μg) of prey expression vector, and 80 μL of distilled water for a protein pair of interest.
2. Add 20 μL of the vector DNA solution per well in four wells of a row (i.e., A1–A4) of a U-bottom 96-well plate.
3. Fill the next four wells with another vector DNA solution to analyze another protein pair.
4. Repeat Step 3 until all other vector DNA solutions are added in the wells of four plates.
5. Store the plates with a plastic cover at room temperature until use.
6. Pour the protoplast suspension (abut 20 mL), PEG/ Ca^{2+} solution (about 35 mL), and the WI buffer (about 50 mL) in solution basins, respectively.
7. Work on the first plate; add 40 μL of the protoplast suspension (1.2×10^4 cells) in each well (*see* **Note 12**). Add the suspension to all wells in a column with one dispense using an 8-channel motorized pipetter (*see* **Note 13**).
8. Immediately after filling all wells in the plate with the protoplast suspension, add 60 μL of the PEG/Ca^{2+} solution to each well (*see* **Note 14**). Add the solution in all wells of a column with one dispense using an 8-channel motorized pipetter (*see* **Note 13**).
9. Put a clear plastic cover on the plate and vortex the plate by a digital vortexer at 800 rpm for 15 s (*see* **Note 15**).

10. Leave the plate with the plastic cover on a working bench for 10 min to incubate the protoplasts suspension.
11. Soon after finishing Step 9, repeat Steps 7–9 with the rest of the plates one-by-one. Take 2.5 min per plate to complete Steps 7–9 (a total 10 min to complete four plates).
12. Work on the first plate: after the 10-min incubation, add 200 μL of the WI buffer in each well in the plate. Add the buffer in all wells of a column with one dispense using the 8-channel motorized pipetter (*see* **Note 13**).
13. Shake the plate using a microplate shaker at 800 rpm for 10 s to mix the WI buffer and the protoplasts suspension (*see* **Note 3**).
14. Place the plate in a microplate swing rotor.
15. Work on the rest of the plates one-by-one: Repeat Steps 12–14. Take 2.5-min intervals between each plate so that the duration of the incubation time with the PEG/Ca^{2+} in each plate will be 10 min.
16. Work four plates side-by-side (no interval is required between each plate) from this step on. Centrifuge the four plates at 100×*g* for 3 min.
17. Remove 220 μL of the supernatant from each well using a microplate washer to leave the protoplasts in 100 μL liquid (*see* **Notes 4 and 16**).
18. Add 200 μL of the WI buffer in the wells. Add the solution to all wells of a column with one dispense using the 8-channel motorized pipetter.
19. Shake the plate using the microplate shaker at 800 rpm for 10 s to mix the WI buffer and the protoplasts suspension.
20. Centrifuge the four plates at 100×*g* for 3 min.
21. Remove 200 μL of the supernatant from each well using the plate washer to leave the protoplasts in 100 μL liquid (*see* **Note 16**).
22. Add 200 μL of the WI buffer in each well. Add the solution to all wells of a column with one dispense using the 8-channel motorized pipetter.
23. Shake the plate using the microplate shaker at 800 rpm for 10 s to mix the WI buffer and the protoplasts suspension.
24. Centrifuge the four plates at 100×*g* for 3 min.
25. Remove 200 μL of the supernatant from each well using a plate washer (*see* **Note 16**).
26. Shake the plate using the microplate shaker at 800 rpm for 5 s and incubate the plate with a plastic cover at 25°C for 16 h.

3.4. Measuring Luminescence

1. Add 1.2 mL of distilled water into a 12 mM VVR stock tube containing 10 μL of the substrate to make the VVR working solution.
2. Add 140 μL of the VVR working solution in each well of a column (a total of 8 wells) of a V-bottom 96-well plate (*see* **Note 13**).
3. Take the VVR working solution from the wells using an 8-channel motorized pipetter and dispense 10 μL in each well of the plate where protoplasts were incubated for 16 h (*see* **Note 17**). Add the solution in all wells of a column with one dispense.
4. Immediately after adding the VVR working solution in the protoplast plate, shake the protoplast plate using a microplate shaker at 800 rpm for 5 s to mix the VVR substrate and the protoplasts suspension.
5. Immediately after shaking the plate, place the plate in a microplate luminometer. Read luminescence intensities with 0.5 s integration time (*see* **Note 18**).

3.5. Sample Preparation and Use of Antibodies to Detect Bait (NLuc Tagged) Protein by Western Blotting

1. Collect the protoplasts precipitated at the bottom of the well with a large-orifice pipette tip. Place the protoplasts collected from the four wells (a total of about 48,000 protoplasts) into a 1.5-mL tube.
2. Centrifuge the tube at 100×*g* for 5 min. After removing the supernatant, wash the pellets (about 10 μL) with 1 mL of TBS buffer.
3. Suspend the pellets in 10 μL of sample buffer and incubate at 95°C for 5 min.
4. Centrifuge at 18,000×*g* for 5 min and take the supernatant (about 20 μL).
5. Load the supernatant (about 20 μL) on a SDS-PAGE gel.
6. Conduct electrophoresis and blotting (*see* **Note 19**).
7. Detect signals using a chemiluminescent detection kit. Use anti-*Renilla* Luciferase antibody as a primary antibody at a 1:2,000 dilution and anti-Mouse-HRP antibody conjugate as a secondary antibody at a 1:30,000 dilution, respectively (*see* **Note 20**).

4. Notes

1. Vector DNAs required for this assay are currently available from the laboratory of Naohiro Kato. The sequence information of the vectors is available

at National Center for Biotechnology Information (http://www.ncbi.nlm.nih.gov). The accession numbers of the vectors are EF565883 for pDuExAn6, EF565884 for pDuExDc6, EF565885 for pDuExD7, GU370778 for pDuExAc6, and GU370779 for pDuExDn6.

2. One of the most important preconditions to successfully detect protein–protein interactions in this assay is to prepare protoplasts from Arabidopsis plants that grow under optimal conditions and have an appropriate age. We grow five plants in a 9 × 9 × 6 cm pot under 100 $\mu m/m^2/s$ photon flux of fluorescence light and use mixed soil; Sunshine® mix 8 (Sun Gro Horticulture), Sunshine® Strong-lite Coarse Vermiculite (Sun Gro Horticulture), and Osmocote (Scotts-Sierra Horticultural Products) at a 3 L: 1 L: 20 mL volume ratio. **Figure 24.1** shows the Arabidopsis plants that we use to prepare the protoplasts. The best leaves for protoplast preparation come from 4-weeks-old plant and a weight of about 0.2 g each. Avoid curled leaves because these leaves are too old or dried to obtain competent protoplasts. All plants in the figure could be used, but many leaves in pot (**A**) are too young and the majority of leaves in pot (**C**) are too old to get the best results.

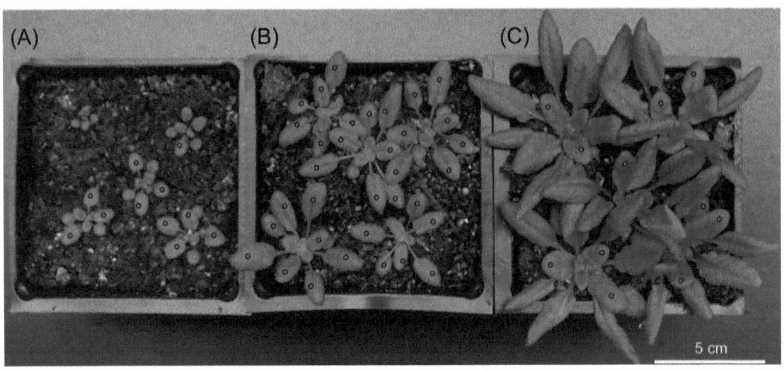

Fig. 24.1. ***Arabidopsis thaliana*** **(ecotype Columbia) plants used for a SLCA.** The plant seeds were germinated on soil in a pot for 7 days and the seedlings were transplanted into individual pots; 5 plants per pot. (**A**) 25 days after germination. (**B**) 28 days after germination. (**C**) 35 days after germination. Dots indicate leaves suitable for the protoplast preparation.

3. To mix fluid in a 96-well plate, orbital shakes in a small diameter are required. A regular orbital shaker would not mix fluid well in a 96-well plate.

4. It is recommended to use a microplate washer that is capable of slowing down the aspiration rate (the rate at which a washer manifold travels down into the microplate while aspirating fluid) to avoid protoplasts from being aspirated with the supernatant.

5. It is recommended to use a microplate luminometer that has a high sensitivity to obtain a high dynamic range of protein–protein interaction values.
6. We use the CompactPrep Plasmid Midi Kit (Qiagen) to save time and get vector DNA of consistent purity in each extraction. We found that with Top10 cells (recommended by Invitrogen to use for Gateway® cloning) and our expression vectors (pDuEx), we get the best yield when we start with 5 mL of the P1 solution for a bacteria pellet collected from 50 mL of an overnight (16 h) culture. Increase the volumes of the rest of the reagents in the kit accordingly.
7. Physically separate plants from their growing location before cutting the leaves to avoid the reuse of plants and the possible induction of a defense mechanism of other plants.
8. It is better not to hold the leaves with excessive forces when using fingers, forceps, or a blade to avoid adding stresses to the cells during the processes. **Figure 24.2** shows piled leaves that are sliced into 1–2 mm strips. It is best to make strips as thin as possible for increased protoplast yield.

Fig. 24.2. **Leaves sliced on a copy paper with a blade.** Five leaves were piled up in one direction and gently held by a middle finger. The piled leaves were sliced by gently stroking the blade with about a 20° angle.

9. **Figure 24.3** shows the leaf strips in the enzyme solution before vacuuming the solution (Step 9) and after shaking the solution (Step 11). Note that the color of the enzyme solution and number of leaf strips are drastically changed.

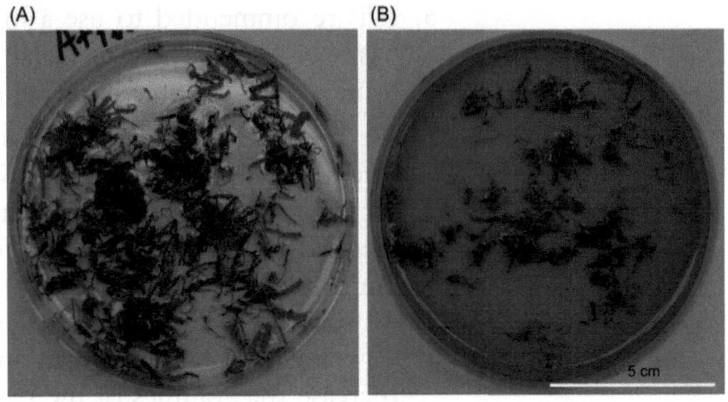

Fig. 24.3. **Leaf strips in the digestion solution.** (**A**) The strips after 30 min vacuum but before the 25°C incubation. (**B**) The strips after the 25°C incubation. The dish was gently shaken 30 times by hand.

10. **Figure 24.4** shows a 50-mL conical tube that is centrifuged at 100×*g* for 3 min in Step 14.
11. **Figure 24.5** shows a 3×10^5 mL^{-1} protoplast suspension in a hemacytometer. Low debris content (i.e., < 10^4 mL^{-1}) gives best results.
12. If the protoplasts precipitate in the basin, mix the solution by repeating aspirations and dispenses five times with a disposable transfer tube. Avoid forming bubbles during mixing.
13. An 8-channel motorized 200-μL pipetter requires aspirating about a 15% extra volume to dispense a given volume of fluid. It can dispense 40 μL of the protoplasts suspension for 4 columns, 60 μL of the PEG/Ca^{2+} solution for 3 columns, 200 μL of the WI solution for 1 column, and 10 μL of the VVR working solution for 12 columns with one aspiration. Set the slowest rates of aspiration and dispense for the protoplasts and PEG/Ca^{2+} solutions. Set the fastest rates for the WI and VVR working solutions.
14. The duration between adding the protoplast suspension and adding the PEG/Ca^{2+} solution should be less than one min. If the duration is too long (i.e., > 5 min), the PEG/Ca^{2+} solution will not be mixed well (*see* **Fig. 24.6**).
15. Mixing the protoplast suspension and PEG/Ca^{2+} solution requires vortex (orbital plus vertical motions). Vortexing should not spill out the protoplast suspension from the wells. If it spills, adjust the rpm.
16. **Figure 24.7** shows a well in a 96-well plate during the transformation. Note changes of protoplast distributions in the well during each step. A mistake made during the transformation (i.e., no centrifugation of the plate before aspiration) can be noticeable by the protoplast distribution.

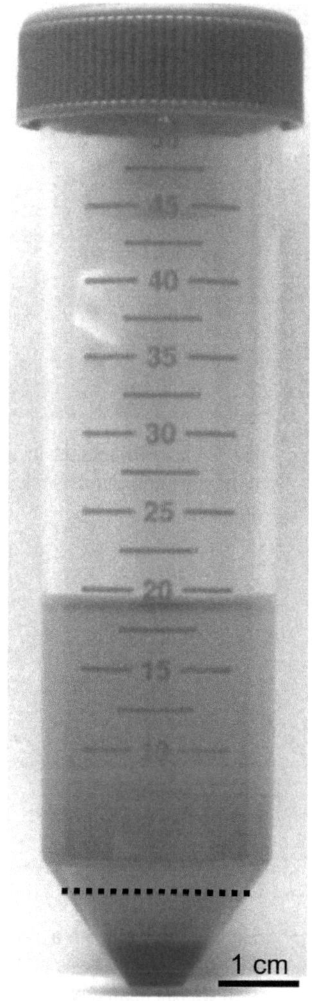

Fig. 24.4. **Protoplasts precipitated in a 50-mL conical tube after the first centrifuge.** A *dashed line* indicates the solution that should remain with the protoplasts in the bottom during the aspiration of the supernatant by a 25-mL pipette.

17. Accumulation levels of recombinant proteins reach a maximum at 16 h after the transformation of the protoplasts in our hands. However, it may be changed based on prey and bait proteins used.
18. **Figure 24.8** shows plots of RLU (relative luminescence unit) against time after the first read by a microplate luminometer.
19. Conduct Western blotting according to a conventional protocol such as published in (14).
20. If signals are not detected, increase vector amounts (i.e., 5 μg each per well) so that transformed protoplasts express more recombinant proteins than that in our default vector amounts (1 μg each per well).

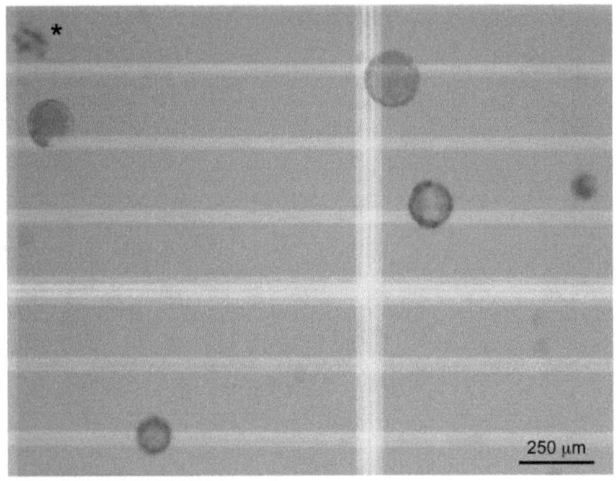

Fig. 24.5. **Protoplasts in a hemacytometer at a 3 × 10^5 mL^{-1} (MMg solution) concentration.** An *asterisk* on the *left top corner* indicates leaf debris presence in the solution.

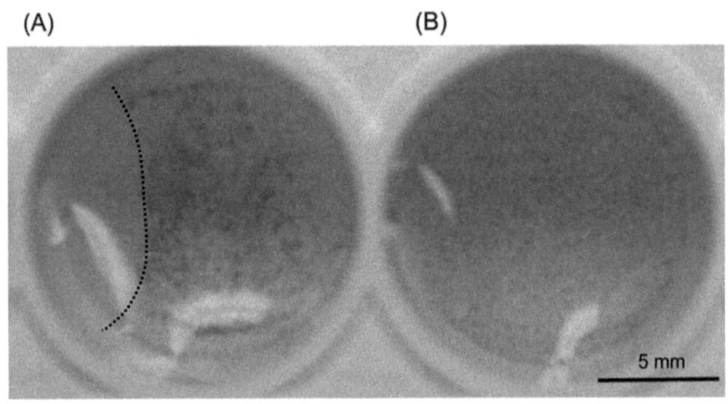

Fig. 24.6. **Wells in a 96-well plate after adding the PEG/Ca^{2+} solution.** The PEG/Ca^{2+} solution was added along the left-side wall of wells. (**A**) Protoplast suspension was incubated with the vector DNA solution for 5 min before adding the PEG/Ca^{2+} solution. The PEG/Ca^{2+} solution remains in the region to the *left* of the *dashed line*. (**B**) Protoplast suspension was incubated with the vector DNA solution for less than 1 min before adding the PEG/Ca^{2+} solution. The entire PEG/Ca^{2+} solution went underneath of the protoplasts.

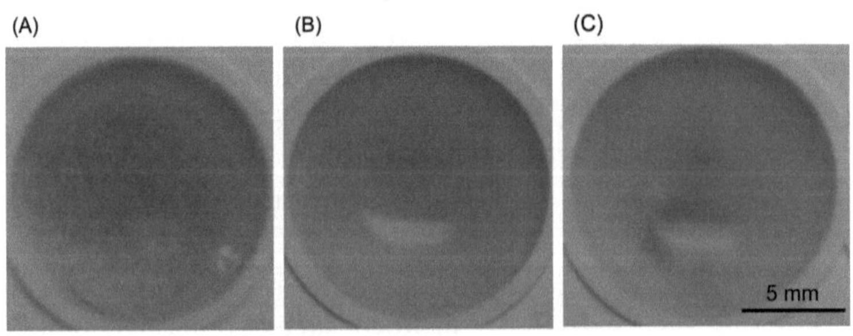

Fig. 24.7. **Changes of protoplast distributions in a well during the transformation.** (**A**) After the first aspiration by the microplate washer. The protoplasts aggregate (look grainy). (**B**) After the second aspiration by the microplate washer. The protoplasts distribute more homogenously. (**C**) After the third aspiration by the microplate washer. The protoplasts tend to precipitate in the *bottom*.

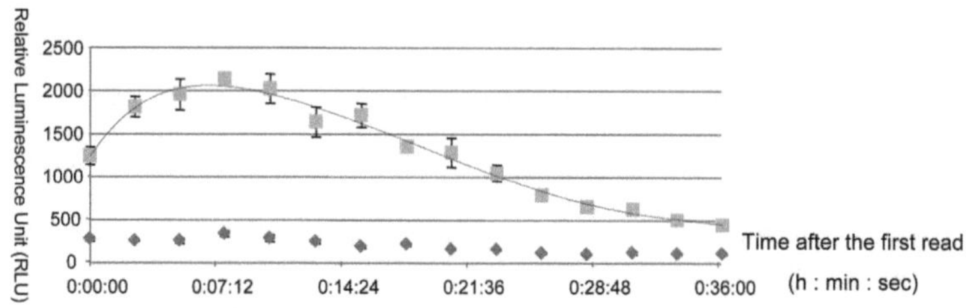

Fig. 24.8. **Luminometer signal plots in the SLCA**. *Squares*: plots of an interacting protein pair (NLuc-A and CLuc-B). *Diamonds*: plots of noninteracting protein pair (NLuc-A and CLuc-C). A, B, and C are proteins of our interest (a family of soluble *N*-ethylmaleimide-sensitive fusion protein attachment protein receptors). *Vertical bars* on each plot indicate standard errors within four wells. RLUs (relative luminescence units) of the interacting protein pair (NLuc-A and CLuc-B) reach the maximum about 8 min after the first reading and gradually decrease over time while RLUs of the noninteracting protein pair (NLuc-A and CLuc-C) stay low.

Acknowledgments

The project was supported by the USDA Cooperative State Research, Education and Extension Service – National Research Initiative – Plant Genome Program, award no. 2006-35604-16627, for N.K. We thank Mr. R. Blake Crochet for his editorial work on the manuscript.

References

1. Ozawa, T., Kaihara, A., Sato, M., Tachihara, K., and Umezawa, Y. (2001) Split luciferase as an optical probe for detecting protein–protein interactions in mammalian cells based on protein splicing. *Anal Chem* **73**, 2516–2521.
2. Paulmurugan, R., Umezawa, Y., and Gambhir, S. S. (2002) Noninvasive imaging of protein–protein interactions in living subjects by using reporter protein complementation and reconstitution strategies. *Proc Natl Acad Sci USA* **99**, 15608–15613.
3. Wilson, T. and Hastings, J. W. (1998) Bioluminescence. *Annu Rev Cell Dev Biol* **14**, 197–230.
4. Fujikawa, Y. and Kato, N. (2007) Split luciferase complementation assay to study protein–protein interactions in Arabidopsis protoplasts. *Plant J* **52**, 185–195.
5. Stefan, E., Aquin, S., Berger, N., Landry, C. R., Nyfeler, B., Bouvier, M., and Michnick, S. W. (2007) Quantification of dynamic protein complexes using *Renilla* luciferase fragment complementation applied to protein kinase A activities in vivo. *Proc Natl Acad Sci USA* **104**, 16916–16921.
6. Kaihara, A., Kawai, Y., Sato, M., Ozawa, T., and Umezawa, Y. (2003) Locating a protein protein interaction in living cells via split *Renilla* luciferase complementation. *Anal Chem* **75**, 4176–4181.
7. Luker, K. E., Smith, M. C., Luker, G. D., Gammon, S. T., Piwnica-Worms, H., and Piwnica-Worms, D. (2004) Kinetics of regulated protein-protein interactions revealed with firefly luciferase complementation imaging in cells and living animals. *Proc Natl Acad Sci USA* **101**, 12288–12293.
8. Paulmurugan, R. and Gambhir, S. S. (2003) Monitoring protein-protein interactions using split synthetic *Renilla* luciferase protein-fragment-assisted complementation. *Anal Chem* **75**, 1584–1589.

9. Paulmurugan, R., Massoud, T. F., Huang, J., and Gambhir, S. S. (2004) Molecular imaging of drug-modulated protein–protein interactions in living subjects. *Cancer Res* **64**, 2113–2119.
10. Kim, S. B., Otani, Y., Umezawa, Y., and Tao, H. (2007) Bioluminescent indicator for determining protein-protein interactions using intramolecular complementation of split click beetle luciferase. *Anal Chem* **79**, 4820–4826.
11. Remy, I. and Michnick, S. W. (2006) A highly sensitive protein–protein interaction assay based on Gaussia luciferase. *Nat Methods* **3**, 977–979.
12. Chen, H., Zou, Y., Shang, Y., Lin, H., Wang, Y., Cai, R., Tang, X., and Zhou, J. M. (2008) Firefly luciferase complementation imaging assay for protein–protein interactions in plants. *Plant Physiol* **146**, 368–376.
13. Yoo, S. D., Cho, Y. H., and Sheen, J. (2007) Arabidopsis mesophyll protoplasts: A versatile cell system for transient gene expression analysis. *Nat Protoc* **2**, 1565–1572.
14. Sambrook, J., and Rusell, D. W. (2001) Molecular Cloning: A Laboratory Manual. Cold Spring Harbor Labortory Press, Cold Spring Harbor, NY.

Chapter 25

Co-immunoprecipitation and Protein Blots

Erika Isono and Claus Schwechheimer

Abstract

Knowledge about the identity of the interacting partners is important for the understanding of the function and the cellular activity of a given protein. Here we describe co-immunoprecipitation and pull-down as methods that are widely used for the identification and characterization of protein–protein interactions. These methods are well suited to find or confirm the interaction among multiple proteins, given the availability of a specific antibody for or a tagged version of the protein of interest.

Key words: Immunoblot, immunoprecipitation, pull-down, western blot.

1. Introduction

Besides molecular biology approaches such as the yeast two-hybrid system (1), fluorescence resonance energy transfer (FRET), bimolecular fluorescence complementation (BiFC) (2, 3), and related techniques, biochemical methods such as co-immunoprecipitations (co-IPs) and pull-down experiments (PDs) are valuable and complementary tools for the identification and characterization of a protein's interacting partners. For a number of reasons, co-IPs and PDs are particularly well suited to identify or examine protein interactions: (i) they can be performed in vivo or in vitro, (ii) they are often unbiased and can lead to the identification of new interacting partners that may provide novel insights into a protein's cellular function, and (iii) they may allow the identification of a more complex protein interaction network.

L. Hennig, C. Köhler (eds.), *Plant Developmental Biology*, Methods in Molecular Biology 655,
DOI 10.1007/978-1-60761-765-5_25, © Springer Science+Business Media, LLC 2010

1.1. Co-immunoprecipitations and Pull-Downs

Co-immunoprecipitations (co-IPs) are useful for the identification or examination of direct or indirect interactions between a protein of interest and others. By using the binding specificity of an antibody against the protein of interest, this method will recover the protein of interest together with its interacting proteins from a total cellular protein extract or a less complex protein mixture. For co-IPs, the protein of interest is captured from the total extract with a specific antibody, which in turn binds to Protein A- or Protein G-coupled matrices (*see* **Fig. 25.1A**). The matrix-bound protein together with its interacting partners can be recovered by centrifugation and the presence or absence of interacting proteins on the matrix can be investigated by western blot following SDS-PAGE or by mass spectrometry. First developed by Kessler in 1975 (4) and later combined with the western blot method developed in 1979 (5), co-IP has become a widely applicable way for testing protein–protein interactions, examining posttranslational modifications, or purifying protein complexes. For co-IPs, either a specific antibody against the protein of interest must be available or a tagged version must be generated that can be recognized by a commercial antibody directed against the protein tag (*see* **Fig. 25.1A, B**). **Table 25.1** lists a range of peptide tags for which commercial antibodies are available.

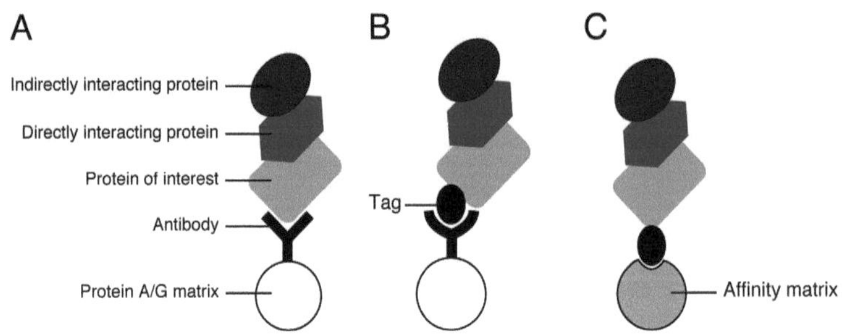

Fig. 25.1. **Co-immunoprecipitation and pull-down.** (**A**) If a specific antibody against the protein of interest is available, it can be used to directly immunoprecipitate the protein together with its interacting proteins. The protein of interest is captured by the antibody, which in turn is captured by the protein A/G matrix. The matrix can be recovered by centrifugation. Non-specifically binding proteins are removed during the washing step. In this way, directly as well as indirectly interacting partners of the protein of interest can be obtained. (**B**) If a specific antibody is not available for the protein of interest, a tagged version of the protein of interest can be used. The procedure of immunoprecipitation is the same as in (**A**); the only difference is that the antibody is directed against the protein tag. (**C**) When using an affinity tag, a specific affinity matrix is used instead of the antibody and protein A/G matrix.

Pull-down experiments (PDs) make use of the same principle as co-IPs, but they employ the highly selective binding specificity of an affinity tag to a small molecule (rather than that of a specific antibody) to purify an affinity-tagged protein of interest together with its associated interacting partners (*see* **Fig. 25.1C**).

Table 25.1
Some commercially available antibodies and antibody-conjugated matrices

Tag	Antibodies	Matrix
FLAG	Anti-FLAG[a]	Anti-FLAG M2 Agarose[a]
HA	Anti-HA[b, c, d]	Anti-HA Affinity Matrix[c]
c-myc	Anti-c-myc[c]	Anti-c-myc Agarose[a, c, d]
GFP	Anti-GFP[c, d, e]	Anti-GFP Agarose[f]
T7	Anti-T7[g]	T7 Tag Antibody Agarose[g, h]

Some suppliers whose products have been successfully used in our laboratory: [a]Sigma-Aldrich, [b]BAbCo, [c]Roche, [d]Santa Cruz, [e]Invitrogen, [f]Vector, [g]Novagen, [h]and Abcam.

Note that other definitions of co-IP and PD are used in the published literature, but the definition given here reflects the dominating view in the community. PDs require the protein of interest to be fused to a tag (an affinity tag), which has a high affinity for a small molecule. Generally, commercially available affinity tags are employed (**Table 25.2**). In rare cases, the protein of interest may itself have an affinity for a small molecule and this feature may be used for protein purification.

Table 25.2
Some commercially available antibodies and antibody-conjugated matrices for affinity tags

Tag	Antibodies	Matrix
His	Anti-His[a, b]	Ni–NTA Agarose[a], Ni–Sepharose[c], TALON beads[d]
GST	Anti-GST[c, e]	Glutathione Sepharose[c]
STREP	Anti-STREP tag[a, f]	Strep Tactin Superflow[a, f], StrepTrap HP[c]

Some suppliers that have been successfully used in our laboratory: [a]Qiagen, [b]Invitrogen, [c]GE Healthcare, [d]Clontech, [e]Sigma-Aldrich, [f]IBA.

When performing a co-IP or PD, the success of the IP or PD can be controlled by western blot with an antibody directed against the protein of interest or against the fusion tag. In turn, the identity of interacting proteins can be confirmed with antibodies directed against candidate interactors or by mass spectrometry of the purified protein extract. (*see* **Fig. 25.2**). The preferred method depends on the available material (e.g., antibodies, knowledge about interacting partners), available technology and expertise (e.g., mass spectrometry), and the specific scientific problem to be answered.

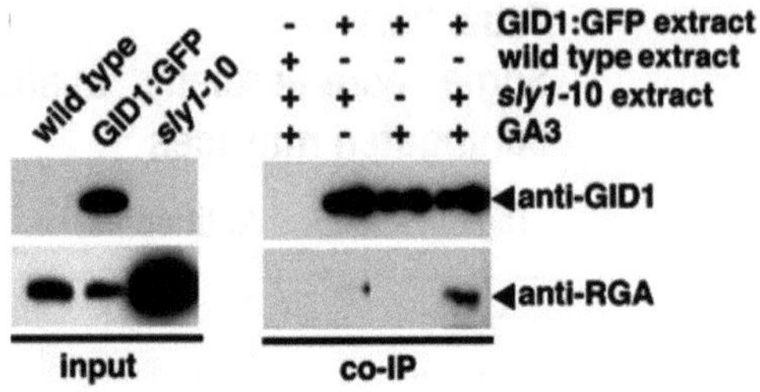

Fig. 25.2. **GA-dependent co-IP of GID1 with RGA**. The Arabidopsis gibberellic acid (GA) receptor GID1 was fused to GFP to obtain GID1:GFP plants (7). GID1:GFP was immunoprecipitated from the total extract using anti-GFP agarose. The purified GID1:GFP was then mixed with total extract of the *sly1-10* mutant, which accumulates the interaction partner RGA, an inhibitor of the GA signaling pathway. The *left panel* shows the protein blot of an input control with anti-GID1 and anti-RGA antibodies and the *right panel* shows the results of the co-IP. Note that GID1 and RGA interact only in the presence of GA (7).

1.2. Notes of Caution

1.2.1. Fusion Protein Functionality

The source of the (fusion) protein for co-IPs as well as for PDs can be proteins expressed in and purified from a heterologous host (e.g., bacteria or yeast) or a protein expressed in plants. Since protein fusions may interfere with protein function or the protein's interactions with other proteins, it is important to note that it has become a standard publication requirement of many scientific journals that proof of the fusion protein's functionality is provided. In the plant field, proof of functionality is generally obtained by the complementation of a mutant phenotype with a transgene expressing the fusion protein. At the same time, such a transgenic line may serve as the source for the fusion protein for co-IP experiments. In other cases, it may be possible to provide proof for protein functionality by demonstrating that the fusion protein has retained its biochemical activity (although it is important to realize that the biochemical activity of a protein and its interactions with other proteins may be distinct activities).

1.2.2. Protein Abundance

When performing a co-IP or PD experiment, it is also important to realize that the ability to detect protein–protein interactions is strongly dependent on the abundance of the proteins of interest.

The interacting protein may be a protein of very low abundance and its detection can be impossible in a complex protein mixture. In such a case, the experimental setup has to be adjusted to enrich the interaction partner. For instance, one can choose tissue that expresses high amounts of the interaction partner or growth conditions under which the interaction partner is enriched. Alternatively, one may decide to overexpress the candidate protein in planta or in a heterologous host to favor protein–protein interactions.

However, the overexpression of a protein in planta or the use of inappropriate amounts of recombinant protein may allow interactions that do not normally take place in a wild-type situation. A good criterion as to whether the protein overexpression alters the behavior of a protein is to see whether the overexpressed protein can complement the mutant (plant) phenotype or whether unexpected phenotypes are observed that can be attributed to non-production or novel protein interactions. Unexpected changes in phenotype could be caused by nonproductive or novel protein interactions; alternatively, they can be caused by rate-limiting amounts of the protein of interest in wild-type cells. To avoid problems to interpret interaction data, the protein of interest is ideally expressed in planta from its endogenous promoter, usually a genomic fragment of the gene of interest that is sufficient to rescue a mutant phenotype.

1.2.3. Nature of Protein Interactions

The nature of the protein interaction will strongly influence the ability to detect the interaction. For example, it is extremely difficult if not impossible to detect protein–protein interactions between an active enzyme and its target, e.g., that of a protein kinase and its phosphorylation target or that of an E3 ubiquitin ligase and its ubiquitylation target. In both cases, the productive interactions will immediately reduce the affinity of the interacting protein to the enzyme, and in the case of ubiquitylation, a productive protein interaction will furthermore lead to the protein's degradation. Thus, these transient interactions are very hard to capture by co-IPs and PDs. In some cases, this problem can be solved by "freezing" the interaction by inactivating the enzyme by mutagenesis such that the interaction can occur, but the biochemical activity of the enzyme is blocked. Alternatively, chemical inhibitors for the enzymatic activity can be used to prevent the dissociation of the interacting proteins. Another indirect solution to this problem is to visualize the biochemical activity of the protein of interest towards its interacting partner rather than the interaction itself, e.g., transfer of radioactive phosphate or gel shift in an SDS-PAGE following protein phosphorylation or ubiquitylation.

2. Materials

2.1. Plant Material and Total Protein Extraction

1. Growth medium: 4.2 g/L Murashige & Skoog medium including Gamborg B5 vitamins, 10 g/L sucrose, 250 mg/L 4-morpholineethanesulfonic acid sodium salt (MES), 0.56% plant agar, pH adjusted to 5.7 with hydrochloric acid (HCl).

2. Protein extraction buffer: 50 mM Tris-HCl pH 7.5, 100 mM NaCl, and 10% (w/v) glycerol. Store at room temperature. Add 1/50 volume of a 50× stock of complete EDTA-free protease inhibitor cocktail (Roche) (prepare the 50× stock by dissolving 1 tablet in 1 mL of buffer, store stock at −20°C) and 1/100 volume of a 10 mM stock of the MG132 26S proteasome inhibitor (Axxor) (prepare the 10 mM stock in DMSO, store at −20°C) immediately prior to use.
3. Motor-driven homogenizer Schuett HomGen with a cooling jacket (Schuett Biotec) with glass tubes and tightly fitting pestles (5 mL, LAT Garbsen).
4. 5×Laemmli buffer (6): 250 mM Tris-HCl pH 6.8, 10% (w/v) SDS, 50% (w/v) glycerol, 0.05% bromophenol-blue (BPB), and 5% β-mercaptoethanol. Store at room temperature.

2.2. Immunoprecipitation

1. Wash buffer: 50 mM Tris-HCl pH 7.5, 100 mM NaCl, and 10% (w/v) glycerol, supplemented with 0.05% Triton X-100 before use.
2. Refrigerated centrifuge.
3. Primary antibody against the protein to be immunoprecipitated (*see* **Note 1**).
4. Rotator with 1.5-mL tube holders (Neolab).
5. Protein A/G PLUS-Agarose.
6. 2 × Laemmli buffer: 100 mM Tris-HCl pH6.8, 4% (w/v) SDS, 20% (w/v) glycerol, 0.02% BPB, and 2% β-mercaptoethanol. Store at room temperature.

2.3. Pull-Down

1. Plant material as described in **Section 2.1**.
2. Buffers as described in the manufacturers' instructions of the respective resin.

2.4. SDS-Polyacrylamide Gel Electrophoresis (PAGE)

1. Separating gel buffer: 1.5 M Tris-HCl pH 8.8 and 0.4% (w/v) SDS. Store at room temperature.
2. Stacking gel buffer: 0.5 M Tris-HCl pH 6.8 and 0.4% (w/v) SDS. Store at room temperature.
3. 30% Acrylamide/bisacrylamide solution (37.5:1), 10% ammonium persulfate (APS), and *N,N,N,N'*-tetramethylethylenediamine (TEMED). Store at 4°C.
4. Running buffer (10×): 250 mM Tris, 1.92 M glycin, and 0.4% (w/v) SDS. Store both 10× and 1× solutions at room temperature.
5. Prestained molecular mass marker, e.g., PageRuler Prestained Protein Ladder Plus.

6. Power supply.
7. Gel apparatus.
8. Syringe and a 20G needle (Terumo).

2.5. Western Blotting

1. Semidry transfer buffer: 25 mM Tris pH 8.3, 192 mM glycine, 20% methanol, and 0.04% SDS. Store at room temperature.
2. Semidry transfer apparatus.
3. Four gel blot papers (1.2 mm) cut in a size 0.5 cm larger than the protein gel.
4. PVDF or nitrocellulose membrane (*see* **Note 2**) cut in the size of the protein gel.
5. Methanol (100%).
6. Tris-buffered saline with Tween-20 (TBST): Prepare a 10× stock with 0.5 M Tris-HCl, pH 7.5, 1.5 M NaCl, 10 mM $MgCl_2$, and 1% Tween-20. Store both 10× and 1× stocks at room temperature.
7. Blocking buffer: 5% (w/v) Nonfat dry milk in TBST.
8. Shaker.
9. Primary antibody against the immunoprecipitated protein.
10. Primary antibodies against the coimmunoprecipitated protein(s).
11. Secondary antibody.
12. Enhanced chemiluminescent (ECL) reagents: ECL Western Blotting Detection Reagents (GE Healthcare) or SuperSignal West Pico chemiluminescent (Thermo Fisher).
13. Development: X-ray films or imaging device such as LAS4000 MINI System (Fuji-Film).

3. Methods

3.1. Plant Material and Total Protein Extraction

1. Grow *Arabidopsis thaliana* seedlings on solid growth medium for 7 days at 22°C under continuous light. Collect 500 mg of seedlings in a 1.5-mL microcentrifuge tube and immediately freeze the sample in liquid nitrogen.
2. Cool the extraction buffer on ice and set the refrigerated centrifuge to 4°C. Place a cooled homogenization glass tube in the ice-filled cooling jacket. Transfer the frozen seedlings into the glass tube and add the extraction buffer (2 mL/g fresh weight material) supplemented with 1× protease inhibitor cocktail and 10 μM MG132 and homogenize

the sample for 1 min at 1,600 U/min. Repeat the homogenization two times, each time allowing the sample to cool for 30 s.

3. Transfer the plant extract to a new 1.5-mL tube and centrifuge the extract for 10 min at 10,000×*g* in the refrigerated centrifuge. Collect the supernatant in a new 1.5-mL tube without touching the pellet.
4. Set 40 μL of extract aside, mix with 10 μL of 5× Laemmli buffer, and boil for 5 min. Quantify the protein concentration using the Bradford assay reagent (BioRad). A typical yield from such a preparation is 1–2 mg of total protein.

3.2. Immunoprecipitation

1. All the solutions to be used should be cooled on ice before use. Add 10 μL of protein A/G agarose (Santa Cruz; 25% slurry) with a cut-off tip to the lysate for preclearing (*see* **Note 3**) and incubate for 15 min at 4°C on a rotator.
2. Centrifuge for 2 min at 2,000×*g* at 4°C and carefully transfer the supernatant to a new 1.5-mL tube. Leave 20 μL of the protein solution in the tube to avoid touching the beads.
3. Add the primary antibody (*see* **Note 4**) to the extract and incubate for 1 h at 4°C on a rotator.
4. Add 20 μL of protein A/G agarose with a cut-off tip and rotate on the rotator for 1 h in the cold room.
5. Centrifuge for 2 min at 2,000×*g* at 4°C and remove the supernatant.
6. Add 1 mL of wash buffer to the beads and wash the beads by inverting the microcentrifuge tube. Centrifuge for 2 min at 2,000×*g* at 4°C and remove the supernatant. Repeat the washing step three times (*see* **Note 5**). After the last wash step, remove as much washing buffer as possible.
7. Add 20 μL of 2× Laemmli buffer to the beads pellet and boil the proteins off the beads for 5 min.
8. Centrifuge for 1 min at 10,000×*g* and transfer the supernatant to a new 1.5-mL microcentrifuge tube.

3.3. SDS-PAGE

1. Use a minigel system such as Mini-PROTEAN 3 (BioRad) with 0.75-mm thick gels for SDS-PAGE. Gel plates have to be thoroughly cleaned with 100% ethanol before use.
2. After assembling the glass plates, prepare a 10% separating gel by mixing 1.7 mL of water, 1 mL of separating buffer, 1.3 mL of 30% acrylamide solution, 25 μL of 10% APS, and finally 3 μL of TEMED (*see* **Note 6**). Pour the gel immediately after mixing the components and leave space (1–1.5 cm) for the stacking gel. Gently overlay the gel with water. Allow 15–30 min for the gel to fully polymerize.

3. Discard the water overlay and thoroughly absorb the remaining water droplets with a paper towel.
4. Prepare the stacking gel by mixing 1.2 mL of water, 0.5 mL of stacking gel buffer, 0.3 mL of 30% acrylamide, 25 μL of 10% APS, and 3 μL of TEMED. Pour the gel after mixing the components and immediately insert the comb. Allow about 15 min for the gel to polymerize. (Gels can be stored at 4°C for several days when wrapped in a wet paper towel and Saran wrap).
5. Once the stacking gel has set, remove the comb carefully and rinse the wells with water. Assemble the gel apparatus and fill the inner and outer chambers with running buffer; flush the wells with a syringe attached to a 20 G needle.
6. Load the samples in the following order: 3 μL of the molecular mass marker mixed with 15 μL of 1× Laemmli buffer, 18 μL of the total extract (*see* **Section 3.1**, Step 4), and 18 μL of the IP-samples (*see* **Section 3.2**, Step 8).
7. Run the gel at 15–20 mA (constant current) until the dye reaches the bottom of the gel (about 60 min, *see* **Note 7**).

3.4. Western Blotting

1. Cut off one corner of the membrane to orient the blot afterwards. Immerse the membrane fully in the semidry blotting buffer for 2–5 min. When using a PVDF membrane, first rehydrate the membrane for 30 s in 100% methanol before immersing it in the semidry buffer (*see* **Note 8**).
2. Soak the filter papers in the semidry blotting buffer and assemble the transfer package in the following order (bottom to top): 2 filter papers – PVDF or nitrocellulose membrane – gel - 2 filter papers. Avoid trapping air bubbles between the different layers during the setup of the transfer cassette; eliminate residual air bubbles by rolling, e.g., a glass tube over the transfer package.
3. Blot the gel for 1 h at a constant current of 1.5 mA/cm^2 membrane (*see* **Note 9**).
4. Prepare the blocking buffer, place the membrane in the blocking solution, and incubate on a shaker for 30 min at room temperature or overnight at 4°C.
5. Dilute the primary antibody in the blocking buffer (*see* **Note 10**) and incubate the membrane on a shaker for more than 1 h at room temperature or overnight at 4°C.
6. Discard primary antibody solution and wash the membrane (*see* **Note 11**) with 15 mL of TBST for 15 min. Repeat at least three times.

7. Add the secondary antibody to 10 mL of TBST and incubate the membrane on a shaker for at least 30 min at room temperature or overnight at 4°C.
8. Wash with 15 mL of TBST at least three times for 15 min.
9. Prepare 2 mL of ECL solution on a large piece of Saran wrap.
10. Take the membrane from the TBST washing solution and allow all excess liquid to drip off the membrane. Place the membrane on the ECL solution with the blotted side facing the solution. Make sure that no air bubbles are trapped between the membrane and the Saran wrap. Incubate for 1 min and expose the membrane either to an X-ray film or a CCD camera system such as a LAS3000 (*see* **Note 12**). Optimal exposure time will vary from experiment to experiment.

4. Notes

1. We have successfully employed commercial anti-c-myc (Roche,) anti-HA (Roche,) anti-GFP (Roche, Invitrogen,) and anti-FLAG M2 (Sigma) antibodies as well as the commercially available anti-HA Affinity Matrix (Roche,) anti-GFP agarose (Vector,) anti-FLAG M2 Agarose (Sigma,) and anti-STREP tactin column (IBA).
2. PVDF membranes are more robust and easier to handle than nitrocellulose membranes. However, depending on the antibodies used, nitrocellulose membranes may give a lower background signal.
3. During this step, proteins that bind nonspecifically to the protein A/G agarose beads will be removed.
4. For IP, use 10 times more of the primary antibody than when using it for a western blot (e.g., western blot dilution 1:1000; co-IP dilution 1:100).
5. The Triton-X100 concentration can be increased to 0.2% if the interaction between the proteins is strong. More washing steps can be performed to reduce background.
6. The appropriate percentage of the separating gel is determined according to the molecular mass of the protein(s) of interest: 50–80 kDa proteins can be well separated on a 10% gel; smaller and larger proteins require a 12.5% or a 7.5% gel, respectively.
7. If a mass spectrometric analysis is going to be performed, the gel should be stained with Coomassie Blue and subjected to an in-gel trypsin digest.

8. Wear gloves when handling methanol. Methanol hydrates the PVDF membrane by removing its hydrophobic surface coating and makes it compatible for protein transfer.
9. The current can be increased to 3 mA/cm^2 and/or the transfer time can be increased to blot proteins with a molecular mass of >100 kDa.
10. To reduce the amount of antibody, hybridization bags can be used, which require as little as 2–3 mL primary antibody solution in the case of a mini-gel.
11. Some primary antibodies are expensive, and a self-raised antibody is precious. In many cases, the primary antibody solution can be recycled. Collect the solution in a 15-mL plastic tube, add 0.02% of NaN_3, and keep it at 4°C or −20°C until use. Depending on the antibody, two to three times of recycling is possible.
12. Cross-reactions of the secondary antibody with the light and heavy chains (25 and 50 kDa, respectively) of IgGs used for the IP are a problem if the protein of interest migrates close to these molecular masses. Antibody-conjugated matrixes in combination with a competitive elution (e.g., HA peptides or FLAG peptides) are a good way to avoid this problem since the IP-antibody does not come off as in Step 7 of **Section 3.2**. Omitting or reducing the amount of ß-mercaptoethanol from the Laemmli buffer is effective since the antibody dissociates from the matrix under reducing conditions. An alternative strategy may be the use of secondary antibodies with reduced reactivity against denatured IgGs (TrueBlot, eBioscience).

References

1. Fields, S. and Song, O.-K. (1989) A novel genetic system to detect protein–protein interactions. *Nature* **340**, 245–246.
2. Hu, C.-D., Chinenov, Y., and Kerppola, T. K. (2002) Visualization of interactions among bZIP and Rel family proteins in living cells using bimolecular fluorescence complementation. *Mol Cell* **9**, 789–798.
3. Walter, M., Chaban, C., Schütze, K., Batistic, O., Weckermann, K., Näke C., Blazevic, D., Grefen, C., Schumacher, K., Oecking, C., Harter, K., and Kudla, J. (2004) Visualization of protein interactions in living plant cells using bimolecular fluorescence complementation. *Plant J* **40**, 428–438.
4. Kessler, S. W. (1975) Rapid isolation of antigens from cells with a staphylococcal protein A-antibody adsorbent: parameters of the interaction of antibody-antigen complexes with protein A. *J Immunology* **115**, 1617–1624.
5. Towbin, H., Staehelin, T., and Gordon, J. (1979) Electrophoretic transfer of proteins from polyacrylamide gels to nitrocellulose sheets: Procedure and some applications. *Proc Natl Acad Sci USA* **76**, 4350–4354.
6. Laemmli, U. K. (1970) Cleavage of structural proteins during the assembly of the head of bacteriophage T4. *Nature* **227**, 680–685.
7. Willige, B. C., et al. (2007) The DELLA domain of GA INSENSITIVE mediates the interaction with the GA INSENSITIVE DWARF1A gibberellin receptor of Arabidopsis. *Plant Cell* **19**, 1209–1220.

Chapter 26

Probing Protein–Protein Interactions with FRET–FLIM

Christoph Bücherl, José Aker, Sacco de Vries, and Jan Willem Borst

Abstract

The quantification of molecular interactions or conformational changes can conveniently be studied by using Förster Resonance Energy Transfer (FRET) as a spectroscopic ruler. The FRET phenomenon describes the transfer of energy from a donor to an acceptor molecule, if they are in close proximity (<10 nm). The most straightforward method to measure FRET is Fluorescence Lifetime Imaging Microscopy (FLIM). In this chapter, we will describe an application of FRET using FLIM to monitor the hexamer formation of CrFP/eYFP-labeled *Arabidopsis thaliana* cell division cycle protein (AtCDC48) expressed in plant protoplasts.

Key words: FRET, FLIM, AtCDC48, CrFP/eYFP.

1. Introduction

The dimensions of proteins vary from 5 to 50 nm covering the experimental range of electron microscopy and, partly, optical microscopy. Although electron microscopy has the highest spatial resolution, it is not applicable in living tissue. In contrast, fluorescence microscopy including the genetically encoded visible protein (VFP) technology, in particular, has all noninvasive capabilities for live cell imaging. However, the spatial resolution of fluorescence microscopy is light-diffraction limited (~250 nm) and no information about molecular (protein) interactions is obtained. In vivo detection of protein interactions has become feasible by combining fluorescence microscopy and Förster Resonance Energy Transfer (FRET). FRET can be measured using

L. Hennig, C. Köhler (eds.), *Plant Developmental Biology*, Methods in Molecular Biology 655,
DOI 10.1007/978-1-60761-765-5_26, © Springer Science+Business Media, LLC 2010

a variety of fluorescence imaging approaches such as FRET- or ratio-imaging and acceptor photobleaching. The most reliable and quantitative method to spatially resolve FRET is Fluorescence Lifetime Imaging Microscopy (FLIM). In this chapter, a protocol will be described to analyze protein interactions in plant protoplasts using a FRET–FLIM approach; hexamer formation of AtCDC48 will be used as an example.

AtCDC48 belongs to the protein family of ATPases associated with various activities (AAA proteins). The characteristic function of AAA proteins is the coupling of ATP hydrolysis to processes like the disassembly or unfolding of protein substrates. Within the *Arabidopsis thaliana* genome, five isoforms of AtCDC48 (A-E) are encoded. Best characterized is AtCDC48A, which shares 77% identity with the mammalian homologue vasolin containing protein (VCP) or p97. The crystal structure of this animal counterpart revealed a homo-hexameric mushroom-like shaped complex (1, 2). Monomeric VCP comprises a *N*-domain important for cofactor and substrate binding, two AAA domains (D1 and D2) connected by a linker, the D1 domain and the linker region being crucial for hexamerization, and the D2 domain responsible for the major ATPase activity, as well as a *C*-terminal tail (3). After complex formation, the *N*- and D1-domains are located at the top of the hexamer, the D2- and C-domains at the bottom. This domain architecture is also assumed for AtCDC48 despite the lack of crystallographic evidence. However, in vitro and in vivo studies showed a hexameric complex for the AtCDC48A isoform (4, 5). Using a FRET–FLIM approach, Aker et al. (5) additionally elucidated oligomerization of AtCDC48C in *Arabidopsis* mesophyll protoplasts. Even though the two mentioned isoforms are predicted to share 95% amino acid identity, their localization pattern is distinct. Whereas AtCDC48A is present at the plasma membrane, the ER, in the cytosol as well as the nucleus, AtCDC48C localizes solely to the nucleus (3). In this chapter, we will focus on the complex formation of the fluorescently tagged AtCDC48 isoforms A and C expressed in *Arabidopsis* mesophyll protoplasts to highlight the potential of FRET–FLIM for investigating in vivo interactions.

2. Materials

2.1. Transient Expression Vectors

1. Plant expression vector: pMON999 (Monsanto, USA).
2. The cDNAs of AtCDC48A and AtCDC48C as well as Cerulean Fluorescent Protein (CrFP) and enhanced Yellow Fluorescent Protein (eYFP) were cloned into pMON999.

3. DNA-Polymerase: PWO-DNA polymerase used in combination with the Mg^{2+}-containing buffer solution provided by the manufacturer.
4. Nucleotides: 25 mM dNTP.
5. Restriction enzymes: Depending on the cloning strategy chosen.
6. Ligase: T4 DNA ligase used in combination with the buffer solution provided by the manufacturer.

2.2. Protoplast Isolation

Protocol adapted from reference (6)

1. Plant material: Rosette leaves of *A. thaliana* plants (ecotype Columbia) grown for 4–5 weeks under long-day conditions (16 h light/8 h dark) and 20–22 C (*see* **Note 1**).
2. Mannitol solution: 0.4 M mannitol, 20 mM KCl, and 20 mM 2-(*N*-morpholino) ethanesulfonic acid (MES) pH 5.7 in Milli-Q water.
3. Calcium chloride solution: 1 M $CaCl_2$ in Milli-Q water.
4. Enzyme solution: 1% (w/v) cellulose R10 (Yakult Honsha Co. LTD, Japan) and 0.2% (w/v) pectinase from Rhizopus sp. (Biochemika/Fulka, Germany) dissolved in mannitol solution; subsequent addition of $CaCl_2$ to a final concentration of 10 mM.
5. Plastic round-bottom tubes (Sarstedt, Germany).

2.3. Protoplast Transfection

Transfection protocol adapted from reference (6)

1. PEG/Ca^{2+} solution: 40% (w/v) Polyethyleneglycol 4000 (PEG, Merck, Germany,) 0.2 M mannitol, and 100 mM $Ca(NO_3)_2$ in Milli-Q water (*see* **Note 2**).
2. W5 solution: 154 mM NaCl, 125 mM $CaCl_2$, 5 mM KCl, and 2 mM MES pH 5.7.
3. W5/Glucose solution: W5 solution containing 1 mM glucose.
4. MMg solution: 0.2 M mannitol, 15 mM $MgCl_2$, and 4 mM MES pH 5.7.
5. Plastic round-bottom tubes (Sarstedt, Germany).
6. Microscope 8-well slides, Lab-tek Nalge Nunc international (Rochester, NY, USA).

2.4. FRET–FLIM

1. Biorad Radiance 2,100 MP system in combination with a Nikon TE 300 inverted microscope (Tokyo, Japan).
2. Hamamatsu R3809U MCP (Hamamatsu, Japan) photomultiplier.
3. B&H SPC 830 module (Becker & Hickl, Germany).

3. Methods

The question, whether a protein acts as an oligomer, can be addressed and answered by several approaches. Phage display and yeast two-hybrid assays are most commonly used (7, 8). However, these methods lack spatial and physiological information, since the proteins expressed are not in their natural environment. The introduction of high-resolution confocal microscopy gave the opportunity to investigate the co-expression of different proteins in their natural habitat. The spatial resolution of a microscope allows detection of fluorescent molecules at subcellular level. At most, co-localization of two proteins equipped with two different fluorophores can be revealed, but physical interactions between proteins on nanometer scale cannot be determined. One possibility to go beyond the optical diffraction limit is using new advanced methods like PALM and STED microscopy, but the resolution is still not at the molecular level (9, 10). However, Förster Resonance Energy Transfer (FRET) microscopy elucidates molecular interactions in living tissue.

The term FRET describes the nonradiative energy transfer from the electronically excited state of a donor fluorophore to an acceptor molecule. This photophysical process based on dipole–dipole coupling was first described by Theodor Förster in 1948 (11). Spectral overlap of donor emission and acceptor absorption, close proximity of the labeled specimen, and adequate orientation of the donor and acceptor transition dipole moments are the requirements for FRET taking place. Exploiting the FRET phenomenon allows the investigation of protein–protein interactions with a distance range of approximately 1–10 nm – depending on the FRET couple used. The energy-transfer efficiency is proportional to the reciprocal of the sixth power of the intermolecular fluorophore distance explaining the high-distance sensitivity of FRET measurements. The so-called critical (or Förster) radius (R_0), a characteristic property of each FRET pair, is the distance between donor and acceptor, at which the energy–transfer efficiency is 50%. The concepts of FRET and relevant equations are summarized in **Fig. 26.1**.

For the quantification of protein interactions by means of FRET, several methods are available. Intensity-based methods such as FRET- or ratio-imaging and acceptor photobleaching have severe disadvantages. The main drawbacks are crosstalk of the emission spectra (donor detected in acceptor window), direct excitation of acceptor by donor laser light, and dependence on differences in donor/acceptor concentrations. An alternative method to spatially resolve FRET is Fluorescence Lifetime Imaging Microscopy (FLIM), which measures the fluorescence lifetime

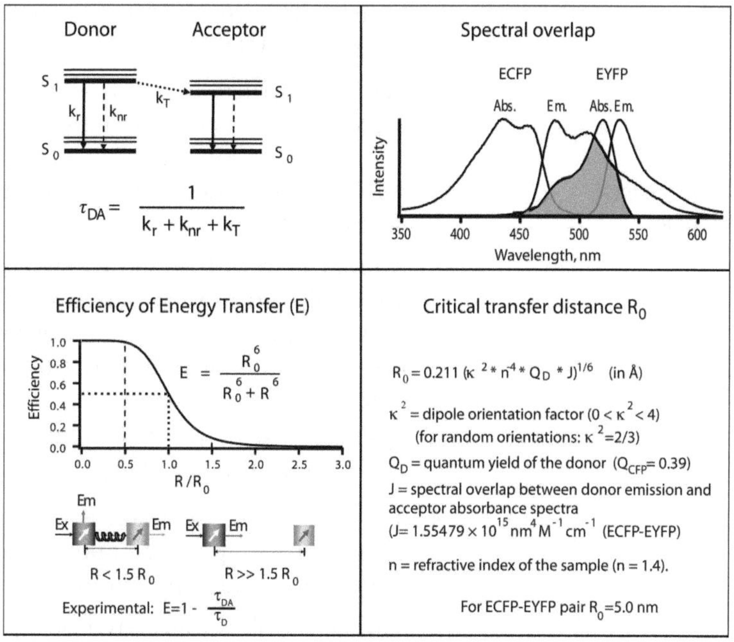

Fig. 26.1. **Summary of FRET principles.** When a donor and acceptor are in close proximity, energy transfer can take place leading to an additional relaxation pathway (k_T) (*top left*). Some prerequisites for FRET are spectral overlap between donor emission and acceptor absorption spectra (*top right*), small distance between donor and acceptor and adequate dipole orientation. These parameters determine the critical transfer distance (R_0), which is characteristic for each FRET pair (*bottom right*). The efficiency of energy transfer can be related to relative distance and experimentally determined from fluorescence lifetime measurements (*bottom left*). **Fig. 26.1** taken from Borst and co-workers (2006) (15).

pixel by pixel. FLIM overcomes problems of intensity-based methods by determining the fluorescence lifetime of the donor molecule only. Molecular interaction between donor and acceptor will result in both quenching of donor fluorescence intensity and consequently a decrease of the donor fluorescence lifetime since energy transfer will introduce an additional relaxation path from the excited state to the ground state (*see* **Fig. 26.1**). The difference of the donor fluorescence lifetime in the absence or presence of acceptor is directly correlated with the FRET efficiency E via

$$E = 1 - \tau_{DA}/\tau_D$$

where τ_{DA} is the fluorescence lifetime of the donor in the presence of acceptor and τ_D is the fluorescence lifetime of the donor alone.

3.1. Transient Expression Vectors

1. Amplify AtCDC48C or AtCDC48A (or the cDNA of your protein-of-interest A) by PCR from an EST using

appropriate primers and fuse to PCR-amplified CrFP and eYFP into pMON999.

2. Verify constructs by sequencing and the size of the fused proteins by Western blotting using anti-GFP antibodies.

3.2. Protoplast Isolation

1. All steps are carried out at room temperature unless otherwise noted.
2. A clean Petri dish (9 cm diameter) is covered with rosette leaves of 4 to 5-week old *A. thaliana* (*see* **Note 1**) plants grown under long-day conditions. This will result in a sufficient amount of protoplasts to perform approximately 6 independent transfections.
3. Slice leaves with a scalpel, add 15 mL of enzyme solution, and swirl to dampen all plant material (*see* **Note 3**).
4. The sliced leaf suspension is placed for 3 min in a vacuum desiccator with powered pump followed by 30 min incubation without pumping, but under vacuum conditions.
5. The Petri dish is transferred on a platform shaker and incubated at 65–80 rpm and 27°C for additional 2 h.
6. Protoplasts are released from the leaf matrix by carefully swirling the Petri dish for 1 min by hand. Subsequently, the suspension is filtered through a 35–100 μm nylon mesh into a clean plastic round-bottom tube.
7. Collect the protoplasts by centrifugation for 3 min at 50×*g* using a tabletop centrifuge; wash once with 5 mL of W5 solution.
8. At this stage, the isolated protoplasts can be kept on ice overnight (*see* **Note 4**).

3.3. Protoplast Transfection

1. All steps are carried out at room temperature, and the protocol is applicable for 6 independent transfections. Besides the double transfections with donor and acceptor constructs encoding the proteins of interest (interaction studies) at least one single transfection with the donor construct only has to be included next to positive and negative controls.
2. Prepare round-bottom tubes and plasmid DNAs. For single transfections, 10–20 μg of the respective DNA is required. For double transfections, 10–15 μg of each construct is combined in one Eppendorf tube. Single transfection is used to obtain protoplasts expressing the donor construct only, whereas the transfections of two plasmids yield protoplasts for the interaction studies (*see* **Note 5**).
3. Collect the isolated and washed protoplasts (*see* **Section 3.2**) by centrifugation for 3 min at 50×*g* in a tabletop centrifuge and remove the supernatant carefully with a pipette.

4. The protoplasts are resuspended in 1.2 mL of MMg (*see* **Note 6**) and aliquots of 200 μL are transferred to round-bottom tubes (*see* **Note** 7).
5. Pipette the respective amount of plasmid DNA into the protoplast suspension and then add 220 μL of PEG/Ca^{2+} solution. Mix well but carefully and incubate for 5 min (*see* **Note** 7).
6. Add 800 μL of W5 solution to stop the transfection process and collect the protoplasts by centrifugation at 50×*g* for 3 min in a tabletop centrifuge.
7. Remove the supernatant with a pipette and wash with 5 mL of W5 solution.
8. Collect protoplasts by centrifugation at 50×*g* for 3 min. Remove the supernatant with a pipette and resuspend the protoplasts in 1 mL of W5 solution containing 1 mM glucose.
9. Transfer the protoplast suspension into a 24-well plate and incubate at 25°C under long-day conditions. In general, measurements should be carried out about 16 h after transfection but depending on the used expression vectors and constructs, earlier or later time points can be chosen. It is recommended to check the expression levels at several time points to figure out the optimal incubation period (*see* **Note** 7).

3.4. FRET-FLIM

1. FLIM is performed on a Biorad Radiance 2 100 MP system in combination with a Nikon TE 300 inverted microscope as described by Russinova et al. (2004) (12). Two photon excitation pulses are generated by a Ti-Sapphire Mira Laser (Coherent,) pumped by a 5 W Verdi Laser, resulting in excitation pulses of 200 fs at a repetition of 76 Mhz. A 60x/1.2 water-immersion objective is used.
2. CrFP emission is selected by a 480DF30 band-pass filter and detected by a Hamamatsu R3809U MCP photomultiplier with a typical time resolution of 50 ps.
3. Fluorescence images of 64 × 64 pixel size are acquired using the B&H SPC 830 module (*see* **Note 8**).
4. The average count rate is around 10^4 photons per second for an acquisition time of 90 s (13).
5. Measurements of single-transfected protoplasts expressing *N*- or *C*-terminally CrFP-tagged AtCDC48A or AtCDC48C result in the donor fluorescence lifetime required as reference. To elucidate the hexamer formation of AtCDC48A/C monomers, protoplasts expressing donor and acceptor constructs are investigated (*see* **Note 9**).

The protoplasts are transferred into an 8-well chamber for imaging (*see* **Note 10**).

3.5. Analysis of FLIM Data with SPCImage 2.9.3

1. SPCImage is the software package included with the B&H acquisition card. The raw data can be imported and analyzed using an exponential model function. The fitted experimental data result in the output of fluorescence lifetime values per pixel depicted as a false-color code. During the fitting process, the chi-square value between model function and data is minimized (14).
2. In general, FRET-FLIM experiments are based on the fluorescence lifetime of donor molecules in the absence and presence of acceptor. The data analysis involves first the determination of the fluorescence lifetime of single transfected protoplasts (donor only). Since one donor population is assumed (no FRET possible), a one-component analysis is carried out (*see* **Note 11**). The average donor fluorescence lifetime subsequently serves as reference value to judge if interactions occur when acceptor molecules are present. Energy transfer from the donor to the acceptor molecule will result in a reduction of the donor fluorescence lifetime and therefore a change in the color code.
3. The double-transfected protoplasts (donor and acceptor) are analyzed based on a two-component model. Thereby the value of one fluorescence lifetime component can be fixed to the mean of the donor-only analysis. This approach assumes two donor populations – one transferring energy to an adjacent acceptor molecule resulting in a reduced fluorescence lifetime, the second showing no FRET and hence exhibiting fluorescence decay kinetics as donor alone.
4. After importing the experimental data, a fluorescence intensity image will be displayed. A blue crosshair allows the selection of single pixels to control the quality of the dataset by surveying the corresponding fluorescence decay histograms. Areas showing high auto-fluorescence (identifiable by the steep fluorescence decay) or less than 200 photon counts for the maximum should be excluded from the fitting process (*see* **Notes 12** and **13**).
5. Before starting the analysis process, several settings have to be chosen. Using the "region of interest" (ROI) tool, specific areas, e.g., plasma membrane or nucleus, of the intensity image can be selected. Additionally, the borders of the fluorescence decay histogram have to be set. Typical values are around 1 ns (before the rising edge) for the left and about 10.5 ns (to prevent TAC (time-to-amplitude converter) noise at long-time scales) for the right border. Depending

on the protoplast analyzed, the number of components for the underlying exponential fit function has to be defined (here one or two). In case of a two-component analysis for double-transfected protoplasts, the τ_2 value is fixed to the mean fluorescence lifetime of the donor fluorophore. All other parameters should remain unfixed independent of the fit model.

6. After performing the calculations, the mean fluorescence lifetime, the distribution of the single pixel values as well as a false-color coded lifetime image will be displayed for the selected ROI (*see* **Fig. 26.2**).

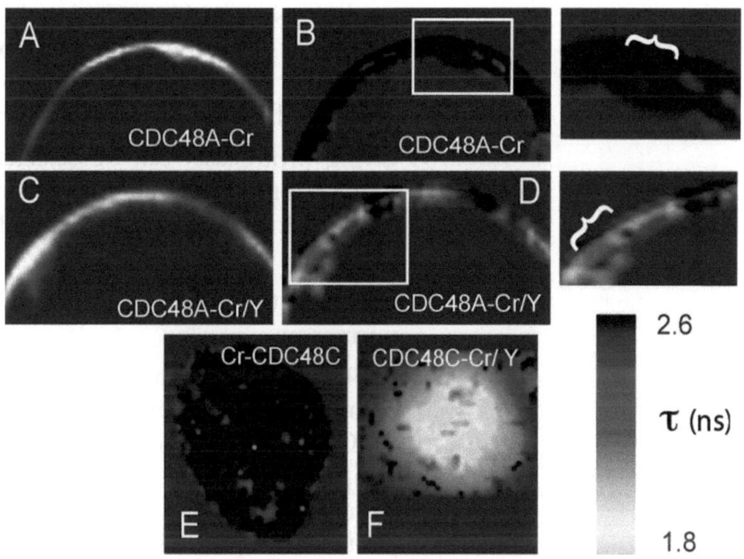

Fig. 26.2. **Interactions between AtCDC48A or AtCDC48C tagged protomers based on FRET measured by FLIM. (A)** Fluorescence intensity image of a section of the plasma membrane of a protoplast expressing the donor molecule AtCDC48A-CrFP alone and **(B)** grey scale or lifetime image. A long lifetime, shown in dark grey, means no interaction; a reduction in donor lifetime generating a shift towards white, means interaction. A combination of AtCDC48A-CrFP and AtCDC48A-eYFP proteins shows a reduction of the fluorescence lifetime at the plasma membrane (**C** and **D**). **(E)** Fluorescence lifetime images of the CrFP-AtCDC48C donor alone and **(F)** the combination of AtCDC48C CrFP with AtCDC48C-eYFP (both C-terminal fusions), showing a reduction in fluorescence lifetime being in line with the mushroom-like shape structure of p97/VCP (5). **Fig. 26.2** taken from Aker et al (2007) (5). Refer online version for color image.

4. Notes

1. Wild-type or transgenic lines possible.
2. Heat twice for 6 s at 300 Watt in a microwave to dissolve PEG.
3. Do not cut the leaves too small because this will lead to a higher ratio of dead protoplasts.

4. Nonetheless, it is recommended to proceed immediately – the quicker the transfection the better.
5. The total volume of plasmid DNA should not exceed 30 μL. Therefore, plasmid solutions with appropriate DNA concentrations of around 1 μg/μL plasmid DNA are recommended.
6. The amount of protoplasts can vary – so adapt the volume of MMg solution.
7. Use tips with enlarged openings to reduce shear forces.
8. Images are taken with a 64 × 64 pixel size with a *xy* pixel resolution of about 200 nm. Higher pixel resolution does not increase spatial information due to the light diffraction limitation. Instead using higher pixel resolution requires a prolonged data acquisition time to ensure the detection of a statistically relevant number of photons.
9. Living protoplasts can be selected by visual inspection. Using a long-pass filter (LP 520 nm), the red fluorescence of the chlorophyll is observed and a good indicator for a living cell.
10. Colocalization has to be verified by confocal laser scanning microscopy.
11. CrFP shows also in absence of an acceptor bi-exponential decay kinetics, but this effect is observed by applying the ADC (analog-digital converter) larger than 64. Therefore, accurate data analysis should make use of two-component analysis even for donor-only expressing protoplasts whereby one component is fixed.
12. In case the number of photons is too low for quantitative analysis, the binning factor can be increased. Binning is a procedure where the selected pixel is analyzed, but the neighboring pixels are included for calculation of the fluorescence lifetime. The binning factor can be calculated according to the following formula:
    Binning factor = $(2n+1)^2$, where n is number of pixels.
    In **Fig. 26.2**, FLIM images are shown where a binning factor of 1 has been applied. In other studies, the effect of binning has been investigated and no significant change was observed.
13. Consider alternative FRET couples such as eGFP/mCherry. First, this FRET couple reduces crosstalk of eGFP fluorescence into the mCherry detection window in intensity-based measurements. Second, eGFP can be excited optimal with a conventional confocal microscope, whereas CrFP is suboptimal excited. Third, direct excitation of mCherry is avoided at the donor

excitation wavelength compared to the CrFP/eYFP situation. Fourth, the fluorescence lifetime of eGFP exhibits a mono-exponential decay profile and therefore quantitative analysis is improved in FRET-FLIM measurements. Fifth, using eGFP as a donor molecule in a FRET-FLIM experiment gives reduced background fluorescence.

References

1. Huyton, T., Pye, V. E., Briggs, L. C., Flynn, T. C., Beuron, F., Kondo, H., Ma, J., Zhang, X., and Freemont, P. S. (2003) The crystal structure of murine p97/VCP at 3.6A. *J Struct Biol* **144**, 337–348.
2. DeLaBarre, B. and Brunger, A. T. (2003) Complete structure of p97/valosin-containing protein reveals communication between nucleotide domains. *Nat Struct Biol* **10**, 856–863.
3. Aker, J., Borst, J. W., Karlova, R., and de Vries, S. (2006) The *Arabidopsis thaliana* AAA protein CDC48A interacts in vivo with the somatic embryogenesis receptor-like kinase 1 receptor at the plasma membrane. *J Struct Biol* **156**, 62–71.
4. Rancour, D. M., Park, S., Knight, S. D., and Bednarek, S. Y. (2004) Plant UBX domain-containing protein 1, PUX1, regulates the oligomeric structure and activity of *Arabidopsis* CDC48. *J Biol Chem* **279**, 54264–54274.
5. Aker, J., Hesselink, R., Engel, R., Karlova, R., Borst, J. W., Visser, A. J., and de Vries, S. C. (2007) In vivo hexamerization and characterization of the *Arabidopsis* AAA ATPase CDC48A-complex using FRET-FLIM and FCS. *Plant Physiol* **145**, 339–350.
6. Sheen, J. (2001) Signal transduction in maize and *Arabidopsis* mesophyll protoplasts. *Plant Physiol* **127**, 1466–1475.
7. Causier, B. and Davies, B. (2002) Analysing protein-protein interactions with the yeast two-hybrid system. *Plant Mol Biol* **50**, 855–870.
8. Burch, L. R., Scott, M., Pohler, E., Meek, D. and Hupp, T. (2004) Phage-peptide display identifies the interferon-responsive, death-activated protein kinase family as a novel modifier of MDM2 and p21WAF1. *J Mol Biol* **337**, 115–128.
9. Betzig, E., Patterson, G. H., Sougrat, R., Lindwasser, O. W., Olenych, S., Bonifacino, J. S., Davidson, M. W., Lippincott-Schwartz, J., and Hess, H. F. (2006) Imaging intracellular fluorescent proteins at nanometer resolution. *Science* **313**, 1642–1645.
10. Willig, K. I., Rizzoli, S. O., Westphal, V., Jahn, R., and Hell, S. W. (2006) STED microscopy reveals that synaptotagmin remains clustered after synaptic vesicle exocytosis. *Nature* **440**, 935–939.
11. Förster, T. (1948) Zwischenmolekulare Energiewanderung und Fluoreszenz. *Ann Phys* **2**, 55–75.
12. Russinova, E., Borst, J. W., Kwaaitaal, M., Cano-Delgado, A., Yin, Y., Chory, J., and de Vries, S. C. (2004) Heterodimerization and endocytosis of Arabidopsis brassinosteroid receptors BRI1 and AtSERK3 (BAK1). *Plant Cell* **16**, 3216–3229.
13. Borst, J. W., Hink, M. A., van Hoek, A., and Visser, A. J. W. G. (2003) Multiphoton microspectroscopy in living plant cells. *Proc SPIE* **4963**, 231–238.
14. Becker, W., Bergmann, A., Hink, M. A., König, K., Benndorf, K., and Biskup, C. (2004) Fluorescence lifetime imaging by time-correlated single-photon counting. *Microsc Res Tech* **63**, 58–66.
15. Borst, J. W., Nougalli-Tonaco, I. A., Hink, M. A., Hoek, A. v., Immink, R. G. H., and Visser, A. J. W. G. (2006) In: Protein–protein interactions in vivo: Use of biosensors based on FRET in reviews in fluorescence 2006, pp 341–355. Geddes, C. D. and Lakowicz, J. R., Eds. Kluwer Academic/Plenum Publishers. New York.

Chapter 27

Plant Chromatin Immunoprecipitation

Corina B.R. Villar and Claudia Köhler

Abstract

Development of multicellular organisms is based on specialized gene expression programs. Because chromatin establishes the environment for transcription, understanding composition and dynamics of chromatin is an important part of developmental biology. The knowledge about chromatin has been greatly advanced by the chromatin immunoprecipitation (ChIP) technique, because ChIP allows to map the position of proteins as well as modifications of DNA and histones to specific genomic regions. Although ChIP has been applied to a wide range of model organisms, including Arabidopsis, it remains a challenging technique, and a careful experimental setup including appropriate positive and negative controls are required to obtain reliable results. Here, we describe a ChIP protocol adapted for material from Arabidopsis, which we routinely apply in our laboratory, and we discuss required controls and methods for data analysis.

Key words: Chromatin, chromatin immunoprecipitation, epigenetics, DNA and histone modifications.

1. Introduction

The eukaryotic genome is packaged into a compact structure called chromatin. The core structure of chromatin is the nucleosome, consisting of approximately 147 base pairs of DNA wrapped around a histone octamer consisting of two copies each of the core histones H2A, H2B, H3, and H4 (1). Chromatin structure is highly dynamic, and opened or closed chromatin conformations are closely connected with specific chromatin functions such as transcription and replication. Chromatin properties that influence gene expression include posttranslational modifications of histones, DNA methylation, nucleosome position, and binding of nonhistone proteins such as transcription factors

L. Hennig, C. Köhler (eds.), *Plant Developmental Biology*, Methods in Molecular Biology 655,
DOI 10.1007/978-1-60761-765-5_27, © Springer Science+Business Media, LLC 2010

(2– 5). To investigate gene regulation, it is required to know when during development and where within the genomic locus histone modification and protein binding occurs. This information can be obtained using chromatin immunoprecipitation (ChIP). In contrast to in vitro assays, ChIP allows the identification of protein–DNA associations in their native chromatin state (6, 7).

In general, ChIP involves the following steps: (i) crosslinking of DNA to protein, (ii) isolating and fragmenting the chromatin, (iii) immunoprecipitation with antibodies recognizing the protein of interest, (iv) elution and purification of the immunoprecipitated DNA, and (v) quantitative PCR with immunoprecipitated material. Our protocol contains several modifications of the original protocols by Colot and Grewal (8, 9) and has been applied successfully by our group (10–12). Some ChIP protocols take advantage of the high binding affinity of histones to DNA, thus eliminating the need for crosslinking of DNA to protein by formaldehyde (native ChIP, 13). In contrast, for studying the association of non-DNA-binding proteins to specific loci, it is advantageous to apply an additional protein–protein crosslinking step before crosslinking protein to DNA by formaldehyde (14–16). There are also different methods to fragment chromatin either by employing sonication or micrococcal nuclease digestion. The protocol described in this chapter involves single or double chromatin crosslinking, chromatin sonication, and classic phenol/chloroform extraction to purify the immunoprecipitated DNA.

Immunoprecipitated DNA can be analyzed in several ways. The basic approach is to perform PCR using primers specific for a certain locus of interest and analyze the products by gel electrophoresis or real-time PCR. However, this method presumes that the protein binds to known sequences, and it cannot detect binding to unknown DNA sequences. Genome-wide approaches such as the combination of ChIP and whole-genome microarrays (ChIP-chip, 17, 18) or the combination of ChIP and direct sequencing (ChIP-seq, 19) should be applied when the identification of novel target genes is the primary focus of the investigation.

2. Materials

2.1. Crosslinking

1. Plant material: Any plant material, whether grown in vitro or on soil, can be used, as long as it is washed thoroughly during harvest to remove debris.
2. Dimethyl adipimate (DMA) solution: Prepare fresh 10 mM DMA (Sigma-Aldrich) in water with 0.25% (v/v) dimethylsulfoxide.

3. Nylon mesh (500-μm pore size, Fisher Scientific).
4. Formaldehyde (1% in water).
5. 2 M glycine.
6. Mortar and pestle.
7. Dessicator linked to vacuum pump.

2.2. Chromatin Preparation

1. Fine nylon mesh for filtration (50-μm pore size, Fisher Scientific).
2. 2 M Sucrose solution in water. Heat and stir continuously to fully dissolve the sucrose.
3. 1 M Tris-HCl (pH 8.0) in water.
4. 1 M $MgCl_2$ in water.
5. 0.2 M phenylmethylsulfonyl fluoride (PMSF) in isopropanol. Can be stored at –20°C.
6. Protease inhibitor solution (25×): Dissolve 1 tablet (complete, Mini, Roche) per 2 mL sterile distilled water. Can be stored at –20°C.
7. 20% (v/v) Triton X-100 in water.
8. 10% (w/v) sodium dodecyl sulfate (SDS) in water.
9. 0.5 M ethylenediaminetetraacetic acid (EDTA) in water.
10. Sonicator: Bioruptor^{TM} 200 (Diagenode).
11. Extraction buffer 1: 0.4 M sucrose, 10 mM Tris-HCl (pH 8.0), 10 mM $MgCl_2$, 0.1 mM PMSF, and 1× protease inhibitors.
12. Extraction buffer 2: 0.25 M sucrose, 10 mM Tris-HCl (pH 8.0), 10 mM $MgCl_2$, 1% Triton X-100, 0.1 mM PMSF, and 1× protease inhibitors.
13. Extraction buffer 3: 1.7 M Sucrose, 10 mM Tris-HCl (pH 8.0), 0.15% Triton X-100, 0.1 mM PMSF, and 1× protease inhibitors.
14. Nuclei lysis buffer: 50 mM Tris-HCl (pH 8.0), 10 mM EDTA, 1% SDS, and 1× protease inhibitors.
15. ChIP dilution buffer: 1.1% Triton X-100, 1.2 mM EDTA, 16.7 mM Tris-HCl (pH 8.0), 167 mM NaCl, and 1× protease inhibitors.

2.3. Immunoprecipitation

1. Protein A agarose beads preblocked with salmon sperm DNA (Upstate). Beads can be stored at 4°C.
2. Antibodies. Prepare aliquots and store at –20°C. Avoid repeated thawing and freezing. Once an aliquot is thawed, keep at 4°C.

2.4. Collection and Washing of Immunoprecipitated DNA

1. 5 M NaCl in water.
2. 10% (w/v) sodium deoxycholate in water.
3. 10% (w/v) octyl phenoxylpolyethoxylethanol (Igepal CA-630, Sigma-Aldrich) in water.
4. 10% (w/v) SDS in water.
5. 1 M Tris-HCl (pH 8.0) in water.
6. 0.5 M EDTA in water.
7. 4 M LiCl in water.
8. Low-salt wash buffer: 150 mM NaCl, 0.1% SDS, 1% Triton X-100, 2 mM EDTA, and 20 mM Tris-HCl (pH 8.0).
9. High-salt wash buffer: 500 mM NaCl, 0.1% SDS, 1% Triton X-100, 2 mM EDTA, and 20 mM Tris-HCl (pH 8.0).
10. LiCl wash buffer: 0.25 M LiCl, 1% Igepal CA-630, 1% sodium deoxycholate, 1 mM EDTA, and 10 mM Tris-HCl (pH 8.0).
11. TE buffer: 10 mM Tris-HCl (pH 8.0) and 1 mM EDTA.
12. Elution buffer: 1% SDS and 0.1 M $NaHCO_3$.
13. 1 M Tris-HCl, pH 6.5.
14. 10 mg/mL Proteinase K.
15. Heating block: Eppendorf™ Thermomixer.

2.5. Purification of Immunoprecipitated DNA

1. Phenol:chloroform (1:1).
2. Ethanol (100%).
3. Ethanol (70%).
4. 3 M sodium acetate (NaOAc), pH 4.8.
5. Glycogen solution (20 mg/mL in water). Store at –20°C.

2.6. Analysis of Immunoprecipitated DNA

1. PCR buffer (10×, without $MgCl_2$): 670 mM Tris-HCl, pH 8.8, 160 mM $(NH_4)_2SO_4$, and 0.1% Tween 20. Store at –20°C.
2. 50 mM $MgCl_2$ in water. Store at – 20°C.
3. 5 mM dNTP mixture in water. Store at –20°C.
4. *Taq* DNA polymerase. Store at –20°C.
5. Quantification software like ImageJ (http://rsbweb.nih.gov/ij/).

3. Methods

3.1. Crosslinking

1. Harvest 1 g of tissue in a 50-mL Falcon tube and rinse thoroughly with water to remove debris. Add 40 mL of freshly

prepared 10 mM DMA solution (*see* **Note 1**). Keep the tissue submerged by using a nylon mesh (500-μm pore size) to push all the material down. This will also make the washing steps easier, with the nylon mesh acting as a sieve.

2. Vacuum-infiltrate on ice for 15 min and then wash twice with distilled water. Pour out the liquid without moving the nylon mesh.
3. Add 37 mL of 1% formaldehyde and then vacuum-infiltrate once more for 20 min.
4. Add 2.5 mL of 2 M glycine to stop crosslinking. Vacuum-infiltrate for another 5 min. Pour out the liquid as before and wash the material twice with distilled water. Take out the plant material and briefly pat dry on tissue paper (*see* **Note 2**).
5. Precool mortar and pestle with liquid nitrogen. After the liquid nitrogen has evaporated, place the dry plant material in the mortar and pour liquid nitrogen over the material. Grind the material quickly but carefully with the pestle to a fine powder. Put the ground tissue in a cold (dipped in liquid nitrogen) 15-mL Falcon tube with the use of a spatula that was also dipped in liquid nitrogen. The sample can be stored at –80°C until needed or used immediately for chromatin preparation.

3.2. Chromatin Preparation

Unless otherwise stated, centrifugation steps are done at 4°C and the material is kept cold on ice at all times.

1. Add 10 mL of extraction buffer 1 to the frozen material and mix thoroughly by vortexing. The material should be kept cold, so put on ice for a few minutes after each vortexing.
2. Filter the solution twice through a double layer of fine nylon mesh into a fresh Falcon tube (*see* **Note 3**).
3. Centrifuge at 3,000×*g* for 20 min and gently remove the supernatant. Resuspend the pellet in 1 mL of extraction buffer 2 by pipetting up and down. Then transfer the solution to a new 1.5-mL tube.
4. Centrifuge at 12,000×*g* for 10 min and completely remove the supernatant. Resuspend the pellet in 400 μL of extraction buffer 3.
5. Add an equal volume (400 μL) of extraction buffer 3 into a fresh 1.5-mL tube and carefully layer the resuspended pellet from the previous step on top.
6. Centrifuge for 1 h at 16,000×*g*. At this stage, ChIP dilution buffer and nuclei lysis buffer should be prepared (*see* **Note 4**).

7. Remove the supernatant and resuspend the pellet in 300 μL of nuclei lysis buffer. Resuspend by pipetting up and down and vortexing while keeping the solution cold. Keep 10 μL of crosslinked chromatin per sample and store at –20°C (*see* **Note 5**).
8. Once resuspended, sonicate the chromatin solution 8 times on a 15-s ON, 1-min-45-s OFF cycle, on a Bioruptor™ 200 sonicator (*see* **Note 6**).
9. Centrifuge at 16,000×*g* for 5 min and place the supernatant in a fresh tube. From this solution, keep another 10 μL to check sonication efficiency (*see* **Note 5**).
10. Measure the remaining solution and bring to 10 times its volume with ChIP dilution buffer. The main point of this step is to dilute the 1% SDS to 0.1%. Split the chromatin solution equally into as many tubes as is required for the immunoprecipitation experiment (one for each antibody plus positive and negative controls). Antibodies against histone H3 may be used as positive control, while nonspecific IgG (or alternatively no antibodies) can be used as negative control. Take note of the amount of chromatin used for each sample for later quantification of the relative amount of immunoprecipitated DNA.

3.3. Immunoprecipitation

1. Transfer 40 μL of protein A agarose beads preblocked with salmon sperm DNA into the same number of fresh tubes as required for the immunoprecipitation experiment. To precisely measure the beads, cut the end of a pipette tip and constantly shake the agarose bead suspension before taking out the beads. Salmon sperm DNA reduces unspecific binding of sample DNA to agarose beads. Rinse the beads three times with 1 mL ChIP dilution buffer. Centrifuge quickly after each wash and remove supernatant.
2. Add the chromatin solution to the tubes containing the rinsed agarose beads. Preclear the samples for 1 h at 4°C with rotation.
3. After preclearing, centrifuge the chromatin solution for 2 min at 12,000×*g*. Carefully transfer the supernatant (no beads included) to a fresh tube. From each solution, take 25 μL as input DNA. Store at –20°C until reversing the crosslinks later.
4. Add the appropriate amount of antibody to each tube. The amount depends on the quality of the antibody and should be determined empirically. Incubate the chromatin plus antibody solutions overnight at 4°C with rotation.

3.4. Collection and Elution of Immunoprecipitated DNA

1. Prepare fresh ChIP dilution buffer. Transfer 50 μL of protein A agarose beads into new labeled tubes and rinse three times with ChIP dilution buffer as before.
2. Collect the immune complexes with the rinsed agarose beads for 1 h at 4°C with rotation. Prepare elution buffer at this point and keep it at 65°C until use.
3. Wash the beads twice (first a quick wash by vortexing and then a second wash for 5 min at 4°C with rotation) using the indicated buffers listed below. Use 1 mL of buffer for each wash. Centrifuge the beads at 4,500×*g* for after each wash.
   a. Low salt wash buffer
   b. High salt wash buffer
   c. LiCl wash buffer
   d. TE buffer
4. Elute immune complexes by adding 250 μL of elution buffer to the pelleted beads. Vortex the solution briefly and then incubate at 65°C for 15 min with gentle agitation on an Eppendorf™ Thermomixer. Centrifuge at 4,000×*g* for 30 s and transfer each eluate to a fresh tube. Repeat the elution and combine the two eluates of each sample into a single tube.
5. Adjust the volumes of input DNA (*see* **Section 3.3**, Step 3,) crosslinked chromatin (*see* **Section 3.2**, Step 7,) and sonicated crosslinked chromatin (*see* **Section 3.2**, Step 9) to 500 μL with elution buffer.
6. Add 20 μL of 5 M NaCl to each sample and incubate at 65°C for at least 6 h.
7. Add 10 μL of 0.5 M EDTA, 20 μL of 1 M Tris-HCl (pH 6.5), and 2 μL of 10 mg/mL proteinase K to each sample and incubate for 2 h at 45°C.

3.5. Purification of Immunoprecipitated DNA

1. Recover immunoprecipitated DNA by phenol/chloroform extraction and ethanol precipitation. To each sample, add an equal volume of phenol:chloroform (1:1). Vortex for 30 s and centrifuge at 16,000×*g* for 15 min. Transfer the aqueous phase to a new tube.
2. Add an equal volume of chloroform. Vortex the solution again for 30 s and centrifuge at 16,000×*g* for 15 min. Split the aqueous phase into two tubes. Add three volumes of 100% ethanol, 0.1 volume of 3 M NaOAc, pH 4.8, and 2 μL of glycogen to each tube and mix by inversion several times. Incubate at –20°C for at least 1 h to precipitate the DNA.

3. Pellet the precipitated DNA by centrifugation at 16,000×*g* for 15 min. Remove the supernatant.
4. Wash the pellet with 1 mL of 70% ethanol and centrifuge again at 16,000×*g* for 15 min. Remove the supernatant and air-dry the pellet for about 15–30 min.
5. Resuspend the pellet in 50 μL of sterile distilled water. Combine the pellets from the same samples to have 100 μL of immunoprecipitated DNA solution for each sample. Store samples at −20°C until use for PCR or any other applications.

3.6. Analysis of Immunoprecipitated DNA

Immunoprecipitated DNA can be analyzed by quantitative or semiquantitative PCR. Whereas the latter will give a good indication whether the ChIP experiment was technically successful, quantitative PCR is recommended to determine enrichment levels. As quantitative PCR is covered in other chapters of this volume, the following section describes ChIP quantification by semiquantitative PCR. Two control samples specific for each ChIP experiment should be included: the input sample and the negative control (e.g., the IgG sample or the "no antibody" sample). Both samples provide essential information about the ChIP experiment and should be included with every primer set used.

1. Design gene-specific primers for standard PCR with an appropriate primer design software. Primers should be 20–30 bp long and should amplify products ranging from 100 to 400 bp (*see* **Note 7**).
2. Prepare standard PCR reactions on ice (1× PCR buffer, 1.5 mM $MgCl_2$, 0.8 mM total dNTPs, and 2 U *Taq* DNA polymerase) with all immunoprecipitated DNA samples and their corresponding input samples, positive PCR control (usually genomic DNA as template), and negative PCR control. Primer concentration should be optimized; 500 nM per reaction is a good starting point.
3. Carry out PCR using the following cycling parameters:

   Initial step: 2 min, 94°C

   35 cycles: 30 s, 94°C (denaturation)

   30 s, 55°C (annealing)

   30 s, 72°C (extension)

   Last step: Hold, 16°C

   These conditions are suitable for most reactions, although number of cycles and annealing temperature should be adjusted. Use dilution curves to make sure that PCR reactions did not reach the plateau phase but were still at the log-linear phase at the last cycle.

4. Analyze samples on a 2% (w/v) agarose gel. Document the image using a gel-documentation system and save the image file. When comparing samples, it is important to run them on the same gel and save them in the same image.
5. Quantify the relative amount of PCR products using ImageJ or other appropriate software.
6. Normalize data by calculating either percent of input or fold enrichment values (*see* **Note 8**). For calculating percent of input, divide the amount of PCR products of the immunoprecipitated sample by the amount of PCR products of the input sample and take into account the ratio of the input sample volume to the immunoprecipitated sample volume (*see* **Note 9** for example). Fold enrichment values can be calculated as enrichment compared to a control sequence. In this case, the ChIP PCR signal derived from the sequence of interest is divided by the signal derived from the control sequence. Alternatively, ChIP signals can also be normalized relative to nucleosome density. In that case, ChIP PCR signals obtained with a specific antibody are divided by the signal obtained with an antibody against a histone, e.g., histone H3.

4. Notes

1. We recommend to use the protein–protein crosslinking agent DMA (or any other protein–protein crosslinking agent) when immunoprecipitating protein complex subunits that do not directly bind to DNA. It is generally not required for ChIP of histones and DNA-binding proteins.
2. If very little plant tissue is available for the ChIP experiment, do not place it directly on the paper to dry. Instead, keep it on the nylon mesh and place the mesh over the paper to dry.
3. We use fine nylon mesh instead of miracloth but either should work fine. Miracloth is absorbent, so before filtration saturate it first with extraction buffer 1 to avoid loss due to absorption into the miracloth.
4. When preparing the nuclei lysis buffer, add the SDS last and store at 4°C, not on ice, to prevent precipitation of the SDS. Each immunoprecipitation (antibody per sample) will require 3 mL of ChIP dilution buffer to wash the agarose beads only, so calculate the amount of ChIP dilution buffer accordingly.
5. An aliquot of the crosslinked chromatin is kept for testing the integrity of the isolated chromatin. After reversing the crosslinks and RNAse A treatment, there should be a clear

band on the agarose gel, without any shearing. On the other hand, the aliquot of sonicated chromatin should yield DNA fragments between 0.3 and 1.5 kb in length after reversing the crosslinks and treating with RNAse A.

6. Make sure to use identical 1.5-mL tubes (brand, material, etc.) for all samples. The Bioruptor^{TM} 200 requires a maximum volume of 300 μL per sample. An hour before use, fill the Bioruptor^{TM} 200 with ice for cooling. Just before use, remove most of the ice except for a thin layer of crushed ice and then add ice water until the level advised by the manufacturer in the user's manual.
7. Standard PCR primers designed for PCR products up to 200 bp in length may also be used for real-time PCR, but the PCR efficiencies have to be determined first. For real-time PCR, the primer concentration will need to be optimized depending on the real-time PCR kit being used.
8. Normalization of ChIP data is very important and care should be taken to select for the most appropriate normalization method. There is no consensus in the literature about ChIP normalization, and several methods are currently used. However, some methods are more likely to result in incorrect data interpretation than others. A relatively unbiased method for data normalization is to calculate the percent of input recovered by the ChIP; however, the disadvantage of this method is bias and/or noise introduced by different handling of input and ChIP samples. Another relatively unbiased way to normalize ChIP data is by calculating fold enrichments over a control sequence. The advantage of this method is that the same ChIP samples are used to analyze sequences of interest and control sequences, eliminating variation caused by sample handling. The disadvantage of this method is to identify a suitable control sequence. For a comprehensive discussion over different normalization methods, their advantages and disadvantages *see* (20).
9. For example, the total chromatin is 3 mL and it is distributed into four tubes, so the approximate amount of chromatin for each sample would be 725 μL. From this, 25 μL is taken as input. Therefore, the amount of chromatin taken for immunoprecipitation is 28 times higher than input chromatin. If input DNA and immunoprecipitated DNA are diluted equally, and 2 μL of immunoprecipitated DNA will yield the same PCR product as 2 μL of input DNA, the immunoprecipitation efficiency is 1/28th or ~3.6%. It is important that the amount precipitated by the specific antibody is higher than that precipitated by the nonspecific IgG or the amount obtained in the "no antibody" control, both corresponding to background signal.

Acknowledgments

Research in the authors' laboratory is supported by grants from the Swiss National Science Foundation [PP00A-106684/1] and ETH [TH-12 06-1].

References

1. Luger, K., Mader, A. W., Richmond, R. K., Sargent, D. F., and Richmond, T. J. (1997) Crystal structure of the nucleosome core particle at 2.8 A resolution. *Nature* **389**, 251–260.
2. Grunstein, M. (1997) Histone acetylation in chromatin structure and transcription. *Nature* **389**, 349–352.
3. Duo, Y., Mizzen, C. A., Abrams, M., Allis, C. D., and Gorovsky, M. A. (1999) Phosphorylation of linker histone H1 regulates gene expression in vivo by mimicking H1 removal. *Mol Cell* **4**, 641–647.
4. Zhang, Y. and Reinberg, D. (2001) Transcription regulation by histone methylation: interplay between different covalent modifications of the core histone tails. *Genes Dev* **15**, 2343–2360.
5. Gehring, M. and Henikoff, S. (2007) DNA methylation dynamics in plant genomes. *Biochem Biophys Acta* **1769**, 276–286.
6. Gilmour, D. S. and Lis, T. J. (1984) Detecting protein-DNA interactions in vivo: Distribution of RNA polymerase on specific bacterial genes. *Proc Natl Acad Sci USA* **81**, 4275–4279.
7. Dedon, P. C., Soults, J. A., Allis, C. D., and Gorovsky, M. A. (1991) A simplified formaldehyde fixation and immunoprecipitation technique for studying protein–DNA interactions. *Anal Biochem* **197**, 83–90.
8. Gendrel, A.-V., Lippman, Z., Yordan, C., Colot, V., and Martienssen, R. A. (2002) Dependence of heterochromatic histone H3 methylation patterns on the Arabidopsis gene *DDM1*. *Science* **297**, 1871–1873.
9. Gendrel, A.-V., Lippman, Z., Martienssen, R. A., and Colot, V. (2005) Profiling histone modification patterns in plants using genomic tiling microarrays. *Nat Methods* **2**, 213–218.
10. Makarevich, G., Leroy, O., Akinci, U., Schubert, D., Clarenz, O., Goodrich, J., Grossniklaus, U., and Köhler, C. (2006) Different Polycomb group complexes regulate common target genes in Arabidopsis. *EMBO Rep* **7**, 947–952.
11. Schönrock, N., Bouveret, R., Leroy, O., Borghi, L., Köhler, C., Gruissem, W., and Hennig, L. (2006) Polycomb-group proteins repress the floral activator AGL19 in the FLC-independent vernalization pathway. *Genes Dev* **20**, 1667–1678.
12. Villar, C. B. R., Erilova, A., Makarevich, G., Trösch, R., and Köhler, C. (2009) Control of *PHERES1* imprinting in *Arabidopsis* by direct tandem repeats. *Mol Plant* **2**, 654–660.
13. O'Neill, L. P. and Turner, B. M. (2003) Immunoprecipitation of native chromatin: NChIP. *Methods* **31**, 76–82.
14. Kurdistani, S. K. and Grunstein, M. (2003) In vivo protein-protein and protein-DNA crosslinking for genome-wide binding microarray. *Methods* **31**, 90–95.
15. Fujita, N., Jaye, D. L., Kajita, M., Geigerman, C., Moreno, C. S., and Wade, P. A. (2003) MTA3, a Mi-2/NuRD complex subunit regulates an invasive growth pathway in breast cancer. *Cell* **113**, 207–219.
16. Nowak, D. E., Tian, B., and Brasier, A. R. (2005) Two-step crosslinking method for the identification of NF-kappa B gene network by chromatin immunoprecipitation. *BioTechniques* **39**, 715–725.
17. Chua, Y. L., Mott, E., Brown, A. P. C., MacLean, D., and Gray, J. C. (2004) Microarray analysis of chromatin-immunoprecipitated DNA identifies specific regions of tobacco genes associated with acetylated histones. *Plant J* **37**, 789–900.
18. Zhang, X., Clarenz, O., Cokus, S., Bernatavichute, Y. V., Pellegrini, M., Goodrich, J., and Jacobsen, S. E. (2007) Whole-genome analysis of histone H3 lysine 27 trimethylation in Arabidopsis. *PLoS Biol* **5**, e129.
19. Jothi, R., Cuddapah, S., Barski, A., Cui, K., and Zhao, K. (2008) Genome-wide identification of in vivo protein–DNA binding sites from ChIP-seq data.*Nucleic Acids Res* **36**, 5221–5231.
20. Haring, M., Offermann, S., Danker, T., Horst, I., Peterhänsel, C., and Stam, M. (2007) Chromatin immunoprecipitation: Optimization, quantitative analysis and data normalization. *Plant Methods* **3**, 11.

Chapter 28

Immunocytological Analysis of Chromatin in Isolated Nuclei

Penka Pavlova, Federico Tessadori, Hans J. de Jong, and Paul Fransz

Abstract

All cells in a multicellular organism have the same genetic constitution, yet their appearance and function may differ enormously, due to differences in the nuclear program. Central in the establishment of this cell diversity are epigenetic marks, which are largely based on covalent modifications of histones and methylated cytosine residues in the DNA sequence. The study of these epigenetic factors in individual cells requires the microscopic visualization of chromatin components. Here we describe a number of protocols to study chromatin in isolated nuclei.

Key words: Immunodetection, chromatin, isolated nuclei, heterochromatin quantification, RHF.

1. Introduction

The inheritance of biological traits involves not only the transfer of genetic information in the form of DNA, but also epigenetic information, which is encrypted in a DNA component, methylated cytosine, and in non-DNA components such as histone modifications, histone variants, nonhistone proteins, and RNA. Chromatin comprises both genetic and epigenetic information. Its structure determines at which stage and in which cell a gene will be expressed. This concept entails a high flexibility of chromatin with respect to molecular composition, biochemical modification, and physical organization. Indeed, chromatin is

L. Hennig, C. Köhler (eds.), *Plant Developmental Biology*, Methods in Molecular Biology 655,
DOI 10.1007/978-1-60761-765-5_28, © Springer Science+Business Media, LLC 2010

highly dynamic, showing continuous changes at different levels of organization. In addition, chromatin occupies various subdomains in the nucleus, each with specific functions (1, 2). The past few years have witnessed major breakthroughs in the elucidation of epigenetic mechanisms controlling gene regulation and heterochromatin formation (3). For example, an increasing number of modified residues have been established including acetylation, methylation, phosphorylation, ubiquitylation, and sumoylation (for reviews *see* 4, 5), all of which contribute to the variation of the epigenetic information laid down at the surface of the nucleosome. The number of modulations is extended by the level of posttranscriptional modifications. For example, lysines can become mono-, di- or tri-methylated, while acetylation can occur up to the tetra-level. It follows that a group of residues in one tail can greatly increase the variation of histone modifications. Indeed, using chromatography and high-resolution tandem mass spectrometry, Garcia et al. (6) characterized over 150 different modified isoforms of the human histone H3.2. The authors further predicted the existence of many hundreds of differently modified forms of histone H3 in human cells. Understanding the dynamics of these modifications will greatly increase our understanding of epigenetic gene regulation. Fluorescence microscopy has proven a valuable tool to study chromatin dynamics. In particular, immunolabeling with specific antibodies or the use of GFP-tagged chromatin proteins are the major techniques. Since posttranscriptionally modified proteins cannot be detected in living cells with fluorescently tagged protein fusions, immunolabeling on fixed nuclei is the recommended technique to monitor modified histones and methylated DNA. This technique requires the recognition of epitopes by specific antibodies via physical contact and therefore a good penetration of the biological material for antibodies. Plant cells in tissue context are held together by a matrix of cell-wall material that is impermeable to macromolecules such as DNA probes or antibodies. Consequently, in order to investigate chromatin by immunolabeling, the antibodies have to get access to the nucleus. This can be achieved by (i) perforation of cell walls, (ii) sectioning of the tissue, and (iii) isolation of nuclei. The latter is the fastest method but the tissue context is lost. Hence, if a rapid examination is required and cellular information is less important, the isolation of nuclei is the method of choice. Here we describe different protocols to examine chromatin of isolated nuclei in the model plant Arabidopsis, including immunolabeling and heterochromatin quantification.

2. Materials

2.1. Isolation of Arabidopsis *Nuclei*

2.1.1. Formaldehyde Fixation of Nuclei for Spreading

1. Razor blade.
2. Glass Petri dishes.
3. Nylon mesh (pore size 100 μm).
4. PBS buffer: 137 mM NaCl, 2.7 mM KCl, 10 mM Na_2HPO_4, and 2 mM KH_2PO_4, pH 7.0.
5. Formaldehyde (4%) in PBS buffer (*see* **Note 1**).
6. Nuclei isolation buffer (NIB): 500 mM sucrose, 100 mM KCl, 10 mM Tris-HCl pH 9.5, 10 mM EDTA, 4 mM spermidine, 1 mM spermine, and 0.1% v/v 2-mercaptoethanol. Add 2-mercaptoethanol just before use. NIB can be stored at –20°C for several months.
7. 4',6-diamidino-2-phenylindole (DAPI) (1 μg/mL) in Vectashield mounting medium (Vector Laboratories, Burlingame, CA, USA).

2.1.2. Formaldehyde Fixation of Nuclei for Squashing

1. Microscopic slides.
2. Dissection needles (Omnilabo International, Breda, The Netherlands).
3. Fine forceps.
4. Small Petri dishes.
5. Moist chamber (lockable box with wet tissues at the bottom).
6. Liquid N_2.
7. Formaldehyde (4%) in PBS buffer (*see* **Note 1**).
8. Cell wall-degrading enzyme mixture: 0.25% (w/v) Cellulase R10 (Yakult, Tokyo, Japan,) 0.25% (w/v) Pectinase (Sigma,) and 0.25% (w/v) Pectolyase (Sigma) Haarlem, the Netherlands) in PBS. A 10× concentrated stock of enzyme mixture can be stored at –20°C.
9. DAPI (1 μg/mL) in Vectashield mounting medium (Vector Laboratories, Burlingame, CA, USA).

2.1.3. Carnoy Fixation of Nuclei for Spreading

1. Microscopic slides.
2. Dissection needles.
3. Small Petri dishes.
4. Fine forceps.

5. Moist chamber for enzymatic digestion.
6. Heating block at 42°C.
7. Freshly prepared ice-cold Carnoy's fixative: 3:1 ratio of ethanol:glacial acetic acid.
8. Citrate buffer: 10 mM sodium citrate/citric acid, pH 4.5.
9. Cell wall-degrading enzyme mixture: 0.6% (w/v) Cellulase R10 (Yakult, Tokyo, Japan) and 0.25% (w/v) Macerozyme R10 (Duchefa, Haarlem, the Netherlands) in citrate buffer. A 10× concentrated stock solution can be stored at –20°C.
10. Acetic acid (50%).
11. DAPI (1 μg/mL) in Vectashield mounting medium (Vector Laboratories, Burlingame, CA, USA).

2.2. Immunolabeling

2.2.1. In Situ Immunolabeling of DNA Modifications (Cytosine Methylation)

1. Heating block at 80°C.
2. Moist chamber.
3. Interphase nuclei preparations from either Carnoy's-fixed or formaldehyde-fixed isolated nuclei on microscopic slides.
4. PBS buffer: 137 mM NaCl, 2.7 mM KCl, 10 mM Na_2HPO_4, and 2 mM KH_2PO_4, pH 7.0.
5. SSC buffer (2×): 300 mM NaCl and 30 mM Na_3 citrate, pH 7.0.
6. Formaldehyde (4%) in PBS (*see* **Note 1**).
7. Bovine serum albumin (BSA) (2%) in PBS.
8. Hybridization buffer (HB50): 50% deionized formamide, 2× SSC, and 50 mM sodium phosphate pH 7.0.
9. Mouse antibody against 5-methyl-cytosine (anti-5 mC, Eurogentec, Seraing, Belgium).
10. Rabbit antimouse~FITC antibody (Jackson Immunoresearch, Soham, UK).
11. DAPI (1 μg/mL) in Vectashield mounting medium (Vector Laboratories, Burlingame, CA, USA).
12. DNAse-free RNAse: 100 μg/mL in 2× SSC.
13. Ethanol series (70%, 90%, 100%).

2.2.2. In Situ Immunolabeling of Chromatin Proteins (Histone H3 Dimethylated at Lysine 9)

1. Moist chamber.
2. Formaldehyde-fixed nuclei on microscopic slides.
3. PBS buffer: 137 mM NaCl, 2.7 mM KCl, 10 mM Na_2HPO_4, and 2 mM KH_2PO_4, pH 7.0.

4. Formaldehyde (2%) in PBS.
5. Primary antibody such as rabbit antihistone H3 dimethylated at lysine 9 (K9) (Millipore; Billerica; MA; USA).
6. Triton X-100.
7. Bovine serum albumin (BSA) (2%) in PBS.
8. Donkey antirabbit~Cy3 (Jackson Immunoresearch) or any other secondary antibody that recognizes rabbit antibodies.
9. DAPI (1 μg/mL) in Vectashield mounting medium (Vector Laboratories, Burlingame, CA, USA).

3. Methods

Like for most cytological methods, the preparation of biological material is crucial for a successful microscopic examination. Here we present different protocols to fix and isolate nuclei for cytogenetic purposes. The first step is the preservation of structures, for which two types of fixation are available: cross-linking and ethanol-acid dehydration (e.g., Carnoy fixative,) Cross-linking is required to investigate protein and RNA targets, while Carnoy fixation is preferred for the analysis of large DNA complexes, such as chromosomes. Depending on the target tissue, different approaches can be applied. We present a spreading protocol for leaf tissue and a squash protocol for meiotic cells. The squashing method provides some preservation of the anther structure, which is convenient for the identification of the meiotic stages (7). The isolation of Carnoy-fixed nuclei is generally used for karyotyping (8, 9) and quantification of heterochromatin content (10, 11).

3.1. Isolation of Nuclei

3.1.1. Spreading Protocol for Formaldehyde-Fixed Nuclei

For the detection of chromatin and nuclear proteins, isolated nuclei need to be fixed in formaldehyde in order to preserve the targeted antigenic epitopes. Additionally, this protocol preserves the 3D structure of the isolated nuclei. To prepare interphase nuclei from adult aerial tissues, grow plants under standard conditions in the greenhouse. Unless specifically interested in large, elongated leaves or other aerial parts, always prefer young healthy leaves (up to 10-mm long), as they are easy to chop and have a higher yield of nuclei (**Fig. 28.1 C, D**).

1. Dissect ~0.5 g of young leaves and fix them in 10 mL of ice-cold 4% formaldehyde in PBS buffer under vacuum for 20 min. (*see* **Note 2**)
2. Wash the sample in ice-cold PBS buffer three times for 5 min.

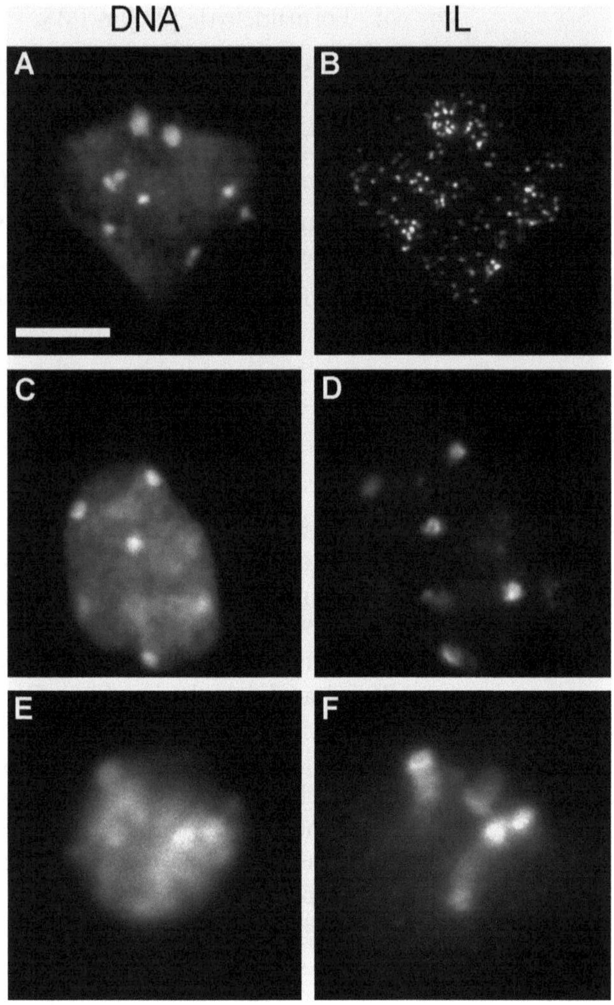

Fig. 28.1. **Examples of Immunolabeling (IL) techniques.** In **A**, **C,** and **E**, DNA is visualized by DAPI staining. **(A–B)** Carnoy-fixed and spread interphase nucleus. Note the intensely stained chromocenters **(A)**. 5-methyl-cytosine immunolabeling (DNA methylation). Note that the fluorescent signals cluster at the periphery of the chromocenters **(B)**. **(C–D)** Formaldehyde-fixed interphase nucleus. DAPI-stained nucleus **(C)** and immunolabeling of histone H3 at dimethylated lysine 9 **(D)**. **(E–F)** Pachytene nucleus prepared by formaldehyde fixation and liquid nitrogen squash. DAPI-stained nucleus **(E)** and immunosignals for H3K9me2 **(F)**. Bar= 5 μM.

3. Transfer the material to a Petri dish and chop it with a razor blade in 500 μL of NIB buffer on ice until a fine suspension is obtained.
4. Filtrate the suspension through nylon mesh (pore size 100 μm) into a new tube (*see* **Note 3**).
5. Centrifuge at ~500×*g* in a standard table-top mini-centrifuge for 3 min at 4°C.
6. Discard the supernatant, add 50 μL of NIB, and gently resupend the pellet.

7. Pipet 2 μL of the nuclear sample onto a clean slide. Add 3 μL of DAPI in Vectashield and mix gently with the pipette tip. Cover with a 24 × 24-mm coverslip.
8. Examine for yield, density of nuclei, and presence of debris under a fluorescence microscope using a UV filter set. The nuclei should appear as bright blue, round, or elongated structures. Chromocenters (intensely stained chromatin regions) and nucleolus (weakly stained region) should be clearly identifiable. Chloroplasts debris appears red under UV excitation. At 100× magnification the field should contain at least up to 5 nuclei. If the density and the purity are satisfactory, proceed to Step 11.
9. To increase the concentration and purification of nuclei, centrifuge at ~500×*g* for 3 min at 4°C.
10. Discard the supernatant and resuspend the pellet with nuclei in ~20 μL of NIB buffer.
11. Pipet 2 μL of the suspension onto a microscopic slide. Spread the droplet of sample with the pipette tip and air-dry at 4°C.
12. The slides can be stored at 4°C for several months.

3.1.2. Squashing Protocol for Formaldehyde-Fixed Nuclei

This protocol makes use of formaldehyde-fixed Arabidopsis inflorescences. The "squashing" step and the use of liquid N_2 are needed to "spread" the meiotic chromosomes for better visualization and to adhere them to the glass slide (**Fig. 28.1 E,F**).

1. Fix young inflorescences in 10 mL of ice-cold 4% formaldehyde in PBS buffer for 20 min (*see* **Note 2**).
2. Wash the sample in ice-cold PBS buffer three times for 5 min.
3. Incubate three inflorescences in ~1 mL of cell wall-degrading enzyme mixture for 1 h in a moist chamber at 37°C (*see* **Note 4**).
4. Take an inflorescence in a Petri dish with water and place it under the stereomicroscope. The yellowish flower buds and the largest white flower bud contain anthers with pollen and microspores or tetrad stages, respectively. In decreasing size order, the second white flower bud contains pollen mother cells (PMCs) in meiosis II. PMCs at prophase I, displaying the diakinesis and pachytene stages, can be found in the third or fourth largest white flower bud.
5. Place the selected flower bud on a clean microscopic slide. Remove excess of fluid (by capillary forces of a fine forceps or a glass capillary tube). Keep ~10 μL of liquid.
6. Dissect the anthers from the flower bud with needles. It is important to have the anthers separated from the rest of

the flower bud to obtain satisfactory visualization of the meiotic stages. The region of the slide where the anthers are located may be marked with a diamond pen, for example, to facilitate subsequent microscopic observation.

7. Cover the liquid droplet with a coverslip and tap it gently with a needle on top of the anthers. This will release the meitoic cells from the tissue and "spread" their content on the slide. Tapping too gently will result in insufficient squashing of the tissue, while tapping too hard will damage the cells.
8. Snap-freeze the slide in liquid N_2 for a few seconds.
9. "Flip" the coverslip off with a razor blade and immediately transfer to PBS at 37°C for ~30 s.
10. Air-dry the slide at RT.
11. The slides are examined by adding 8 μL of DAPI in Vectashield and covering with a 24 × 24-mm coverslip. They can be stored in a clean box at 4°C. Allow stored slides adapting to room temperature before further use (*see* **Note 5**).

3.1.3. Spreading Protocol for Carnoy's Fixed Nuclei

In this protocol, the tissue is fixed using Carnoy's fixative. This results in spreading of the nuclei (**Fig. 28.1 A, B**). The 2D samples prepared with this protocol are particularly useful for heterochromatin quantification and karyotyping. However, since the antigenic epitopes are not well preserved, Carnoy fixation is not advisable for immunolabeling of chromatin proteins. This protocol can be used to prepare nuclear spreads of leaf mesophyll, root tissue and for pachytene chromosome preparation from immature flower buds. The example given here is for leaf mesophyll nuclei.

1. Fix tissue in freshly prepared ice-cold Carnoy's fixative at least three times for 30 min or until the tissue becomes white and the fixative remains clear. For leaf mesophyll nuclei spreads, young leaves are preferred (up to 10-mm long). Fixed material can be stored in fixative or 70% ethanol at 4°C or –20°C for several months.
2. Wash fixed leaves in distilled water two times for 10 min in a small Petri dish.
3. Replace water by citrate buffer and wash three times for 5 min.
4. Incubate 2–3 leaves in ~500 μL of cell-wall degrading enzyme mixture for 30 min in a moist chamber at 37°C (*see* **Note 4**).
5. Take a leaf with a needle and place it on some parafilm under the stereomicroscope.

6. Remove excess liquid with a pipette and add ~5 μL of distilled water. The water on the parafilm should form a "bead", with the leaf floating inside.
7. Tap the leaf with a dissection needle until a fine, milky suspension has formed. If large leaf fragments subsist, extend the enzyme incubation time. Be careful not to pierce the parafilm with the dissection needle. The obtained fine suspension can be used right away (Step 8) or pipetted into a microcentrifuge tube and stored on ice or at 4°C overnight.
8. Pipet 1–5 μL (depending on the concentration of the suspension) of the suspension to a clean microscopic slide. Add 20 μL of 50% acetic acid to the suspension. Place the slide on a heating block at 42°C and spread the drop by careful stirring with a needle for 30 s without touching the slide surface. Add ~20 μL of 50% acetic acid and continue stirring carefully for an additional 30 s. During this step, the acid-soluble cytoplasmic components are removed; it is important to work carefully to obtain "clean" nuclei.
9. Precipitate the nuclei by carefully pipetting ~100 μL of Carnoy's fixative around the sample drop and wait until the fixative covers the whole slide. While doing this, gently tilt the slide to keep the sample droplet in the middle of the slide. When the whole slide is covered, discard the liquid in excess by tilting.
10. Air-dry the preparation with a hairdryer. Alternatively, after Step 9, the slide can be rinsed in PBS (2 min), postfixed in 1% formaldehyde in PBS (2 min,) and rinsed again in PBS (2 min) before air-drying (*see* **Note 6**).
11. The slides may be examined by adding 8 μL of DAPI in Vectashield and covering with a 24 × 24-mm coverslip. The nuclei should appear as bright blue, round or elongated structures. Inside the nuclei, the chromocenters (compact chromatin regions, intensely stained) and the nucleolus (weaker staining) should be clearly identifiable. Preparations in which nuclei appear to be covered by cytoplasm are not suitable for FISH or DNA immunolabeling experiments, as penetration of the nuclei by the probes and the detection antibodies will be hindered.
12. Store the slides in a dust-free box at 4°C (*see* **Note 5**).

3.2. Immunolabeling

Immunolabeling of isolated nuclei can be established on nuclei that are fixed on a slide or on a nuclei suspension in a microcentrifuge tube. Immunodetection of nuclei fixed on a slide requires fewer manipulation steps, and the slide preparations can be stored for several months before immunodetection. However, when the nuclei are spread on a slide and dried, they become slightly flattened. The loss of 3D structure can be prevented by performing the immunodetection in a tube. Both protocols are described.

3.2.1. Immunolabeling of Methylated DNA on a Slide

The methylated cytosine, which is located in the major groove of the DNA helix, needs to be accessible for antibodies. Therefore, immunolabeling of methylated DNA requires a denaturation step. The epitope is a DNA constituent, not a peptide, so the plant material can be fixed in either Carnoy's fixative or formaldehyde.

1. Bake the slides (*see* **Section 3.1**) at 60°C for 30 min on a heating block.
2. Rinse in PBS two times for 5 min.
3. Postfix in 1% formaldehyde in PBS for 10 min at room temperature.
4. Rinse in PBS two times for 5 min.
5. Dehydrate successively in 70, 90, and 100% ethanol, each step for 1–2 min, and air-dry.
6. Pipet 20 μL of HB50 onto the slide, cover with a 24 × 24-mm coverslip and denature at 80°C for 2 min on a heating block.
7. Rinse the slides in ice-cold 70% ethanol for 2 min in a coplin jar. The coverslip should come off during the rinse.
8. Rinse the slides successively in ice-cold 90 and 100% ethanol, each step for 1–2 min, and air-dry.
9. Add 100 μL of 2% BSA, cover with a 24 50-mm coverslip and incubate at 37°C for 60 min in a moist chamber.
10. Rinse the slides three times for 5 min in PBS in a coplin jar.
11. Prepare a 1:50 dilution of mouse-anti-5 mC antibody in PBS.
12. Pipet 50 μL of the antibody solution onto the slide, cover with a 24 × 32-mm coverslip, and incubate in a moist chamber at 37°C for 30 min.
13. Remove coverslip carefully and rinse three times for 5 min in PBS.
14. Prepare a 1:500 dilution of rabbit antimouse~FITC antibody conjugate in TNB buffer.
15. Pipet 50 μL of the antibody solution onto the slide, cover with a 24 × 32-mm coverslip, and incubate in a moist chamber at 37°C for 30 min (*see* **Note 7**).
16. Rinse three times for 5 min in PBS.
17. Mount in Vectashield with 1 μg/mL DAPI.

3.2.2. Immunolabeling of Methylated DNA in a Tube

1. Carry out Steps 1–4 of the nuclei isolation protocol (*see* **Section 3.1.1**).

2. Collect the nuclei at ~500×*g* in a standard table-top mini-centrifuge for 2 min at 4°C.
3. Discard the supernatant and resuspend the nuclei in 20 μL of PBS.
4. Collect the nuclei at ~400×*g* for 2 min at 4°C.
5. Discard the supernatant and resuspend in 20 μL of HB50.
6. Repeat Steps 4 and 5.
7. Denature at 80°C for 4 min in a water bath.
8. Cool the suspension on ice and collect the nuclei at ~400×*g* for 2 min.
9. Discard the supernatant and resuspend the pellet in 20 μL of ice-cold PBS.
10. Repeat Step 8 and collect the nuclei at ~400×*g* for 2 min.
11. Discard the supernatant and resuspend in 40 μL of 1% BSA.
12. Incubate at 37°C for 30 min.
13. Add 40 μL of mouse anti-5 mC antibody at a 1:100 dilution in 2% BSA and incubate at 37°C for 30 min.
14. Collect the nuclei at ~400×*g* for 2 min.
15. Discard the supernatant and resuspend the nuclei in PBS.
16. Add 40 μL of donkey antimouse~Cy3 at a 1:500 dilution in 2% BSA and incubate at 37°C for 30 min
17. Collect the nuclei at ~400×*g* for 2 min.
18. Discard the supernatant and resuspend in 40 μL of 1:500 diluted donkey antimouse~Cy3 antibody conjugate in 1% BSA.
19. Incubate at 37°C for 30 min.
20. Collect the nuclei at ~400×*g* for 2 min.
21. Discard the supernatant and resuspend in 1× PBS
22. Mount 2 μL of the suspension on a microscopic slide and add DAPI in Vectashield.
23. Store the tube with the rest of the suspension at 4°C for not longer than one week.

3.2.3. Immunolabeling of Chromatin Proteins (e.g., H3K9me2) on a Slide

1. Postfix the slide preparations (*see* **Sections 3.1.1** or **3.1.2**) in 2% formaldehyde in PBS at room temperature for 30 min.
2. Rinse in PBS two times for 3 min.
3. Wash in freshly prepared 0.5% Triton X-100 PBS in a coplin jar.
4. Rinse in PBS two times for 3 min.

5. Incubate the slides in 1% BSA in PBS at 37°C for 30 min in a moist chamber. Use 100 μL per slide and cover with a 24 × 50-mm coverslip.
6. Rinse in PBS at room temperature for 5 min.
7. Prepare a 1:50 dilution of anti-H3K9me2 antibody in 1% BSA in PBS, pipet 50 μL onto each slide, cover with a 24 × 32-mm coverslip, and incubate in a moist chamber at 37°C for 2 h or at 4°C overnight.
8. Carefully remove the coverslip and rinse in PBS at room temperature three times for 20 min.
9. Prepare a 1:100 dilution of Donkey antirabbit~Cy3 antibody conjugate in 1% BSA in PBS, pipet 50 μL onto each slide, cover with a 24 × 32 mm coverslip, and incubate at 37°C for 30 min (*see* **Note 7**).
10. Rinse in PBS at room temperature three times for 5 min.
11. Mount in Vectashield with 1–2 μg/mL DAPI.

3.2.4. Immunolabeling of Chromatin Proteins (e.g., H3K9me2) in a Tube

1. Carry out Steps 1–4 of the nuclei isolation protocol (*see* **Section 3.1.1**).
2. Collect the nuclei at ~400×*g* for 2 min.
3. Discard the supernatant and resuspend the nuclei in 40 μL of 1% BSA.
4. Incubate at 37°C for at least 1 h.
5. Add 1 μL of undiluted rabbit anti-H3K9me2 antibody to the suspension and incubate at 37°C for 2 h.
6. Collect the nuclei at ~400×*g* for 2 min.
7. Discard the supernatant and resuspend in 40 μL of PBS
8. Repeat Step 6.
9. Discard the supernatant and resuspend in 40 μL of 1% BSA
10. Add 1 μL of undiluted goat antirabbit~Alexa^{488} antibody conjugate and incubate at 37°C for 30 min.
11. Collect the nuclei at ~400×*g* for 2 min.
12. Discard the supernatant and resuspend in PBS.
13. Mount 2 μL of the suspension on a microscopic slide and add DAPI in Vectashield.
14. Store the tube with the rest of the suspension at 4°C for not longer than one week.

3.3. Quantification of Heterochromatin in Arabidopsis Nuclei

In Arabidopsis, virtually the entire heterochromatic fraction of the genome – pericentromeric tandem repeats, inactive rDNA genes, and transposable elements – is accommodated in chromocenters. We have based the determination of the relative heterochromatin fraction (RHF) on this morphological property of the Arabidopsis nucleus. The RHF represents the fraction of a nucleus' total chromatin observed as heterochromatin (10, 12). Note that the RHF is a relative quantitative parameter and does not give information about the actual heterochromatin content. The RHF is only biologically relevant when different conditions, genetic backgrounds, or cell types are compared. The analysis is performed on flattened Carnoy-fixed nuclei, stained with DAPI and imaged with a CCD camera. Semiautomated measurement of the RHF can be achieved with interactive, freely available image analysis software such as *Object Image* (http://simon.bio.uva.nl/Object-Image/object-image.html) and *CHIAS* (http://www.bio.eng.osaka-u.ac.jp/cl/E_top.html). *See* **Note 8** for a protocol for RHF measurement with *Image Pro Plus* (Media Cybernetics Inc., Bethesda, USA) using a macro for (semi)automated analysis.

3.3.1. Acquisition of Images

1. Prepare slides with Carnoy's-fixed spread nuclei (*see* **Section 3.1.3**).
2. Stain slides with DAPI.(*see* **Note 9**)
3. Examine slides with the fluorescent microscope.
4. Photograph DAPI-stained nuclei with a CCD camera. Avoid overexposure of the nucleus.
5. Photograph at least 30 nuclei per sample.

3.3.2. Principle Steps for RHF Measurement

The measurement of the relative heterochromatin fraction is based on the fluorescence intensity of the DNA stain per area. Here we describe a standard protocol for the RHF measurement (*see* **Note 8**).

1. Determine the average fluorescence intensity of the nucleus (I_n).
2. Determine the background fluorescence intensity (I_b).
3. Subtract the background fluorescence intensity from the nucleus fluorescence intensity.
4. Determine the area of the nucleus (A_n).
5. Calculate the total nucleus fluorescence intensity: $A_n \times (I_n - I_b)$.
6. Repeat this procedure for each chromocenter present in the measured nucleus and sum up the values to obtain the average fluorescence intensity of the heterochromatin fraction (I_c).
7. Subtract the background fluorescence intensity from the chromocenter intensity.
8. Determine the total area of all chromocenters in the nucleus (A_c).

9. Calculate the total heterochromatin fluorescence intensity: $A_c \times (I_c - I_b)$.
10. Calculate RHF:

$$\mathrm{RHF} = \frac{A_c \times (I_c - I_b)}{A_n \times (I_n - I_b)}$$

4. Notes

1. 4% Formaldehyde is prepared by dissolving 4% paraformaldehyde (w/v) in PBS. Stir for ~30 min on a heating plate (~60°C) under a fume hood. Addition of a few droplets of NaOH can speed up the process. Paraformaldehyde can be stored at -80°C, but should be used quickly (within a couple of days) once thawed.
2. Formaldehyde fixation is accelerated under vacuum. In general, the fixation is established, when the tissue sinks.
3. Filters for microcentrifuge tubes are easily made by cutting the narrow end of a P1000 tip, heating it mildly on the flame of a bench top burner, and then pressing the hot tip onto the nylon mesh. Once cooled down, check that the mesh sticks to the tip.
4. The incubation time and the composition of the cell-wall degrading mixtures are an approximation. When using younger or older tissue, or tissue for which the composition of the cell wall is different (e.g., root tissue, flower buds, cultured cells), the incubation time and the composition of the enzymatic mixture should be adapted.
5. Slides may be stored at 4°C already mounted with Vectashield and DAPI and a coverslip. Before starting with Step 1 of any of the immunolabeling protocols described here, slide preparations should be rinsed twice for 5 min in PBS.
6. The standard spreading protocol may sometimes yield faint chromosome/nuclei preparations for unknown reasons. This becomes visible on slides only after denaturation at high temperature. This can be prevented by formaldehyde postfixation immediately after spreading in Carnoy's fixative.
7. The secondary antibody given here is an example. To enhance the intensity of the immunolabeling signal, an additional detection step may be added with an antibody raised against the secondary antibody.

8. Protocol to measure the RHF with *Image Pro Plus* (Media Cybernetics Inc., Bethesda, USA) using a macro for (semi)automated analysis.

   Program steps

   1. Open *Image Pro Plus* and run the RHF-macro (*see* below).
   2. Import a picture of a nucleus.
   3. Open automatically MS Excel.
   4. Open a file.
   5. Turn the image to grey scale.
   6. Duplicate the image.
   7. Mask the chromocenters with the Equilize command.
   8. Choose the Best Fit option to enhance the contrast and dynamic range of the active image, by stretching the histogram between the minimal value black (0) and maximal value white (255).
   9. Apply LUT (Lookup Table) command to interpret the image with enhanced contrast.
   10. Blur the chromocenters with a median filter (7 × 7 kernels).
   11. Use threshold and segmentation to find the region of interest (the nucleus in this step) and outline it.
   12. Save the outline.
   13. Measure within the outline selected parameters such as:
       - area (area of object)
       - area polygon (area included in the polygon defining the perimeter of the object's outline)
       - count "adjusted" (size weighted object for nuclei with irregular form)
       - density mean (average intensity of the object)
       - density sum (sum of intensity inside the object)
       - heterogeneity (fraction of pixels that deviate more than 10% from the average intensity
       - perimeter (length of the object's outline)
       - roundness ({perimeter^2}/{4*pi*area})
   14. Threshold and find the edges of the chromocenters (CCs).
   15. Perform a segmentation step and draw the outlines around the chromocenters and number them from the top left corner to the bottom right corner.
   16. Save the outline.

17. Measure the same parameters as for the nucleus on the CCs.
18. Export automatically the data to Microsoft Excel or the clipboard together with the names of all analyzed files within a folder.
19. To measure the RHF, only the density sum parameter of the nucleus and the CCs is required.
20. Sum the density sum of all the CCs and divide by the density sum of the nucleus.
21. The rest of the parameters are informative in a more detailed analysis of the nuclei and their chromocenters.

Macro script

1. Option Explicit
2. Sub nucleifast’ ()
3. Dim filename3 As String*100
4. Dim imname As String
5. Dim imdir As String
6. Dim lenname As Integer
7. Dim outlinename As String
8. Dim i As Integer
9. Dim infname As String*255
10. Dim iname2 As String
11. Dim more As Integer
12. Dim imdir2 As String
13. Dim file As Integer
14. Beginning:
15. ret= ipappcloseall()
16. file = IpStGetName(“Find the file you want to process”,“c:\ ”, “*.tif”, Filename3)
17. If file = 0 Then
18. MsgBox (“please select a file to analyze”)
19. GoTo Beginning
20. End If
21. imdir = CurDir(filename3)
22. imdir2 = imdir
23. i=0
24. retry:
25. more = ipstsearchdir(imdir2, “*.tif”,i, infname)
26. Debug.Print more
27. If more = 2 Then

28. i=i+1
29. GoTo retry
30. End If
31. ret = IpAppRun("C:\Program Files\Microsoft Office\Office11\excel.exe", RUN_MINIMIZED, RUN_AUTOCLOSE)
32. ret = IpDde (DDE_OPEN, "excel", "sheet1")
33. ret = IpDde(DDE_SET, "row", "1")
34. ret = IpDde(DDE_SET, "col", "1")
35. Do While more = 1
36. imname = Dir (infname)
37. imdir = "c:\Documents and Settings\Penka Pavlova\"
38. lenname =Len(imname)
39. Debug.Print lenname
40. outlinename = Left(imname,lenname-4)
41. ret = IpWsLoad(INFNAME,"tif")
42. ret = IpWsConvertImage(IMC_GRAY, CONV_SCALE, 0, 0, 0, 0)
43. ret = IpWsDuplicate()
44. ret = IpHstEqualize(EQ_BESTFIT)
45. ret = IpLutApply()
46. ret = IpFltMedian(7, 5)
47. ret = IpBlbShow(1)
48. ret = IpBlbSetAttr(BLOB_AUTORANGE, 0)
49. ret = IpBlbEnableMeas(BLBM_ALL, 0)
50. ret = IpBlbSetAttr(BLOB_CLEANBORDER,1)
51. ret = IpBlbEnableMeas(BLBM_AREA, 1)
52. ret = IpBlbEnableMeas(BLBM_AREAPOLY, 1)
53. ret = IpBlbEnableMeas(BLBM_SIZECOUNT, 1)
54. ret = IpBlbEnableMeas(BLBM_DENSITY, 1)
55. ret = IpBlbEnableMeas(BLBM_DENSSUM, 1)
56. ret = IpBlbEnableMeas(BLBM_ HETEROGENEITY, 1)
57. ret = IpBlbEnableMeas(BLBM_PERIMETER, 1)
58. ret = IpBlbEnableMeas(BLBM_ROUNDNESS, 1)
59. ret = IpTemplateMode(1)
60. ret = IpSegShow(1)
61. ret = IpSegSetRange(0,100, 255)

62. ret = IpTemplateMode(0)
63. ret = IpBlbCount()
64. ret = IpBlbUpdate(0)
65. ret = IpTemplateMode(1)
66. ret = IpBlbHideObject(0,0,0)
67. ret = IpTemplateMode(0)
68. ret = IpBlbUpdate(4)
69. ret = IpBlbSaveOutline(imdir2 & "\" & outlinename & "nuclei.scl")
70. ret = IpBlbSaveOutline(imdir2 & "\" & outlinename & "nuclei.out")
71. ret = IpSegCreateMask(2,0,1)
72. ret = IpSegShow(0)
73. ret = IpBlbLoadOutline(imdir2 & "\" & outlinename & "nuclei.scl")
74. ret = IpAppSelectDoc(1)
75. ret = IpOpImageLogic(3, OPL_AND, 1)
76. ret = IpBlbLoadOutline(imdir2 & "\" & outlinename & "nuclei.scl")
77. ret = IpBlbMeasure()
78. ret = IpDde (DDE_OPEN, "excel", "sheet1")
79. ret = IpBlbSaveData ("", S_APPEND+S_HEADER+S_Y_AXIS+S_DDE)
80. ret = IpAppSelectDoc(4)
81. ret = IpWsDuplicate()
82. ret = ipfltsobel()
83. ret = IpFltShow(0)
84. ret = IpFltClose(MORPHO_2 × 2SQUARE, 10)
85. ret = IpFltErode(MORPHO_2 × 2SQUARE, 3)
86. ret = IpBlbShow(1)
87. ret = IpBlbSetAttr(BLOB_AUTORANGE, 1)
88. ret = IpBlbSetAttr(BLOB_BRIGHTOBJ, 1)
89. ret = IpAppSelectDoc(3)
90. ret = IpBlbCount()
91. ret = IpTemplateMode(1)
92. ret = IpBlbHideObject(0,0,0)
93. ret = IpTemplateMode(0)
94. ret = IpBlbUpdate(4)
95. ret = ipappmenuselect(1, 5 , "", DLG_Menu_coord)

96. ret = Ipmacrostop(“Draw unfound chromocentres and press OK”, 0)
97. ret = IpBlbUpdate(4)
98. ret = IpBlbSaveOutline(imdir2 & “\” & outlinename & “chromocentre.scl”)
99. ret = IpBlbSaveOutline(imdir2 & “\” & outlinename & “chromocentre.out”)
100. ret = IpBlbMeasure()
101. ret = IpDde (DDE_OPEN, “excel”, “sheet1”)
102. ret = IpBlbSaveData (“”,S_APPEND+S_Y_AXIS+S_DDE)
103. ret = IpDde(DDE_SET, “row_inc”, “1”)
104. ret = IpDde(DDE_PUT, “r[-4]c"”, Chr$(34) & INFname & Chr$(34))
105. ret = IpDde(DDE_SET, “append”, “1”)
106. i = i+1
107. more = ipstsearchdir(imdir2, “*.tif”,i, infname)
108. ret= ipappcloseall()
109. Loop
110. ret = IpAppRestore()
111. End Sub

9. Although DAPI has a binding preference for AT-rich DNA, we found no significant differences in RHF values between DAPI-stained nuclei, propidium iodide-stained nuclei or Sytox Green-stained nuclei.

References

1. Fransz, P., ten Hoopen, R., and Tessadori, F. (2006) Composition and formation of heterochromatin in Arabidopsis thaliana. *Chromosome Res* **14**, 71–82.
2. Lamond, A. I. and Earnshaw, W. C. (1998) Structure and function in the nucleus. *Science* **280**, 547–553.
3. Mager, J. and Bartolomei, M. S. (2005) Strategies for dissecting epigenetic mechanisms in the mouse. *Nat Genet* **37**, 1194–1200.
4. Berger, S. L. (2007) The complex language of chromatin regulation during transcription. *Nature* **447**, 407–412.
5. Margueron, R., Trojer, P., and Reinberg, D. (2005) The key to development: Interpreting the histone code? *Curr Opin Genet Dev* **15**, 163–176.
6. Garcia, B. A., Shabanowitz, J., and Hunt, D. F. (2007) Characterization of histones and their post-translational modifications by mass spectrometry. *Curr Opin Chem Biol* **11**, 66–73.
7. Mylne, J. S., Barrett, L., Tessadori, F., Mesnage, S., Johnson, L., Bernatavichute, Y. V., Jacobsen, S. E., Fransz, P., and Dean, C. (2006) LHP1, the Arabidopsis homologue of HETEROCHROMATIN PROTEIN1, is required for epigenetic silencing of FLC. *Proc Natl Acad Sci USA* **103**, 5012–5017.
8. Fransz, P., Armstrong, S., Alonso-Blanco, C., Fischer, T. C., Torres-Ruiz, R. A., and Jones, G. (1998) Cytogenetics for the model system *Arabidopsis thaliana*. *Plant J* **13**, 867–876.

9. Lysak, M. A., Fransz, P. F., Ali, H. B., and Schubert, I. (2001) Chromosome painting in *Arabidopsis thaliana*. *Plant J* **28**, 689–697.
10. Soppe, W. J., Jasencakova, Z., Houben, A., Kakutani, T., Meister, A., Huang, M. S., Jacobsen, S. E., Schubert, I., and Fransz, P. F. (2002) DNA methylation controls histone H3 lysine 9 methylation and heterochromatin assembly in Arabidopsis. *EMBO J* **21**, 6549–6559.
11. Tessadori, F., van Driel, R., and Fransz, P. (2004) Cytogenetics as a tool to study gene regulation. *Trends Plant Sci* **9**, 147–153.
12. Tessadori, F., Schulkes, R. K., van Driel, R., and Fransz, P. (2007) Light-regulated large-scale reorganization of chromatin during the floral transition in Arabidopsis. *Plant J* **50**, 848–857.

Chapter 29

Bisulphite Sequencing of Plant Genomic DNA

Ernst Aichinger and Claudia Köhler

Abstract

DNA methylation is a prominent epigenetic mark and extensively found within plant genomes. It has two major roles– first, defending the genome against invasive DNA and second, regulation of gene expression. Since the first report of 5-methylcytosine found in wheat germ, many improvements in detection of methylated cytosine residues have been made and genome-wide methylation maps for *Arabidopsis thaliana* are now available. The development of fast, reproducible, and single-base pair resolving detection methods for DNA methylation at defined loci advanced our understanding of the establishment and maintenance of DNA methylation patterns. Bisulphite conversion of unmethylated cytosine residues followed by detection methods such as sequencing of distinct loci has become accepted as the gold standard for detecting 5-methylcytosines. Treatment of single-stranded DNA with bisulphite under acidic conditions leads to the conversion of cytosine residues to uracil whereas 5-methylcytosine is less sensitive and remains unchanged. Here, a detailed protocol for bisulphite conversion, primer design, and optimization of PCR conditions is given. Specific requirements for plant DNA are discussed.

Key words: DNA methylation, 5-methylcytosine, bisulphite conversion, bisulphite sequencing, plant epigenetics.

1. Introduction

The first detection of 5-methylcytosine in DNA has been reported from tubercle bacillus in 1925, and in plants, from wheat germ in 1950 (1, 2). In the meantime, DNA methylation has been found in many prokaryotes and most eukaryotes. In prokaryotes, DNA methylation is found at cytosine and adenine residues and is involved in host defense against invasive DNA such as viral DNA and DNA mismatch repair (3). In contrast to prokaryotes, DNA methylation in eukaryotes is almost exclusively found at the

L. Hennig, C. Köhler (eds.), *Plant Developmental Biology*, Methods in Molecular Biology 655,
DOI 10.1007/978-1-60761-765-5_29, © Springer Science+Business Media, LLC 2010

carbon-5 position of cytosine residues and is usually associated with transcriptional regulation (4). In plants, DNA methylation is involved in the control of gene expression and the defense against invasive DNA such as transposable elements (5). More recently, a role in repression of cryptic transcription has been proposed (6). While DNA methylation in mammals typically occurs at symmetric CG-residues, in plants methylation occurs symmetrically at CG and CNG (N is any residue) and asymmetrically at CHH (H is A, C, or T) sequences. Consequently, plants have more potential sites of DNA methylation. Methyltransferases are enzymes involved in the establishment or maintenance of methylation, and mutations within the encoding genes cause developmental abnormalities in animals as well as in plants (7). Signals generated from the RNAi machinery or histone marks are necessary for the dynamic regulation of DNA methylation (8, 9). In plants, DNA glycosylases are responsible for active demethylation using a mechanism for DNA base-excision repair (10). Mutants defective in the DNA glycosylase DEMETER (DME) exhibit hypermethylation of imprinted loci (11) and related proteins of the DME family protect genes against deleterious hypermethylation (12).

In order to understand the dynamics of establishment, maintenance, and removal of DNA methylation as well as the biological function of DNA methylation in different cell types, during development and in response to environmental changes, fast, reproducible, accurate as well high-resolving detection methods are required. Detection of DNA methylation by bisulphite conversion followed by DNA sequencing, generally termed bisulphite sequencing, is the gold standard method and widely used within genome-wide profiling approaches or for sequencing of distinct loci.

Bisulphite mediates deamination of cytosine to form uracil. Since the methyl substitution at position 5 of cytosines makes the amino group at position 4 resistant to the bisulphite deamination, this reaction can be used to discriminate 5-methylcytosine residues in DNA from those of cytosine (13). The deamination of cytosine by sodium bisulphite involves four critical steps: (i) denaturation of the DNA into single strands; (ii) reaction of bisulphite with the 5–6 double bond of cytosine to give a cytosine sulphonate derivative; (iii) hydrolytic deamination of the resulting cytosine–bisulphite derivative to give a uracil sulphonate derivative; and (iv) removal of the sulphonate group by a subsequent alkali treatment to give uracil (14). Such treated DNA can be used for further analysis by different methods, e.g., deep sequencing, DNA microarray hybridization, or target-specific PCR amplification. Because the conversion of cytosines to uracils creates noncomplementary strands (i.e., uracils opposite guanines), DNA must be amplified with separate pairs of primers that are specific for either the top or bottom DNA strands. Following PCR amplification, the uracils are amplified as thymines, whereas methylated cytosine residues are amplified as cytosines (*see* **Fig. 29.1**).

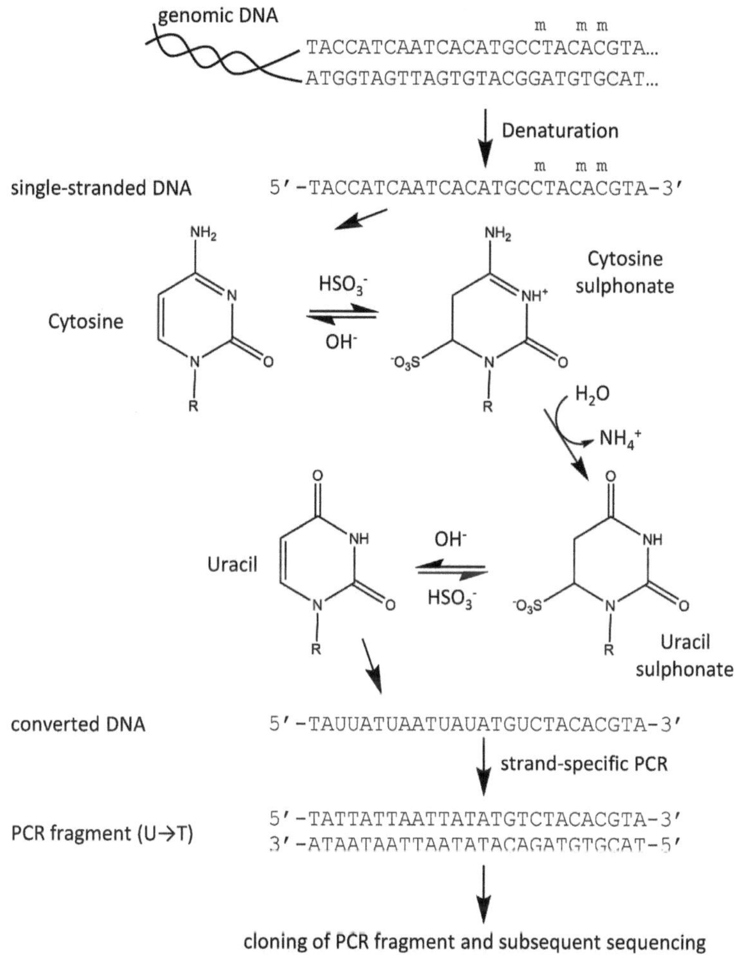

Fig. 29.1. **Scheme of bisulphite conversion.** Unmethylated cytosine residues of single-stranded DNA are sulphonated, followed by deamination to uracil sulphonate and desulphonation to uracil. Methylated cytosine residues are protected from this reaction. As the converted DNA is not complementary any more, the PCR has to be strand specific.

Due to the high number of potential methylated sites in plant DNA compared with mammals and the high variability in DNA methylation, the design of primers and selection of PCR conditions are critical steps and will be described in a separate section below. For a detailed and statistical analysis, the PCR product has to be cloned and multiple individual clones have to be sequenced.

2. Materials

2.1. DNA Isolation

Use plant DNA extraction kit (e.g., Nucleon PhytoPure, GE Healthcare, or MasterPure Plant Leaf DNA purification kit, Biorad).

2.2. Pretreatment of DNA

1. Restriction enzyme with appropriate buffer, e.g., EcoRI.
2. Phenol/chloroform: Mix 1:1 aquaphenol with chloroform. Phenol is toxic and mutagenic, avoid inhalation and skin contact. Chloroform is irritant and harmful, avoid inhalation and skin contact.
3. Glycogen solution (20 μg/μL).
4. Ammonium acetate solution (10 mM).
5. Ethanol (100%, 70%).

2.3. Bisulphite Conversion

1. NaOH (3 M): Prepare fresh the same day.
2. Hydroquinone (20 mM): Dissolve 220 mg of hydroquinone in 100 mL of water. Hydroquinone is toxic, avoid inhalation and skin contact.
3. Bisulphite solution: Dissolve 10.1 g of disodium bisulphite (sodium metabisulphite) in 20 mL of water with slow stirring to avoid aeration (takes around 20 min). Adjust pH to 5.1 with 10 M NaOH (around 500 μL). Add 825 μL of 20 mM hydrochinone. Adjust volume to 25 mL. Prepare always fresh immediately before use or on the same day and keep in the dark. For complete dissolving of bisulphite, add NaOH. Disodium bisulphite is toxic, avoid inhalation and skin contact.
4. Wizard® DNA clean up (Promega).
5. TE buffer: 10 mM Tris-HCl and 1 mM EDTA, pH 8.0. For 500 mM EDTA, dissolve EDTA in water by vigorous stirring and pH adjustment to pH 8.0 by addition of NaOH pellets. EDTA is irritant, avoid skin contact.

2.4. PCR Amplification of Bisulphite-Treated DNA

1. Taq polymerase, PCR buffer, and 50 mM $MgCl_2$ solution.
2. dNTP (5 mM each of dATP, dCTP, dGTP, dTTP).
3. Primers (10 μM).
4. Agarose.

3. Methods

3.1. DNA Isolation

Ensure high quality of genomic DNA by using an appropriate isolation method, e.g., Qiagen DNeasy® Plant Mini kit or

Nucleon™ Phytopure kit. The sample DNA should be free of proteins to ensure complete denaturation and should show a ratio A_{260}/A_{280} between 1.7 and 2.0. For bisulphite conversion, use 50 ng to 2 μg of genomic DNA (can be upscaled for species with large genomes to 50 μg) (*see* **Note 1**).

3.2. Pretreatment of DNA

1. Digest sample DNA with an appropriate restriction enzyme that is not cutting in the region of interest in a suitable sample buffer (e.g., 2 μg of DNA incubated with 20–40 units enzyme in 200 μL water overnight at the appropriate temperature) (*see* **Note 2**).
2. Add 200 μL of phenol/chloroform, vortex rigorously, and centrifuge for 1 min at 10,000×*g*. If not indicated otherwise, steps are carried out at room temperature.
3. Transfer the upper aqueous phase to a new reaction tube. Add 200 μL of chloroform, vortex rigorously, and centrifuge for 1 min at 10,000×*g*.
4. Transfer the upper aqueous phase to a new reaction tube. Add 1.5 μL of glycogen, 38.6 μL of 10 M ammonium acetate, and 385 μL of ice-cold absolute ethanol, vortex, and centrifuge for 15 min at 4°C at maximum speed (<12,000×*g*) (*see* **Note 3**).
5. Remove the supernatant and wash twice with 500 μL of 70% ethanol.
6. Vacuum dry the pellet at room temperature.
7. Resuspend the pellet in 40 μL of water.

3.3. Bisulphite Conversion

1. Divide the digested DNA between two PCR tubes and add 1 μL of freshly prepared 3 M NaOH. Incubate for 30 min at 39°C, followed by incubation at 90°C for 2 min. Place directly on ice and centrifuge briefly.
2. Add 208 μL of freshly prepared bisulphite solution (pH 5.0), mix, centrifuge briefly, and transfer 114 μL of the solution to a new tube.
3. Incubate in a thermocycler for 16 h at 55°C with a 5-min jolt every 3 h to 95°C (*see* **Note 4**). Afterwards proceed immediately with desalting as prolonged incubation in bisulphite solution can lead to DNA degradation (15).
4. Use an appropriate kit for desalting such as Wizard® DNA clean up (Promega) (*see* **Note 5**).
5. Measure the exact volume of recovered DNA in TE buffer and add 6.3 M NaOH to a final concentration of 0.3 M (for 50 μL sample, use 2.5 μL of 6.3 M NaOH). Mix briefly, centrifuge, and incubate at 37°C for 15 min.

6. Add 10 M ammonium acetate to a final concentration of 3 M, add 2 μL of glycogen, and 3 volumes of ice-cold absolute ethanol (for 52.5 μL sample, use 22.5 μL of 10 M ammonium acetate and 225 μL of absolute ethanol) (*see* **Note 3**).
7. Mix by inversion and centrifuge at 14,000×*g* for 15 min at 4°C.
8. Remove the supernatant and wash the pellet once with 500 μL of 70% ethanol.
9. Remove the supernatant and air dry the pellet (around 20 min).
10. Resuspend the pellet in 100 μL of TE buffer. The bisulphite-treated DNA can be stored at –20°C.

3.4. PCR Amplification of Bisulphite-Treated DNA

PCR primer design for bisulphite-treated plant DNA is a critical step and different from the design for mammalian DNA (*see* **Note 6**). After bisulphite treatment of the DNA, the two strands are not complementary anymore and have to be analyzed separately. Therefore, primer pair(s) have to be strand specific. For amplifying the sense DNA strand, the 3′ primer is complementary to the target strand and should be G poor and C rich (*see* **Note** 7), whereas the 5′ primer should be G rich and C poor. Conversely, for amplifying the antisense strand, the 5′ primer is complementary to the target strand and should be C rich and G poor, whereas the 3′ primer should be G rich and C poor. Maximum amplicon length is around 500 bp; shorter fragments will enhance PCR yield. Primer design for bisulphite-treated plant DNA program can also be done using the Kismeth program (http://katahdin.mssm.edu/kismeth/revpage.pl) (16).

1. 2 μL of bisulphite-treated DNA is used for PCR amplification with bisulphite conversion specific primers (*see* **Note 8**):

2 μL	Template DNA
2 μL	10 × Taq PCR buffer
0.8 μL	$MgCl_2$
1 μL	dNTPs
1 μL	5′ Primer
1 μL	3′ Primer
0.1 μL	Taq polymerase
12.1 μL	Water

2. The reaction conditions have to be optimized for each primer pair. Below is an example of PCR conditions (*see* **Note 9**):

94 °C	3 min
94 °C	30 s
anneal.temp.	30 s
72 °C	1 min
Go to Step 2, 40 cycles	
72 °C	5 min
Hold at 16 °C	

Freeze 5 μL of PCR product for optional follow-up usage (*see* **Note 10**) and separate the remaining PCR product by agarose gel electrophoresis (use a 1.5–2.5% agarose gel depending on expected size). If a defined band with the expected size is visible, purify the PCR product from the gel and clone it into a suitable vector using a PCR cloning kit such as pJET (CloneJet, Fermentas) or pDrive (Qiagen).

3. For reliable quantitative conclusions, sequence at least 8–10 individual clones.

3.5. Data Processing

The sequence data have to be tabulated to visualize the methylation state of the individual clones and calculated for statistical information. This can be done either by manual comparison or the use of available online software like Kismeth and CyMATE (*see* **Note 11** and **Fig. 29.2**) (16, 17). The following section describes manual comparison:

1. Extract the sequence results into FASTA format starting and ending within the primer sequences.
2. Align sequences including the native sequence using ClustalW2 that is freely available (or alternative alignment programs). In ClustalW2, arrange the native sequence on top by changing the settings "Output order" to "Input."
3. Each C to T conversion in the bisulphite-treated sequence indicates an unmethylated C in the original genomic sequence. Preserved C residues signify methylated cytosines. Mark all converted and unconverted C residues with two different colors, sum up, and visualize with the help of statistic software such as SigmaPlot (Systat Software Inc.) or a spreadsheet application like Excel (Microsoft) or OpenOffice.org Calc (Openoffice). The different methylation sites (CG, CNGm and CHH) should be labeled differently as they are targeted by different pathways (*see* **Note 12** and **Fig. 29.2**).

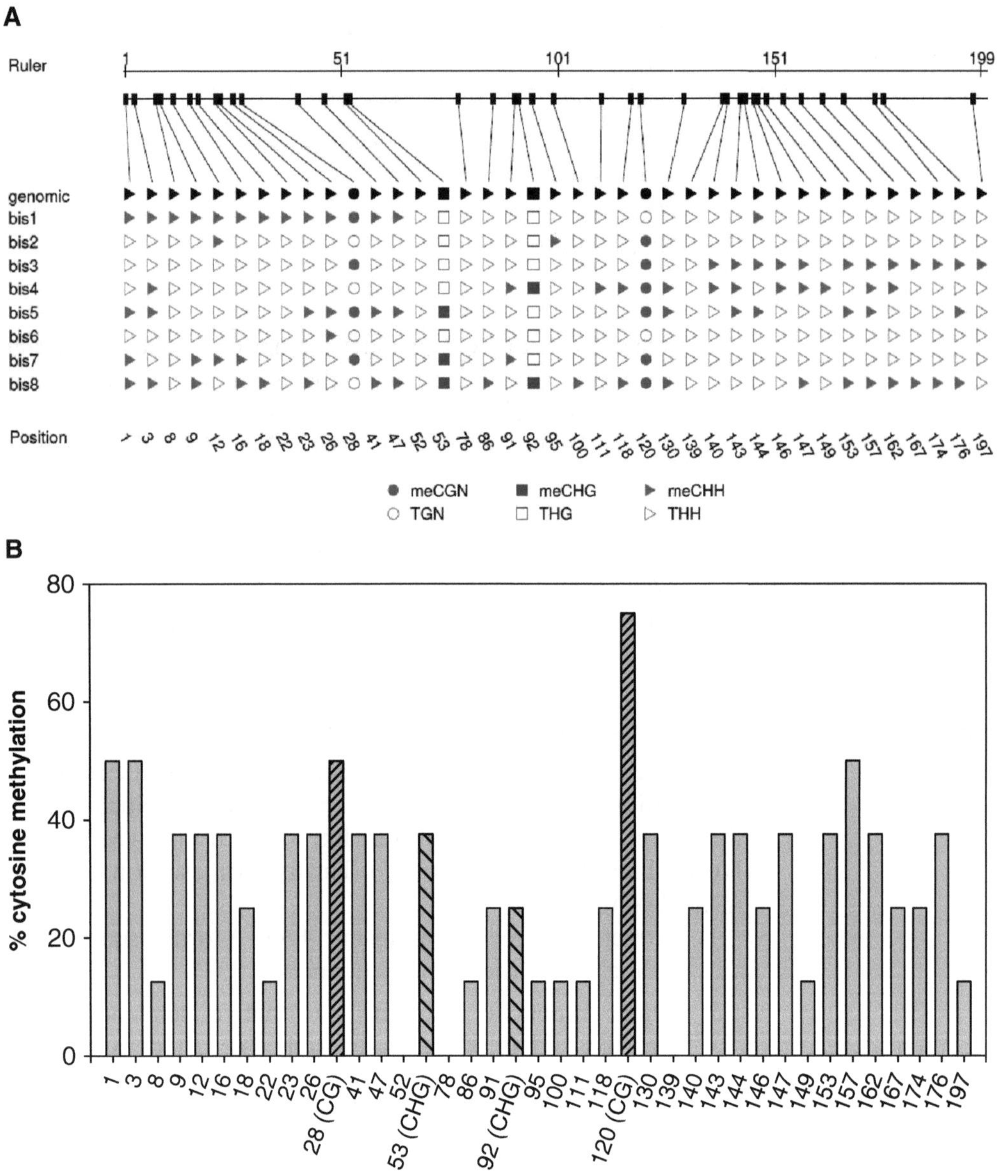

Fig. 29.2. **Graphical representation of bisulphite sequencing results.** Report of a 200-bp fragment analyzed by CyMATE (**A**) or manually (diagram prepared in SigmaPlot) (**B**).

4. Notes

1. For increased amounts of genomic DNA, the total volumes have to be adapted. Too high concentrations can cause DNA-strand reannealing after denaturation and incomplete bisulphite conversion (18).

2. High molecular weight DNA will be fragmented to allow complete denaturation. In general, six-base cutters that do not target the regions of interest are used. Alternatively or in case no suitable restriction enzymes can be found, sonication of DNA or shearing through narrow needles can be used to produce smaller fragments.
3. Addition of glycogen as an inert carrier is not mandatory but increases the recovery rate of DNA by alcohol precipitation. Alternatively, samples can be incubated for 1 h at –20°C.
4. The length of treatment depends on the fragment size and amount of sample DNA. High molecular weight DNA and high amounts of DNA need prolonged incubation times (5 h for 100 ng DNA, 16 h for 2 μg DNA) (18). If the reaction is not carried out in a thermocycler, ensure that the reaction takes place in the dark as light causes oxidation. If a thermocycler without a heated lid and mineral oil is used instead, remove mineral oil after bisulphite treatment by freezing the sample and pipetting off the mineral oil.
5. Alternative desalting procedures suitable for single-stranded DNA can be used. Other commercially available kits are Microcon® centrifuge clean up (Millipore) or QIAquick® (Qiagen).
6. In plants, cytosines in all sequence contexts can potentially be targeted for methylation, including asymmetric CHH residues. Therefore, differential methylation of both strands has to be considered (19).
7. To avoid biased amplification of methylated or unmethylated DNA, degenerated bases should be used within the primer pairs (Y instead of C/T for 5′ primer and R instead of G/A for 3′ primer for amplifying the sense strand). Always prefer a low quantity of potential methylation sites within your primer annealing sequence. Alternatively inosine, which can pair with all nucleotides, can be used. In contrast to bisulphite primer design for mammalian DNA, avoid potential conversion/methylation sites at the last two bases of the primer.
8. PCR amplification for each primer pair has to be optimized. Different DNA polymerases can work differentially for each primer pair and should be tested if no or only faint bands are visible. Furthermore, the PCR master mix can be adapted by varying the Mg^{2+} concentrations (do not elevate Mn^{2+} concentrations as it can lead to lower fidelity). Several proof-reading DNA polymerases (e.g., Pfu) cannot amplify uracil-containing templates and should therefore not be used in the first cycles of PCR amplification.

9. Testing annealing temperatures by gradient PCR or extended elongation steps in the first 20 cycles (1 min 30 s per elongation step) can help to find optimal PCR conditions.
10. If there is no PCR product visible after gel electrophoresis, use some of the backup DNA for nested or seminested PCR.
11. CyMATE (http://www.gmi.oeaw.ac.at/en/cymate-index/cymate-v2/) is a freely available Web-based program for graphical and quantitative evaluation of bisulphite data (16). The input sequences can be uploaded in FASTA or CLUSTAL formats. It was developed especially for plant DNA and classifies cytosine residues due to their sequence context. A similar program represents Kismeth (http://katahdin.mssm.edu/kismeth/revpage.pl) with an additional feature of primer design for bisulphite-treated plant DNA (17).
12. The different steps of bisulphite sequencing harbor potential for the generation of artifacts (20). First, bisulphite conversion could be incomplete, e.g., due to proteins bound to DNA or strand reannealing. This can be checked by bisulphite sequencing of regions with known methylation states or use of an independent method, e.g., by the use of methylation-sensitive enzymes combined with Southern probes. Second, PCR amplification can be biased, favoring either converted, unconverted, or methylated regions. The use of degenerated primers can help to eliminate this problem. For mammalian systems, a simple check for primer pairs exists. A 1:1 mixture of bisulphite-converted DNA of commercially available unmethylated and methylated human DNA is used for PCR. Subsequent digestion with restriction enzymes cutting the methylated or unmethylated DNA (e.g., Taq1 cuts TCGA) should give an approximate ratio of 1:1 and PCR conditions are optimized to reach this ratio. Methylated plant DNA is not commercially available thus far; however, it could be enzymatically synthesized. Furthermore, cloning bias can be tested by directly sequencing the PCR products. All partially methylated residues found in the alignment of individual clones should show a double peak of C and T in the chromatogram of the sequence reaction.

References

1. Johnson, T. B. and Coghill, R. D. (1925) Research on pyrimidines. C111. The discovery of 5-methyl-cytosine in tuberculinic acid, the nucleic acid of the tubercle bacillus. *J Am Chem Soc* **47**, 2838–2844.
2. Wyatt, G. R. (1950) Occurrence of 5-methyl-cytosine in nucleic acids. *Nature* **166**, 237–238.
3. Arbor, W. and Linn, S. (1969) DNA modification and restriction. *Annu Rev Biochem* **38**, 467–500.
4. Bird, A. (2002) DNA methylation patterns and epigenetic memory. *Genes Dev* **16**, 6–21.
5. Chan, S. W., Henderson, I. R., and Jacobsen, S. E. (2005) Gardening the genome: DNA methylation in *Arabidopsis thaliana*. *Nat Rev Genet* **6**, 351–360.
6. Tran, R. K., Henikoff, J. G., Zilberman, D., Ditt, R. F., Jacobsen, S. E., and Henikoff, S. (2005) DNA methylation profiling identifies CG methylation clusters in Arabidopsis genes. *Curr Biol* **15**, 154–163.
7. Goll, M. G. and Bestor, T. H. (2005) Eukaryotic cytosine methyltransferases. *Annu Rev Biochem* **74**, 481–514.
8. Huettel, B., Kanno, T., Daxinger, L., Bucher, E., van der Winden, J., Matzke, A. J., et al. (2007) RNA-directed DNA methylation mediated by DRD1 and Pol IVb: A versatile pathway for transcriptional gene silencing in plants. *Biochim Biophys Acta* **1769**, 358–374.
9. Gehring, M. and Henikoff, S. (2007) DNA methylation dynamics in plant genomes. *Biochim Biophys Acta* **1769**, 276–286.
10. Ikeda, Y. and Kinoshita, T. (2009) DNA demethylation: A lesson from the garden. *Chromosoma* **118**, 37 41.
11. Gehring, M., Huh, J. H., Hsieh, T. F., Penterman, J., Choi, Y., Harada, J. J., et al. (2006) DEMETER DNA glycosylase establishes MEDEA polycomb gene self-imprinting by allele-specific demethylation. *Cell* **124**, 495–506.
12. Penterman, J., Zilberman, D., Huh, J. H., Ballinger, T., Henikoff, S., and Fischer, R. L. (2007) DNA demethylation in the Arabidopsis genome. *Proc Natl Acad Sci USA* **104**, 6752–6757.
13. Frommer, M., McDonald, L. E., Millar, D. S., Collis, C. M., Watt, F., Grigg, G. W., et al. (1992) A genomic sequencing protocol that yields a positive display of 5-methylcytosine residues in individual DNA strands. *Proc Natl Acad Sci USA* **89**, 1827–1831.
14. Clark, S. J., Statham, A., Stirzaker, C., Molloy P. L., and Frommer, M. (2006) DNA methylation: Bisulphite modification and analysis. *Nat Protoc* **1**, 2353–2364.
15. Grunau, C., Clark, S. J., and Rosenthal, A. (2001) Bisulfite genomic sequencing: systematic investigation of critical experimental parameters. *Nucleic Acids Res* **29**, E65–E65.
16. Gruntman, E., Qi, Y., Slotkin, R. K., Roeder, T., Martienssen, R. A., and Sachidanandam, R. (2008) Kismeth: Analyzer of plant methylation states through bisulfite sequencing. *BMC Bioinformatics* **9**, 371–385.
17. Hetzl, J., Foerster, A. M., Raidl, G., and Mittelsten Scheid, O. (2007) CyMATE: A new tool for methylation analysis of plant genomic DNA after bisulphite sequencing. *Plant J* **51**, 526–536.
18. Clark, S. J., Harrison, J., Paul, C. L., and Frommer M., (1994) High sensitivity mapping of methylated cytosines. *Nucleic Acids Res* **22**, 2990–2997.
19. Luo, S. and Preuss, D. (2003) Strand-biased DNA methylation associated with centromeric regions in Arabidopsis. *Proc Natl Acad Sci USA* **100**, 11133–11138.
20. Warnecke, P. M., Stirzaker, C., Song, J., Grunau, C., Melki, J. R., and Clark, S. J. (2002) Identification and resolution of artifacts in bisulfite sequencing. *Methods* **27**, 101–107.

Subject Index

L. Hennig, C. Köhler (eds.), *Plant Developmental Biology*, Methods in Molecular Biology 655,
DOI 10.1007/978-1-60761-765-5, © Springer Science+Business Media, LLC 2010

S

T

V

W

Y

Z

MIX
Papier aus verantwortungsvollen Quellen
Paper from responsible sources
FSC® C105338

If you have any concerns about our products,
you can contact us on
ProductSafety@springernature.com

In case Publisher is established outside the EU,
the EU authorized representative is:
Springer Nature Customer Service Center GmbH
Europaplatz 3, 69115 Heidelberg, Germany

Printed by Libri Plureos GmbH
in Hamburg, Germany